Reviews on Analytical Chemistry—Euroanalysis VIII

Reviews on Analytical Chemistry—Euroanalysis VIII

Edited by

D. Littlejohn

Department of Pure and Applied Chemistry, University of Strathclyde, Glasgow

D. Thorburn Burns

Department of Analytical Chemistry, The Queen's University of Belfast, Belfast

THE ROYAL
SOCIETY OF
CHEMISTRY

This volume contains invited lectures from the Proceedings of Euroanalysis VIII, held at Edinburgh, UK on 5–11 September 1993.

Special Publication No. 154

ISBN 0-85186-982-3

A catalogue record of this book is available from the British Library.

Published by The Royal Society of Chemistry,
Thomas Graham House, Science Park, Cambridge CB4 4WF

Typeset by Vision Typesetting, Manchester
Printed and bound by Bookcraft (Bath) Ltd.

Preface

This book contains texts of most of the invited lectures presented at the Euroanalysis VIII European Conference on Analytical Chemistry held in Edinburgh, Scotland on 5–11 September 1993. Like earlier conferences in this series, Euroanalysis VIII was a broad-spectrum meeting covering many of the most important techniques and applications in modern analytical chemistry. In particular, the conference emphasized developments in biomedical and pharmaceutical analysis, sensors, and environmental analysis. Special sessions were organized on Computer-based Analytical Chemistry (COBAC) and the Validity of Analytical Measurements (VAM). These and other subjects are featured in the reviews and overviews presented in this volume.

The many facets of analytical chemistry were covered by the invited and contributed lectures (176) and invited and contributed posters (350), which formed the main body of the Scientific Programme. Open, structured discussions were held on educational matters, the validity of forensic analysis, and separation science. The conference attracted 649 registered scientific participants from 48 countries; the majority of participants (495) came from outside the UK.

Arranging a major international conference is always a collective effort. We appreciate the support provided by The Royal Society of Chemistry, which assumed financial and organizational responsibility. Special thanks are due to Miss P. E. Hutchinson, Secretary of the Analytical Division. We also wish to record our thanks to the Organizing Committee, the Conference Praesidium, and Professor L. Niinistö, in particular, who is Chairman of the Working Party on Analytical Chemistry, Federation of European Chemical Societies. The participation of a number of lecturers and delegates was made possible through generous donations from various sponsors, too many to mention on this occasion but we are grateful none-the-less for their support. Special thanks are reserved for the members of the Scientific Programme Committee (D. Littlejohn, R. I. Aylott, K. D. Bartle, D. Thorburn Burns, J. Marshall, J. N. Miller, M. R. Smyth, and A. M. Ure) each of whom acted as Session Conveners and liaised with the invited lecturers concerning the submission of manuscripts for this book. Finally, we wish to record our gratitude to the authors of the various chapters for their co-operation during the editing process.

D. Thorburn Burns
Chairman—Organizing Committee

D. Littlejohn
Chairman—Scientific Programme Committee

Engraved by J. Posselwhite.

BLACK.

*From a Print by Ja.ˢ Heath, after
a Picture by Raeburn.*

Under the Superintendence of the Society for the Diffusion of Useful Knowledge.

London, Published by Charles Knight & Cᵒ Ludgate Street.

Contents

Dark Ages to Enlightenment, Alchemy to Analysis: Progress in Scottish Chemistry Within a European Framework

D. Thorburn Burns

DEPARTMENT OF ANALYTICAL CHEMISTRY, THE QUEEN'S UNIVERSITY OF BELFAST, BELFAST BT9 5AG, UK

Summary

This review outlines aspects of the three phases of Scottish Chemistry, from the advent of alchemy into Western Europe to the birth of scientific chemistry.

The alchemical period (1200–1650) is illustrated by the travels and contributions of *Michael Scot* (1175–1235), *John Damian* (*fl.* 1501–13), and *Alexander Seton* (died 1604), otherwise known as the 'Cosmopolite'.

Chemistry *per se* emerged from alchemy in the post Baconian 'scientific revolution' (1600–1700). Much seventeenth century chemistry retained alchemical overtones and terminology due to the then imperfect understanding of the nature of the elements. *Matthew Mackaile* (*fl.* 1657–96) and *William Davidson* (1593–1669) are typical of such 'transition period' chemists.

The period of the so called 'enlightenment' (1700–1800) and the establishment of chemistry as an independent science was followed by the period to date, that of scientific chemistry. *William Cullen* (1710–90) gave Scottish chemistry a social profile it was to retain for the rest of the eighteenth century, as John Robison put it, 'Cullen made chemistry a liberal science and a study of gentlemen'. *Joseph Black* (1728–99) completed the transition of chemistry from an empirical to a quantitative science by his studies of *magnesia alba*. Black's paper 'Analysis of the Waters of Some Hot Springs in Iceland' exhibited fine innovative analytical work of a quality far ahead of his time. He was a founder member of the Royal Society of Edinburgh (1783) as well as instituting the world's first Chemical Society in Edinburgh in 1785.

It is shown that, as in the early periods, Scottish chemistry had European dimensions, links can also be traced to all the countries of previous Euroanalysis conferences and indeed to the next, Bologna in 1996.

1 Introduction

Scotland's impressive contributions to the development of chemistry as a science are well known,[1] but it is not often appreciated that, from the advent of alchemy into Western Europe until the birth of scientific chemistry, Scotland fostered a succession of outstanding exponents of the 'divine art' of alchemy.[2-9] The history of chemistry in Scotland thus involves four phases: an alchemical period (1200–1650), one of the emergence of chemistry in the late alchemical period, in the Bacon 'scientific revolution' (1600–1700), the period of the 'enlightenment' and the establishment of chemistry as an independent science (1700–1800), followed by the period to date, that of scientific chemistry.

2 The Alchemical Period

Michael Scot (1175–1235)

Michael Scot[10-15] was one of the leading European intellectuals in the early thirteenth century although he appears shadowy and intriguing to later generations. He was born about 1175 in the Scottish border country. He studied in Oxford and then in Paris before moving further south in Europe. The earliest certain date in his cosmopolitan academic career is 18 August 1217 on which day, in Toledo, he completed his translation of al-Bitrogi's 'De Spharea'. He was known to be in Bologna in the Autumn of 1220. Scot's eminence as a master of Latin learning, of Hebrew, and of Arabic was recognized by the Popes, between 1224 and 1227 he was given permission to hold benefices in England, Scotland, and was even offered the Irish archbishopric of Cashel. By 1227 he had reached Sicily and became attached to the Court of Frederick II, Holy Roman Emperor and King of Sicily, at whose request many of his works were written. 'Liber particularis' included a section on the then known metals. The composition of metals was referred to as the Sulfur–Mercury theory of Jabir.[14] Several alchemical manuscripts have been attributed to Scot.[11,12] His medical text 'Physionomia' was an early 'best seller' and enjoyed a wide circulation in manuscript form prior to its first printing in 1477.[16] Some of his formulations remained in pharmacopoeias and books of secrets such as those of Alexis of Piedmont well into the seventeenth century.[17] He died in the Emperor's service about 1235. Michael Scot is now seen to be of the greatest importance as a translator, reforming the translations of Aristotelian metaphysics and natural philosophy and also acquainting the Latin world with the recent Arab thought.[13] He was the most notable Briton to figure in the history of medieval Italy.

John Damian (fl. 1501–13)

Some three hundred years later James IV of Scotland founded Scotland's first research laboratory in Stirling Castle. James' excursions into alchemy were prompted by his interest in medicine because of the potent medicinal virtues ascribed at that time to the 'philosophers stone'. As the 'stone' perfected metals so it was an unfallible cure for the ills of imperfect man. James IV's chief assistant was a Lombard, John Damian.[2-9] In order to provide him with 'an emolument and the necessary leisure for research', the King created him Abbot of Tongland in Galloway.[18] The Lord High Treasurer's accounts 1501–08[19] show large sums expended on chemicals, equipment, and on wages for assistants including the 'boy that kepit the furnes fire'. In 1501 the Abbot's attempts to achieve the stone were interrupted by another spectacular activity—an attempted flight to Paris. Failure was ascribed to the presence of hens' feathers in the Abbot's wings which 'yearned and covet the midden'. He recovered and chemical work continued up to the death of the King of Flodden, 9 September 1513.

Alexander Seton (died 1604)

Alchemy reached its zenith in the seventeenth century when it was a popular pursuit of many noble families. Alexander Seton[2-9] a nebulous and enigmatic figure, the mysterious 'Cosmopolite', is said to have left Scotland for the Continent in 1602. He shoots across alchemy 'like a streamer of the northern morn'[3] from 1601 to 1604. He is recorded as having achieved a dramatic series of transmutations of base metals into gold at Enkhuysen, Basel, Strasbourg, Frankfurt, Cologne, Hamburg, Helmstedt, and Dresden.[5,20] In Munich, he married a Burgher's fair daughter. He joined the court of Christian II, Elector of Saxony, who on failing to extract Seton's great secret, imprisoned and tortured him. He was rescued by a Polish nobleman Michael Sendivogius who took him with his wife to Cracow, where he died in 1604 leaving the remains of his much coveted red tincture to Sendivogius. Sendivogius married Seton's widow who possessed two manuscripts said to have been written by Seton, these were later published by Sendivogius as 'Novum Lumen Chymicum',[21] who also assumed the cognomen 'Cosmopolite'. 'New Light of Alchemy' was a best seller throughout the seventeenth and eighteenth centuries;[22,23] Sir Isaac Newton treated the work with great respect and made a 16000 word manuscript transcript from it.[5,24]

3 The Emergence of Chemistry

In the long ages of alchemy there were many who carried out practical operations such as the assaying of metals, chemical based pharmacy, examination of mineral waters, *etc.* and recorded their work devoid of the theoretical, allegorical, or religious aspects of alchemy. It is largely through this work and its records that chemistry arose, phoenix like, from the ashes of alchemy. The transition to chemistry was slow, at times incomplete, and much seventeenth century chemistry retains alchemical terminology due in part to imperfect understanding of the nature of an element. Indeed, alchemical or alchemical type symbols remained in use, as in Dalton's symbolic notation,[25] until the adoption of Berzelius symbols as used today after their introduction in 1813.[26,27] Matthew Mackaile and William Davidson are two such Scottish transitional period chemists of interest.

Matthew Mackaile (fl. 1657–96)

Mackaile trained as an apothecary in Edinburgh and later practised medicine in Aberdeen.[28,29] He wrote the first book on Scottish mineral waters with a detailed chemical content, 'Fontium Mineralium Moffetensium' (1659)[30] later translated into English by the author (1664).[31] The preface contained a 26 page account of the 'elements of chymie' so that the spagyricall (*i.e.* chemical) description of the wells may be better understood. He was against alchemy *per se* and later wrote a medical tract 'Scurvie Alchymie discovered'.[32] Despite being against alchemy, Mackaile used the Aristotelian four qualities[6] (hot, dry, cold, and wet) and two (earth, phlegm) of the four elements (fire, water, and air) coupled with the Paracelsian *tria prima*, namely mercury, sulfur, and salt, this last being sub-divided into fixed and volatile. He did, however, describe the chemistry of materials such as iron and sulfur which we now know as elements. 'Moffet Wells' demonstrates Mackaile was well read, he makes detailed reference to the Italian Andreas Baccius'[33] book on 'Hot Baths' (1571) and Zwelfer's revision of the 'Pharmacopoeia Augustana' published in Vienna (1652).[34]

The oil well at Libberton, two miles south from Edinburgh, mentioned on the title page of 'Fons. Moffetensis' was known much earlier than 1659 since it had been visited by James VI in 1617. It is interesting to speculate as to why it took until the 1860s for the Scottish shale oil industry to develop.

William Davidson (1593–1669)

Davidson[35-37] graduated Master of Arts from Marischal College Aberdeen in 1617. Shortly afterwards Davidson migrated to France where he appears to have graduated in medicine. He formed a lifelong friendship with Jean Baptiste Morin and studied chemistry in the household of Morin's patron, Claude Dormy bishop of Boulogne. Later he moved to Paris, in due course Gallicized his name to d'Avissone, and in 1644 he became physician to Louis XIII. In 1648 Davidson was appointed first professor at the Jardin du Roi in Paris.[36] The lectures were delivered in Latin and accompanied by practical demonstrations. His lectures there, as for earlier private courses, attracted pupils from many countries. It was for this reason Davidson published a text, 'Philosophia Pyrotechnica seu Curriculus Chymiatricus' in four parts (1633–5).[38] This, complete with an allegorical illustration of the 'whole work', was issued in several editions first in Latin and then in French.[39]

Following a legal wrangle concerning his position as *intendant* of the Jardin du Roi he resigned in 1651 to become physician to the wife of John Casimir, King of Poland, and also director of the Royal Botanic Garden in Warsaw. On the death of the Queen in 1667 he returned to Paris where he died in 1669.

The first two parts of Davidson's text are devoted to theoretical and speculative considerations following Paracelcian concepts, the third deals with chemical terms, and the fourth the apparatus and operations of chemistry. This later part contains some of the earliest contributions to crystallography—'a new subject which none before has elaborated'. The perfect bodies being associated with the old elements earth, fire, air, and water. Despite being an alchemist in theory he was a chemist in practice.

4 The 'Enlightenment' and Establishment of Chemistry as an Independent Science

By the start of the eighteenth century, 'the century of the enlightenment',[40,41] chemistry had emerged from the ashes of alchemy but was largely associated in the universities with medicine.

James Crawford, a graduate of Leyden, was appointed Professor of Physick (Medicine) and Chymistry at Edinburgh University in 1713 and taught chemistry intermittently from 1714. The first lectureship in chemistry (within the Faculty of Medicine) at Glasgow University was held by William Cullen from 1746 to 1755.

William Cullen (1710–90)

William Cullen[42–44] wrote very little on chemistry and made no major fundamental discoveries but contributed significantly to chemical aspects of *materia medica*. His single paper on chemistry, *Of the Cold Produced by Evaporating Fluids*,[45] was in due time followed up in the studies of Black, Watt, and Irvine at Glasgow. Cullen's contribution to chemisty was via his teaching[46,47] which involved novel pictorial chemical equations and his pioneering introduction of voluntary practical work. He was however to remark,[46] 'the laboratory has been open to you but I am sorry to find that so few of you have frequented it'. Joseph Black, who succeeded Cullen in the Glasgow lectureship after Cullen's move to Edinburgh in 1755, was one who did attend the laboratory and benefited. Cullen gave 18 years to chemistry, 8 in Glasgow and 10 in Edinburgh. His Edinburgh lectures on the history of chemistry[42] were based on Borrichius's 'Conspectus' (1696) and on Boerhaave's 'Elements of Chemistry' (1741) translated by Shaw. Cullen gave Scottish chemistry a social profile it was to retain for the rest of the eighteenth century. As Robison put it[43a] 'Cullen made chemistry a liberal science and a study of gentlemen'. When Cullen took up the Chair of Physick he was again succeeded by his pupil, Joseph Black. Cullen's last 30 years were devoted to the practice and teaching of medicine. His reputation was spread by his many popular texts. Cullen played a major role in the founding of the Royal Society of Edinburgh in 1783.[48]

Joseph Black (1728–1799), Belfast, Glasgow, and Edinburgh

Joseph Black,[42,49–52] was born on 16 April 1728 in Bordeaux where his father, a Belfast man of Scots descent, was in the wine trade. He was educated by his Scottish Mother until the age of twelve when he was sent to Belfast to become a pupil in the old Latin School. This school, established by the Earl of Donegal in 1666, stood at the corner of Ann Street and Church Lane, which was then called School-house Lane.[53] It remained in existence until the Belfast Academy was established in 1786. In 1744 Joseph Black entered the University of Glasgow.

His matriculation entry reads 'Josephus Black filius natu quartus Joannis Black Mecatoris in Urbe Bordeaux in Gallia, ex urbe de Belfast in Hibernia'. He read the general arts curriculum for three years, then, after being pressed by his father to choose a profession, studied medicine and chemistry under William Cullen. Black says he attended Cullen's lectures, studied the publications of Marggraf and Reaumur, and worked in Cullen's laboratory.[49]

In 1752 Black went to Edinburgh to complete his medical studies and graduated MD in 1754. His now famous thesis 'On the acid humour arising from food, and magnesia alba'[54] deals mainly with the acidity of chemical experiments and an explanation and proof of Black's doctrine of the relationship between mild and caustic alkalis.

In June 1755 he read a paper to the Philosophical Society of Edinburgh, a revised account of the chemical experiments in his thesis together with further work. His paper was published the following year,[55] and reprinted 1779,[56] 1782,[45] 1898, and reissued in 1963.[57] This was Black's most significant chemical publication.[58] He showed that the change from chalk to lime consists of the withdrawal of fixed air (carbon dioxide).

'A piece of perfect quicklime, made from two drams of chalk, and which weighed one dram and eight grains, was reduced to a fine powder, and thrown into a filtrated mixture of an ounce of fixed alkaline salt, and two ounces of water. After a slight digestion, the powder being well washed and dried, weighed one dram and fifty-eight grains. It was similar in every detail to fine powder of ordinary chalk, *i.e.*[55]

$$CaCO_3 \rightarrow CaO + CO_2$$

$$CaO + H_2O + K_2CO_3 \rightarrow CaCO_3 + 2KOH'$$

Black's theory was opposed by J. F. Meyer who explained the production of causticity by absorption of a 'phlogiston like' substance, oily *acidium pigue*, from the fire.[59] Some support to Black's work provided by the Glasgow trained surgeon, D. Macbridge of Dublin using apparatus[60,61] devised by Black and communicated to Hutcheson, lecturer in chemistry at Trinity College, Dublin. Detailed independent confirmation was given by the work of N. J. E. Jacquin in a small book published in Vienna.[62] Jacquin was the first professor of metal production and chemistry at the Hungarian Mining Academy at Selmecbanya (now Banska Stiavnica, Slovakia).[63,64] Later Jacquin became professor of chemistry in Vienna. A detailed account of the controversy was given by Lavoisier in his 'Essays Physical and Chemical' (1774).[65,66]

Black, as recognized by Lavoisier, thus laid a foundation stone of the revolution in chemistry based on quantitation. The letters from Lavoisier to Black were published at the behest of Thomas Andrews in the Report of the 41st Meeting of the British Association held in Edinburh in 1871.[67] His equipment was simple, the balance and other apparatus used in this and other early work are in the Playfair Collection housed in the Royal Museum of Scotland.[68]

Black succeeded Cullen in the Glasgow lectureship in 1756.[42,69] Whilst at Glasgow Black developed his second important line of research, that on latent heats and specific heats. This important work was never published independently but discussed in his chemical lectures[70] and recorded in manuscript form by students[71,72] and in the edited edition of his lectures, published posthumously by Robison.[73] James Watt was, at the time of Black's lectureship, instrument maker to the University of Glasgow and his experiments on the improvement of the steam engine were greatly assisted by Black's discovery of latent heat. Watt owed much to Black's advice and their extensive correspondence is extant.[74] The letters show Watt's considerable knowledge of contemporary chemisty which was sufficient for Robison to recommend to Black's Trustees and Executors that Watt edit Black's Lectures.[73] In the event he was too busy and the task fell to Robison who dedicated the works to Watt, 'Black's most illustrious pupil'. Watt was also one of the links of Black to the Lunar Society[75] of Birmingham.

Black edited an edition of Martine's 'Essays on the Construction of Thermometers' to which he added tables of the different scales of heat as exhibited in his annual lecture course.[76]

Black left Glasgow in 1766[77,78] and was followed in the chemical lectureship by John Robison and then by William Irvine. Irvine's pupil Adair Crawford, an Ulster Scot, who worked on 'animal heat', published (1779)[79] the earliest book based on research conducted in the Glasgow Chemistry Department. The second edition of 1788 contains much new material, *e.g.* experiments to determine the heat of combustion of hydrogen and oxygen.[80]

Crawford was for a time physician at St Thomas's Hospital London then Professor of Chemistry at Woolwich Arsenal where he was assisted by Cruickshank in the discovery of strontium. They examined a mineral from a lead mine in Strontian, Argyllshire and showed quite decisively it contained a peculiar earth different from baryta.[81]

Crawford provided a link with Scandinavian chemistry. He was visited by Johan Gadolin of Abo[82] during his tour of Germany, Holland, England, and Ireland (1786–8). Gadolin was, like Crawford, also particularly interested in thermochemistry and mineral analysis.

It has been stated that Black did no significant research work after being appointed to be Edinburgh Chair in 1766.[51,68,78] This is not true. Rancke-Madsen[83] and Szabadváry[84] have both commented that Black's paper 'Analysis of the Waters of Some Hot Springs in Iceland'[85] exhibited fine analytical work of a quality far ahead of his time. It was also innovative. Black was the first to observe the inference of carbon dioxide in alkali titration and an indicator error, he corrected the latter with a blank and was the first to use a back titration, and titration by weight. One must assume other commentators have not read the paper or are dismissive of progress in analytical chemistry.

After 1766 Black took an increasing interest in the rapidly developing chemical based industries in Scotland and above all in teaching. Black achieved his most widespread fame in his lifetime for his teaching and specifically for his use of lecture demonstrations. The lecture contents were kept up to date as can be judged from the contemporary student lecture notes,[71] over 90 sets survive over the session 1766–7 through to 1796–7. Several lectures were devoted to developing Cullen's concepts of exchange reactions based on Geoffrey's table of relative affinity.[47] Black's method of expressing reactions was to write alchemical symbols for the parts of a simple salt or binary compound in a circle with a line drawn through the circle separating the two symbols, the second compound in the reaction was similarly represented. The symbols were arranged so that the two parts in the upper semi-circles combined, and the two parts in the lower semi-circles formed another compound. Equilibrium was indicated by the point of touching of the circles. Black also used an explicit 'lever' form giving chemical names. Black was for most of his career a phlogistonist but ultimately abandoned the phlogiston theory and adopted Lavoisier's new system of elements.

The new French chemistry came to Britain via Edinburgh through, amongst others, Robert Kerr's translations[86] of works by Lavoisier[87] and Berthellot.[88] Black was a Founder Fellow of the Royal Society of Edinburgh[48] and the world's first Chemical Society.

5 The World's First Chemical Society

Having just celebrated the sesquicentenary of the Royal Society of Chemistry it is appropriate to note that its first President, Thomas Graham, held the Chair of Chemistry at Andersons' University (now the University of Strathclyde), Glasgow from 1830–37. It is perhaps even more important to note that the world's first Chemical Society met in Edinburgh.[89] The list of members, dated 1785, in Black's handwriting, contained 59 members of which 58 were students attending Black's classes 1783–7. It included Thomas Beddoes (founder of the Pneumatic Institute). The first volume of the proceedings 'Dissertations Read Before the Chemical Society Instituted in the Beginning of 1785', was presented to the Royal Irish Academy by Sir William Betham, 26th January 1846, and lay unnoticed in its library for more than a century.[90] It attracted the attention of Professor P. J. McLaughlin of St Patrick's College, Maynooth, Ireland. Subsequently this volume was presented to the Edinburgh University Chemical Society by the Council of the Royal Irish Academy.

6 Conclusion

Concluding an account of 'Progress in Scottish Chemistry' at the time of Joseph Black of necessity excludes all work done in the period of scientific chemistry. That in analytical chemistry has been surveyed earlier, up to 1880,[29] as has that on 'Early Optics and Spectroscopy'.[91] Following Thomas Melvill's first description in 1752 of the sharp sodium, yellow line, spectroscopy has since been developed and practised assiduously in Scotland.

From the account of the early periods of Scottish chemistry it is clear that it had European dimensions and that the valuable concepts within 'Euroanalysis', those of mutual support and exchange of information, had their exponents as far back as the thirteenth century. The early regard and state support for chemistry and the 'provision of suitable emolument and necessary leisure for research' for its practitioners has regrettably been eroded in recent times.

References

1. J. C. Irvine, *J. Chem. Educ.*, 1930, **7**, 2808.
2. J Read, *Chymia*, 1948, **1**, 139.
3. J. Read, *Chem. Drug.*, 25th June 1938, 742.
4. J. Read, *Ambix*, 1938, **2**, 60.
5. J. Read, 'Humour and Humanism in Chemistry', Bell, London, 1947.
6. J. Read, 'Through Alchemy to Chemistry', Bell, London, 1957.
7. J. Read, 'Prelude to Chemistry', Bell, London, 1936, 2nd edn., 1939. Reprinted Scientific Book Guild, 1961.
8. C. J. S. Thompson, 'The Lure and Romance of Alchemy', Harrap, London, 1932.
9. E. J. Holmyard, 'Alchemy', Penguin, Harmondsworth, 1957.
10. J. Wood Brown, 'An Enquiry into the Life and Legend of Michael Scot', D. Douglas, Edinburgh, 1897.
11. D. H. Haskins, *Isis*, 1928, **10**, 250.
12. S. Harrison Thomson, *Osiris*, 1938, **5**, 523.
13. L. Thorndike, 'Michael Scot', Nelson, London, 1945.
14. J. Read, *Scientia*, 1938, **32**, 190.
15. L. Minio-Paluello, 'Michael Scot' in 'Dictionary of Scientific Biography', ed. C. C. Gillispie, Vol. IX, Scribner's Sons, New York, 1974.
16. M. Scoti, '-Phisionomia-', Jalobus de Fivizzano, Venitii, 1477.
17. 'Les Secrets du Seigneur Alexis Piedmontois', P. Rigaud, Lyon, 1620 [1st edn. Venice, 1556]. See p. 105 for details of 'Pillules de M. Michel d'Escossois ...'.
18. C. H. Dick, 'Highways and Byways in Galloway and Carrick', Macmillan, London, 1927.
19. 'Compota Thesaurarium Regum Scotorum: Accounts of the Lord High Treasurer of Scotland', 13 Vols., HM General Register House, Edinburgh, 1879–1978, Vols. 1–4 cover the period 1473–1513.
20. D. G. Morhof, 'De Metallorum Transmutatione', Janssonium a Waesberge, Hamburgi, 1673.
21. Divi Leschi Genus Amat (anagram Michael Sendivogius), 'Novum Lumen Chymicum ...' [np], 1604 [earliest known printing, Library of Congress].
22. Michael Sendivogius (trans J. F.), 'New Light of Alchymie ...', R. Cotes for T. Williams, London, 1650.
23. Le Cosmopolite (trans F. Guiraud), 'Traicte du Soulphre second principe de Nature', A. Pacard, Paris, 1618.
24. B. J. T. Dobbs, 'The Foundation of Newton's Alchemy or "The Hunting of the Greene Lyon"', Cambridge University Press, Cambridge, 1975.
25. D. Thorburn Burns, *Fresenius' Z. Anal. Chem.*, 1990, **337**, 205.
26. J. J. Berzelius, *Ann. Philos.*, 1813, **2**, 259.
27. M. P. Crosland, 'Historical Studies in the Language of Chemistry', Heinemann, London, 1962. Reprinted Dover, 1978.
28. M. Mackaile, in 'Dictionary of National Biography', Vol. 35, ed. S. Lee, Smith, Elder, and Co., London, 1893, p. 115.
29. D. Thorburn Burns, *Anal. Proc.*, 1990, **27**, 202.

30. M. Mackaile, '... Fontium Mineralium Moffatensium', Higgins (for R. Brown), Edinburgh, 1659.

31. M. Mackaile, 'Moffat Well Trans ... as also the Oyly-Well ... at St Catherines Chapel ... to those in subjoyned, a character of Mr Culpeper and his writings', R. Brown, Edinburgh, 1664.

32. M. Mackaile, 'The Diversitie of Salts and Spirits Maintained ... Animadversions upon Dr C. his papers communicated to the RS ... as also Scurvie Alchymie discovered ...' printed J. Forbes, Aberdeen, 1683.

33. A. Bacci, 'De Thermis Libri Septem ...', V. Valgrisium, Venitiis, 1571.

34. I. Zwelfer, 'Pharmacopoeia Augustana ...', [Vienna], 1652.

35. J. Read, *Ambix*, 1961, **9**, 70.

36. C. de Milt, *J. Chem. Educ.*, 1941, **18**, 503.

37. J. Small, *Proc. Soc. Antiq. Scot.*, 1875, **10**, 265.

38. W. D'Avissoni, 'Philosophia Pyrotechnica seu Curriculus Chymiatricus', J. Bessin, Paris, 1633–35, 2nd edn., 1641.

39. Daviffone (trans J. Hellot), 'Les Elemens de la Philosophie de l'Art du Feu ou Chemie', F. Piot, Paris, 1651, 2nd edn., 1657.

40. A. L. Donovan, 'Philosophical Chemistry in the Scottish Enlightenment', University Press, Edinburgh, 1975.

41. J. Golinski, 'Science as Public Culture', Cambridge University Press, Cambridge, 1992.

42. A. Kent (ed.) 'An Eighteenth Century Lectureship in Chemistry', Jackson, Glasgow, 1950.

43. J. Thompson, 'An Account of the Life, Lectures and Writings of William Cullen', 2 Volumes, Blackwood, Edinburgh, 1832–59.

43a. J. Thompson, 'An Account of the Life, Lectures, and Writings of William Cullen', Vol. 1, Blackwood, Edinburgh, 1832–59, p. 46.

44. W. Cullen, in T. Thompson 'A Biographical Dictionary of Eminent Scotsmen', Chambers, Edinburgh, 1870, reprinted, G. Olms, Hildesheim, 1971.

45. W. Cullen, 'Of the Cold Produced by Evaporating Fluids, and of Some Other Means of Producing Cold', Essays and Observations II, 145, Edinburgh, 1756 [reprinted with Black's 'Experiments on Magnesia Alba', W. Creech, Edinburgh, 1782].

46. W. P. D. Wightman, *Ann. Sci.*, 1955, **11**, 154; *Ann. Sci.*, 1956, **12**, 192.

47. M. P. Crosland, *Ann. Sci.*, 1959, **15**, 75.

48. N. Campbell and R. M. S. Smellie, 'The Royal Society of Edinburgh (1783–1983)', Royal Society of Edinburgh, Edinburgh, 1983.

49. W. Ramsay, 'The Life and Letters of Joseph Black, MD', Constable, London, 1918.

50. J. G. Fyffe and R. G. W. Anderson, 'Joseph Black a Bibliography', Science Museum, London, 1992.

51. H. Riddell, *Proc. Rep. Belfast Nat. Hist. Philos. Soc.*, 1919–20, **3**, 49.

52. A. D. C. Simpson (ed.), 'Joseph Black 1728–1799. A Commemorative Symposium', Royal Scottish Museum, Edinburgh, 1982.

53. W. Garvin and D. O'Rawe, 'Northern Ireland's Scientists and Inventors', Blackstaff Press, Belfast, 1993.

54. J. Black, 'De Humore Acido a Cibis orto, et Magnesia Alba ...', G. Hamilton and J. Balfour, Edinburgh, 1754.

55. J. Black, in 'Essays and Observations, Physical and Literary, Read before a Society in Edinburgh, and published by them', Vol. 2, ed. G. Hamilton and J. Balfour, Edinburgh, 1756, p. 157.

56. J. Black, 'Experiments upon Magnesia Alba, Quicklime and other Alkaline Substances', W. Creech, Edinburgh, 1777.

57. J. Black, 'Experiments ...', Alembic Club Reprints, No. 1, Alembic Club, Edinburgh, 1898, re-issued 1963.

58. H. Guerlac, *Isis*, 1957, **48**, 124.

59. J. F. Meyer, 'Chymischen Versuche zur nahren Erkenntniss des ungeloschten Kalchs ...', J. W. Schmidt, Hanover, 1764.

60. D. Macbride, 'Experimental Essays ...', A. Miller, London, 1764.

61. D. Macbride, 'Essais D'Experience ...', P. G. Cavelier, Paris, 1766.

62. N. J. E. von Jacquin, 'Examen Chemicum Doctrinæ Meyerianæ de Acido Pingui et Blackianæ de aere Fixo respectus calcis', J. P. Kraus, Vindobonæ, [Wein], 1769.

63. E. Szabadváry and Z. Szökefalvi-Nagy, 'A Kémia Története Magyarországon', Akad. Kiadó, Budapest, 1972.

64. F. Szabadváry, 'History of Analytical Chemistry in Hungary', in 'Reviews on Analytical Chemistry presented at Euroanalysis II, Budapest', ed. W. Fresenius, Masson, Paris, 1977.

65. A. L. Lavoisier, 'Opuscules Physiques et Chimiques', Durand, Paris, 1774.

66. A. L. Lavoisier, (trans. with notes and an appendix T. Henry), 'Essays Physical and Chemical', J. Johnson, London, 1776.

67. M. Lavoisier to Dr Black, in 'Report of the 41st Meeting of the British Association for the Advancement of Science, held in Edinburgh, 1871', J. Murray, London, 1872.

68. R. G. W. Anderson, 'The Playfair Collection and the Teaching of Chemistry at the University of Edinburgh, 1713–1858', Royal Scottish Museum, Edinburgh, 1978.

69. J. W. Cook, *J. R. Inst. Chem.*, 1953, **77**, 361.

70. H. Guerlac, 'Joseph Black's Work on Heat', in 'Joseph Black (1728–1799)', ed. A. D. C. Simpson, Royal Scottish Museum, 1982.

71. W. A. Cole, 'Manuscripts of Joseph Black's Lectures on Chemists', in 'Joseph Black (1728–1799)', ed. A. D. C. Simpson, Royal Scottish Museum, 1982.

72. J. R. Partington, *Chymia*, 1960, **6**, 27.

73. J. Black, 'Lectures on the Elements of Chemistry delivered in the University of Edinburgh ... published from his manuscripts by J Robison ...', Mundell, Edinburgh, 1803. [American Edn., M. Carey *et al.*, 3 Vols., 1807, 1806, 1806. The printing was not completed until 1807 and to make the work appear up-to-date the title page of Vol. 1 was replaced by a new one dated 1807.]

74. E. Robinson and D. Mackie, 'Partners in Science, Letters of James Watt and Joseph Black', Constable, London, 1970.

75. D. Thorburn Burns, *Anal. Proc.*, 1991, **28**, 402.

76. J. Black, 'Essays on the Construction and Graduation of Thermometers ... A new edition ... with tables of the Different Scales of Heat, exhibited by Dr Black in his Annual Course of Chemistry', ed. G. Martine, W. Creech, Edinburgh, 1792.

77. E. L. Hirst and M. Ritchie, *J. R. Inst. Chem.*, 1953, **77**, 505.

78. W. P. Doyle, 'Chemistry in the University of Edinburgh—Joseph Black', University of Edinburgh, reprinted from University of Edinburgh Journal, summer issue, 1983.

79. A. Crawford, 'Experiments and Observations on Animal Heat, and the Inflamation of Combustible Bodies. Being an Attempt to Resolve these Phenomena with a General Law of Nature', Murray and Sewell, London, 1779.

80. A. Crawford, 'Experiments and Observations on Animal Heat and the Inflamation of Combustible Bodies', 2nd edn., J. Johnson, London, 1788.

81. A. Crawford, *Med. Commun.* (of the Society for Promoting Medical Knowledge), 1790, **2**, 301.

82. O. Mackitie, 'Johan Gadolin and his Contribution to Analytical Chemistry', in 'Euroanalysis IV, Reviews on Analytical Chemistry', ed. L. Niinisto, Akad Kiadó, Budapest, 1982.
83. E. Rancke-Madsen, 'The Development of Titrimetric Analysis till 1806', G. E. C. Gad, Copenhagen, 1958.
84. F. Szabadváry, 'History of Analytical Chemistry', Pergamon, Oxford, 1966.
85. J. Black, *Trans. R. Soc. Edinburgh*, 1794, **3**, 95.
86. Robert Kerr in 'A Biographical Dictionary of Eminent Scotsmen', ed. R. Chambers, new edition by Thomas Thomson, 1870, G. Olms Verlag, Hildesheim, 1971, p. 441.
87. A. L. Lavoisier, 'Elements of Chemistry ... translated by Robert Kerr ... First edition in English', W. Creech, Edinburgh, 1790.
88. C. L. Berthollet, 'Essay on the New Method of Bleaching ... translated by Robert Kerr ...', W. Creech, Edinburgh, 1790.
89. J. Kendall, *Endeavour*, 1942, **1**, 106.
90. J. Kendall, *Nature (London)*, 1947, **159**, 869.
91. D. Thorburn Burns, *Anal. Proc.*, 1988, **25**, 253.

Traceability to the Mole: A New Initiative by CIPM

J. D. Fassett and R. L. Watters, Jr.

CHEMICAL SCIENCE AND TECHNOLOGY LABORATORY, NATIONAL INSTITUTE OF STANDARDS AND TECHNOLOGY, GAITHERSBURG, MARYLAND, USA 20899

1 Introduction

The Bureau International des Poids et Mesures (BIPM) was set up by the Convention du Metre, signed by seventeen States on 20 May 1875. There are now over 40 nations which are members of this Convention. The BIPM operates under the supervision of the Comité International des Poids et Mesures (CIPM). BIPM is chartered to ensure world-wide unification of physical measurements, including carrying out comparisons of national and international standards and ensuring the co-ordination of corresponding measurement techniques. The BIPM is principally noted for the establishment of fundamental standards and scales for measurement of physical quantities and maintaining the international prototypes.

An *ad hoc* Working Group of the CIPM was established in September 1990 'to advise the CIPM on whether or not the BIPM should have a significant role in addressing the problem of providing uniformity and traceability in chemical and physico-chemical measurements'.[1] This action was taken in response to concern that increased international uniformity and accuracy are required for the progress of advanced technology and the recognition that a significant proportion of industrial production and international trade is dependent on analytical chemical measurement. In addition, an increased international concern for the state of the environment (land, water, and air pollution) was recognized as a further driving force for improvement of international uniformity for chemical measurements.

The Working Group has proposed an exploratory programme of co-operative work to be undertaken among leading national chemical metrology laboratories. This programme is designed to test the hypothesis that co-ordinated activity on the analysis of a few key reference materials, using one or two reference methods of wide application, will provide such laboratories with a base from which to extend international comparability to a wider range of methods and reference materials. The Working Group has proposed that the use of primary reference materials of pure elements and internationally accepted atomic weights will provide not only the basis for assessment of comparability, but also will achieve traceability to the mole, the primary SI unit in chemistry.

One of the initial projects is an interlaboratory comparison of a relatively simple matrix using isotope dilution mass spectrometry (IDMS) as the reference method. The experiment was designed so that potential errors due to sample inhomogeneity, chemical matrix effects, chemical blanks, and sample treatment would be minimized at this initial stage. Solutions of several inorganic elements dissolved in dilute acid were prepared as samples. The reference concentrations for the elements in these solutions are tied to the primary standards used in their preparation.

The technique of IDMS was chosen as the reference method because its inaccuracies and imprecisions have been shown to be relatively small, the few interferences are well-characterized and documented, and IDMS is used currently within some national traceability networks to evaluate the accuracy of other chemical analysis methods. Two international intercomparisons of IDMS have been conducted in the nuclear industry where the method has widespread application[2,3] and where the need for uniformity and accuracy in the safeguarding of special nuclear materials is of significant international concern.

The purpose of this report is to describe the status of this CIPM initiative in chemical measurement. The technique of IDMS will be outlined and its basis relative to the kilogram, mole, and atomic weights will be described. Initial results have been submitted, but have not yet been officially evaluated.

2 Isotope Dilution Mass Spectrometry (IDMS)

IDMS[4-6] is based on addition of a known amount of enriched isotope (called the 'spike') to a sample. After equilibration of the spike isotope with the natural element in the sample, mass spectrometry is used to measure the altered isotopic ratio(s). The concentration is directly derived from this ratio. A major advantage of the technique is that chemical separations, if required for accurate ratio measurement, need not be quantitative. In addition, ratios can be measured very reproducibly and, thus, concentrations can be determined very precisely.

The IDMS process is shown diagrammatically in its most general form in Figure 1. It can be seen from this figure that analytical results are traceable only to the primary standards, and the processes of weighing and mass spectrometric isotope ratio measurement. The weighing process ties the technique to the fundamental unit, the kilogram. The mass spectrometric isotope ratio measurement process ties the technique to the atomic weights of the elements, linking mass to the number of atoms of a substance, or mole, the fundamental unit of chemistry.

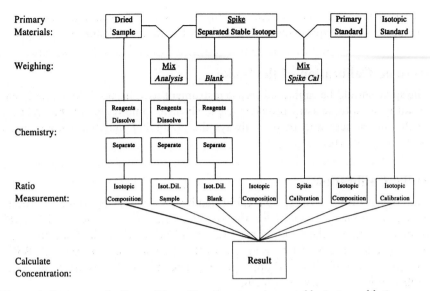

Figure 1 *Summary of all possible calibration steps in a stable isotope dilution mass spectrometric analysis*

3 Calibration Steps in IDMS

Each of the seven vertical lines in Figure 1 represents a possible calibration step and are discussed in order, from left to right, below.

Sample Isotopic Composition

The determination of the isotopic composition of the analyte element in the sample can be omitted for elements which are known not to vary significantly in nature.[7] However, for isotope dilution inductively coupled plasma mass spectrometric (ID-ICP-MS) analysis of complex matrices, we routinely include this step to assess the extent of isobaric interferences and the need for chemical separations to achieve the requisite accuracy.[8]

Isotope Dilution Process

This process includes the mixing of the calibrated spike with the sample and equilibration. Chemical separations or other chemical processing of the sample are not required but may be done to facilitate accurate mass spectrometric measurement.

Blank Calibration

Contamination of the sample during the analysis is possible from the reagents used and from the sample handling steps. An estimate of the blank can be made by repeating all steps of the isotope dilution process, but without the sample.

The blank is often the factor that limits the sensitivity of the technique. Ultratrace chemical analysis requires the use of chemical clean rooms and high purity reagents.

Isotopic Calibration of the Spike

The spike should be calibrated on the instrument used in the IDMS experiment because the isotopic assay provided will likely have some systematic bias relative to this instrument and there is always the possibility that the isotopic assay provided is in error.

Spike Calibration

The concentration of the spike is calibrated by a reverse isotope dilution process against a primary standard for the analyte. From the atomic weight of the primary standard, the molar concentrations of primary standard solutions are defined. These are mixed by weight with the spike for highest accuracy. Spike solutions are stable if stored appropriately, and it is not necessary to do spike calibrations with every experiment.

Isotopic Assay of the Primary Standard

A primary standard by definition should have its isotopic composition established. For the few elements that have a variable isotopic composition in nature, of which lithium and lead are the most notable examples, it is important that the primary standard be isotopically characterized.

Isotope Ratio Calibration

All mass spectrometers have some degree of isotopic discrimination or fractionation that results from the ionization process, ion extraction process, and/or the ion detection process. An isotopic standard can be used to calibrate this fractionation. However, this calibration is not strictly necessary—if the same fractionation occurs for all ratio measurements in the above calibrations, systematic errors due to this effect will cancel, to the first approximation.

4 Isotopic Spikes

The IDMS technique is dependent upon the availability of spike materials. The quality of the spike is characterized by its enrichment relative to the naturally occurring isotopic abundance ratio. This enrichment does not have to be very high to be useful in IDMS. The quality of the spike is reflected in the magnification of the ratio measurement uncertainty to the concentration measurement uncertainty. This magnification increases when the ID ratio approaches either the sample or spike ratio. Thus, when the spike and natural ratios are not considerably different, the range for minimization of this magnification factor is smaller and optimum spiking of the samples becomes more important. This consideration also becomes more important for mass spectrometric methods with reduced ratio measurement precision. When the spike is highly enriched, or not present naturally (such as ^{233}U), the magnification factor is minimal for spike-to-natural ratios spanning several orders of magnitude.

5 Inorganic Mass Spectrometric Techniques

The traditional inorganic mass spectrometric methods for isotope ratio measurement were developed using thermal ionization. This technique is noted for its high precision and also its need for extensive sample preparation. The successful development of ICP-MS, first introduced commercially in 1983, is making mass spectrometry much more widely available to general analytical laboratories. Thus, it was recognized in this interlaboratory exercise that ICP-MS would be the technique likely to be used.

Currently, ICP-MS coupled with IDMS is used extensively at NIST for certification of reference materials.[8,9] A compilation of precisions achieved in spike calibration carried out at NIST during reference material certifications is shown in Table 1. A spike calibration typically represents mixture of the spike with at least two standard solutions prepared from one or more primary standard or high purity material and sampling these solutions in duplicate. The stated precision is the relative standard deviation (1s) for these independent measurements of the spike concentration. It can be seen from the data in Table 1 that measurement precisions of 0.25% relative are routine. The elemental concentrations of these spike calibrations and the relatively simple matrix should reflect the samples prepared in the interlaboratory experiment, described below. Furthermore, the samples provided in this exercise are prepared in a manner very much similar to spectrometric solution standards, typically certified with a total uncertainty of 0.3% relative. A number of these solutions have been examined by IDMS during spike calibration experiments, and in no case has there been

Table 1 *Measurement Precisions for Spike Calibration*

Isotopic Spike	Concentration/$\mu mol\ g^{-1}$	Precision* RSD
^{62}Ni	0.88518	0.06%
	0.008561	0.19%
^{65}Cu	3.15557	0.08%
	0.52213	0.23%
^{67}Zn	1.59907	0.36%
^{97}Mo	0.07604	0.21%
^{109}Mg	1.12712	0.08%
^{111}Cd	0.5589	0.14%
	0.0009799	0.33%
^{135}Ba	0.80244	0.17%
^{206}Pb	32.8477	0.11%
	0.00239	0.07%
^{235}U	0.016584	0.09%

*The relative standard deviation (RSD), 1s, for a minimum of four independent mixes of the spike with primary standards.

a disagreement *versus* the independently prepared primary standard solutions. Thus, it was felt that the experimental design of this initial exercise was extremely robust.

6 Interlaboratory Measurement Experiment

Two solutions were initially prepared for the interlaboratory experiment and each contained three elements. The analyte concentrations were made at the $\mu g\,g^{-1}$ level in these solutions. They were prepared from pure metals or primary standards, and dissolved in high-purity acids and water. The gravimetrically calculated concentrations were taken as the target values. Isotopic spikes were included with the samples to facilitate the IDMS measurements. The appropriate concentrations for the spikes were provided as well as the supplier's isotopic assays. The intent of the experiment was for each laboratory to calibrate the spike *versus* its own primary standard.

The first solution consists of a mixture of Mg, Pb, and Cd. The second solution consists of a mixture of Li, Mo, and Fe. Each solution presents certain measurement challenges. For instance, Fe is not readily measured by ICP-MS because of a large interference from ArO. The natural abundances of Li and Pb isotopes can vary considerably in nature. Isotopic standards for these elements were also included to allow calibration of the atomic weights of these elements. Further, Mo has potential stability and equilibration problems and Li is a light element, which could result in MS fractionation problems. These measurement challenges should not present problems for an experienced mass spectrometry laboratory.

7 Participating Laboratories

Many of the laboratories represented by the members of the CIPM have agreed to participate in the interlaboratory comparison, as well as some additional laboratories which expressed an interest to participate (between 15 and 20 laboratories). Laboratories not equipped to perform IDMS were allowed to participate by using other methods of analysis, however this data will be evaluated separately. The proposed target was 1% relative agreement with the known concentration values of the prepared solutions.

8 Proposed Future Activities

This initial experiment was considered as a first in a hierarchy of chemical measurements. It is recognized that the problem of traceability in chemical measurement derives from the diversity of substances to be analysed and the range and complexity of analytical methods available. It also is recognized that a number of chemical traceability systems have been developed at the national level based on reference materials and reference or validated methods. The present experiment should provide the framework for future work as well as for assessment of the present state-of-the-art for measurement comparability, at least by the technique of isotope dilution mass spectrometry. It is expected that co-operative action by the national metrology laboratories will result in improvement of the state-of-the-art and lead to better levels of comparability for increasingly demanding chemical measurements.

References

1. Comité International des Poids et Mesures, Report of the 79th Meeting, 1990, p. 100.
2. W. Beyrich and E. Drosselmeyer, 'The Interlaboratory Experiment IDA-72 on Mass Spectrometric Isotope Dilution Analysis', KfK 1905/Eur5203e, 1975.
3. W. Beyrich, W. Golly, G. Spannagel, P. DeBievre, W. H. Wolters, and W. Lycke, *Nucl. Technol.*, 1986, **75**, 73.
4. K. G. Heumann, in 'Inorganic Mass Spectrometry', Wiley, New York, 1988, Chapter 7, pp. 301–376.
5. K. G. Heumann, *Int. J. Mass Spectrom. Ion Phys.*, 1982, **45**, 87.
6. J. D. Fassett and P. J. Paulsen, *Anal. Chem.*, 1989, **61**, 643A.
7. *Pure Appl. Chem.*, 1991, **63**, 991.
8. E. S. Beary and P. J. Paulsen, *Anal. Chem.*, 1993, **65**, 1602.
9. E. S. Beary, P. J. Paulsen, and J. D. Fassett, *J. Anal. Atom. Spectrom.*, 1994, in press.

Is there a Tower in Ransdorp? Harmonization and Optimization of the Quality of Analytical Methods and Inspection Procedures

W. G. de Ruig[1] and H. van der Voet[1,2]

[1]DLO—STATE INSTITUTE FOR QUALITY CONTROL OF AGRICULTURAL PRODUCTS (RIKILT–DLO), PO BOX 230, 6700 AE WAGENINGEN, THE NETHERLANDS
[2]DLO—AGRICULTURAL MATHEMATICS GROUP (GLW–DLO), PO BOX 100, 6700 AC WAGENINGEN, THE NETHERLANDS

Summary

The reliability of analytical methods has become a focus of interest in the past decade. A number of international bodies are offering guidelines, protocols, standards, *etc.*, which can be helpful in optimizing and maintaining the quality of analyses. It may be profitable, to consider laboratory activities as a production process, producing relevant information, instead of a series of isolated actions. Errors are inherent in the analytical process. They are not 'mistakes'; however, in relation to effort and costs, they have to be known about. The quality and cost of a method have to be tuned to the requirements of the client. Suitable measures of quality are the risks of false positive and false negative results. New highly sophisticated analytical methods have come into practice; and also simple screening methods are available. The question arises as to how to balance the use of high-tech and simple analytical methods. Under some circumstances, the use of generally described, objective criteria, as opposed to a detailed description of methods, may be preferred. For demonstrating the quality of a method of analysis, the receiver operating characteristic (ROC) curve can be a helpful tool.

1 Introduction

When we consider the plethora of current and draft international standardization documents several questions come to mind:

- How detailed should standardization protocols be?
- How can we deal with the existing situation where there is a high degree of standardization for simple concepts (*e.g.* detection limits in a univariate

setting, within-run and between-lab variability), but little standardization for more complex situations (*e.g.* multivariate detection limits, non-constant covariance matrices with many sources of variability)?

- In what detail should the statistical analysis of results be prescribed?
- How can we formulate regulations that stimulate rather than inhibit improvements in analytical chemical procedures? Inhibition of progress can easily occur if regulations are too specific and method changes are not allowed.
- What is the client role in many of the documents? Many protocols seem to be written in complete isolation from the diverse needs and wants of analytical chemistry clients, *e.g.* legal institutions which need to take binary decisions (guilty/not guilty), scientists who need to compare the measurements on certain groups of samples as precisely as possible, or farmers who get paid proportionally to certain measurement results. It should be realized that a given analytical chemical method may be used to answer very different questions, implying varying requirements and the capability of analytical methodology.

Two main questions arise: (a) Do we do the things properly? (b) Do we do the proper things?

Answers to questions like these can be based on developments described recently.[1,2]

We have formulated the following statements which can be helpful in discussion of these questions.

(i) Consider laboratory activities as a production process.
(ii) Errors are not mistakes!
(iii) The risks of false results are a suitable measure of quality.
(iv) Listen to your clients.
(v) The quality of methods can only be measured after defining quality measures.
(vi) Use generally accepted rules (if applicable).
(vii) Use of quality criteria can be helpful.
(viii) Use ROC curves to demonstrate the quality of an inspection procedure.

2 Consider Laboratory Activities as a Production Process

Basically, it is now recognized that an analytical chemical laboratory is like any other production unit: there are products (analytical data) and customers (those in need of analytical results, either quantitative or qualitative), who pay for these results. Manufacturers of more physical products like, *e.g.* automobiles, have long become aware of key elements for insuring the high quality of their products: products should be manufactured in a clearly defined and stable *production process*, which is subjected to stringent procedures of *quality control* (guaranteeing desired quality to the producer) and *quality assurance* (guaranteeing desired quality to the consumer). Quality has become an important topic in industry where more and more consumers require producers to adopt procedures in accordance to the ISO 9000 series.

Analytical chemical laboratories are also an industry. Our product is 'information', that is the difference in knowledge before and after the chemical analysis. Decision makers in industry and government base their strategies on analytical results. The consequences can be immense.

Thus we have to transfer the ideas of general quality control and assurance to analytical chemistry. One of the main factors which will change many of the existing operating procedures is that analytical data should be produced in a process, not as an isolated activity. In practical terms, this means that a certain minimum volume of samples per unit of time have to be analysed in order to permit quality control and assurance procedures to operate. The traditional order in operating procedures: (a) prepare apparatus and chemicals; (b) analyse; (c) calculate results; is better replaced by something like: (a) maintain analytical process in a state of control; (b) while permanently checking this with quality control samples, analyse customer samples and quality assurance samples using a well-defined time scheme.

3 Errors are Not Mistakes!

Not because the method, or because the operator, is invalid but inherent to all analytical procedures is that we make 'errors'. We have not to blame ourselves for making such errors, for they are due to the uncertainty in the method. Absolute certainty does not exist. However, it is our duty to give insight into these uncertainties. The question: 'what is the real amount μ of an analyte in a sample?' is answered by two figures: an amount X, which is the best estimate of μ, and a statistical quantity to express the uncertainty of X. It is common practice to use the standard deviation, S for the latter.

A useful starting point is a statistical model in which the observed value X is seen as the sum of the true value μ, a possible systematic error (bias), and a number of random errors

$$X \;=\; \mu \;+\; \delta \;+\; \varepsilon \tag{1}$$

$$
\begin{array}{llll}
\text{test} & \text{true} & \text{systematic} & \text{random} \\
\text{result} & \text{value} & \text{error} & \text{error} \\
 & \text{(bias)} & & \\
 & & \text{trueness} & \text{precision} \\
 & & \llcorner\!\!\text{---}\text{ accuracy }\text{---}\!\!\lrcorner &
\end{array}
$$

The terms trueness, precision, and accuracy are defined in the new ISO 5725-1 (for references, see Section 7). In its predecessor ISO 5725-1986 no term for the systematic error was provided, but it was described as 'accuracy of the mean'.

The disribution of ε is often approximated by a normal distribution. Assuming such distribution, it is easy to make calculations, *e.g.* on confidence intervals. If the real situation does not differ too much from the mathematical formula, the conclusion will not differ too much from reality.

The 'errors', or may be better 'deviations', have causes. Some sources of variability are: time, operator, equipment, calibration, laboratory. Bias can be caused by matrix effects. Basic methods for estimating the *trueness* of a test method are presented in ISO 5725-4. The *precision* can be described in terms of repeatability and reproducibility; this is done in ISO 5725-2.

Repeatability conditions are defined as conditions where mutually independent test results are obtained with the same method on identical test material in the same laboratory by the same operator using the same equipment within short intervals of time.

Reproducibility conditions are conditions where test results are obtained with the same method on identical test material in different laboratories with different operators using different equipment. In fact, these are two extreme conditions: the repeatability describes the minimum variability due to random errors and the reproducibility the maximum variability. In laboratories there is a need for inbetween estimates of the random errors. Therefore, in the new ISO 5725-3 intermediate measures on the precision of a test method are introduced. It establishes intermediate precision measures to take account of changes of observation conditions (time, operator, equipment, recalibration) within one laboratory.

4 The Risks of False Results are a Suitable Measure of Quality

In a population to be inspected, a major part (hopefully) is in reality fulfilling the requirements; this is denoted as the *real negative* (RN). The part that does not fulfil the requirements is the *real positive* (RP). However, due to the uncertainty of the analysis, there is a probability that part of the real negative is found to be positive and part of the real positive is found negative. Therefore, there is not a one-to-one correspondence between the inspection result: 'rejected' and 'accepted' and the real situation, Figure 1.[3]

The quality of the inspection procedure can be expressed in the probability of false positive (FP) in the acceptable part, $\alpha = \mathrm{FP/RN}$ and the probability of false negative (FN) in the rejectable part, $\beta = \mathrm{FN/RP}$. In Section 9 we will explain, how these parameters can be used in a practical way.

Traditionally analytical methods have been characterized by, *e.g.* limit of detection, limit of quantification, and repeatability and reproducibility at some specified levels. These concepts are not relevant if the customer only wants to know if a substance is present above the maximum permissible value—and the probabilities of false positive or false negative associated with such a decision. In the case of analytical data produced for their use in inspection procedures it is better to characterize the analytical method by its ability to prevent inspection errors of both kinds. Thus important changes have to take place by adapting quality assessments which correspond to the form of the analytical data as needed by the customer.

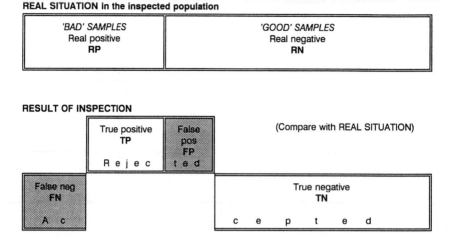

Figure 1 *Relation between the real situation and result of inspection.* $\alpha = FP/RN$; $\beta = FN/RP$. *Shaded area = false results*

5 Listen to Your Clients

Analytical chemists have to be more flexible with regard to what constitutes a 'good' performance. Typical criteria from the analytical viewpoint (How low can we go? How precise can we be?) may be of no interest for a specific client. In many cases the question of 'good enough' performance should be left to the customers, although analytical chemists will be obliged by their clients to provide adequate data in an adequate form to allow proper decision making by the client.

For example, we can detect the presence of hormones by a GC–MS method, with a capacity of 10 samples a day and costs of 2000 ecu per sample. A rapid, cheap immuno assay or receptor assay, with a throughput of hundreds of samples per day, can cost 50 ecu per sample. For the same price, 400 samples analysed by a simple assay may control the whole population of cows much better than 10 samples analysed by GC–MS.

Another example is that of the scientist–customer who has performed a randomized block experiment (with, *e.g.* tomato plants), and is interested in any chemical differences between several types (*e.g.* control samples *versus* samples after genetic manipulation). Their interests are best served if the samples from the tomato plants are analysed using the same randomized block scheme, so that tomato types can be compared within blocks, where analyses in the same block are performed under repeatability conditions. Note that repeatability is a key criterion for this scientist, rather than reproducibility or even trueness.

The previous examples may look unattractive to analytical chemists because responsibility is transferred to the customer. However, on other points the flow is the reverse. Clients of analytical data should not in general become involved with the technical aspects of the measurement process. They purchase analytical data on their samples at an agreed level of assured quality; it is the responsibility of analytical chemists to accomplish this. It is the individual analytical chemist in each laboratory who is responsible for selecting the method and conducting the analysis, the sample preparation method, the calibration scheme, the number of replicates, *etc.* This is a very important aspect of the problem: firstly, it allows for a considerable reduction in the size of international standard protocols; secondly, it presents opportunities for improvements to be made without violating existing protocols.

Analytical chemists should also be concerned about which samples they analyse. What guarantee can they give for the analytical result from a sample about which they know absolutely nothing? Can any kind of interference be excluded? In practice any claim about performance will always be linked to a population of samples, *e.g.* veal calf urine samples from a certain geographic origin. This population should be clearly defined for any validated method. Population variability (in the example calf urine matrix effect variability) should be properly represented in the quality control and assurance samples, for example by a random sampling mechanism.

The main conclusion is that the interaction between analytical chemists (as producers) and their customers will result in contracts which include not only the cost of the analyses, but also the quality that will be assured for a clearly defined population of samples. Quality can be formulated in many ways, and choosing certain criteria that are relevant for the customer will be part of the contract. The analytical chemical laboratory will be responsible for the whole analytical process including quality control and the analysis of quality assurance samples. Standard operating procedures will be an intralaboratory matter, and quality will be a property of a certain laboratory and its chemists rather than of an analytical method *per se*.

6 The Quality of Methods Can Only Be Measured after Defining Quality Measures

When you want to know 'what the quality of our method is', *e.g.* what is its limit of detection; in a ringtest, what outliers are, *etc.*, do not hope that mathematicians and statisticians can present the solution, they cannot! First, clear definitions should be agreed upon, that are applicable to your problem; thereafter calculations can be made. When generally accepted definitions exist, it is wise to use them, particularly when they are implemented in harmonized protocols and these protocols reflect your aim. However, they give no absolute answer to how good the analysis is but, only to 'what is the result according to the stated, arbitrarily fixed definition?'

Therefore you have not to be astonished, when different tests give different results, *e.g.* when a result in a ringtest is denoted as an outlier according to the Dixon test and it is not an outlier according to the Grubbs test. In general it is not justified to reject an outlier without more information.

Concepts like 'limit of detection' and 'limit of determination' have been defined in numerous ways in the literature, so what is discussed appears as the limit of confusion! The practical meaning of all these definitions is: when the real concentration is very low, the precision becomes too low, so that test results are no longer acceptable for making inspection decisions.[4] What is 'too low', has to be specified in terms of a maximum acceptable standard deviation of the test results, as agreed with your customer. A good review of detection limit definitions has been written by Currie.[5]

Notice that 'limit of detection', *etc.* refers to an inspection or a control procedure. In the case of testing a method by a collaborative study, it is best that all primary values are reported, without regard to any assumed limit of detection. On the contrary, one goal of such a study may be to determine a limit of detection!

In modern (ISO) standards a clause on the precision of the method, giving values for repeatability, r and reproducibility, R is obligatory.[6] However, we have to be aware, that variability is a dynamic process rather than a static value. Thus, the reported values for r and R are not of eternal value. They have been found once, in a particular collaborative study. Another collaborative study on the same method, whether or not with the same laboratories, will result in other values of r and R. We advise, that the precision of test methods is monitored, *e.g.* by quality control charts.

Another misleading idea is, that a *significant* difference is the same as a *real* difference. In Figure 2a two levels, μ_1 and μ_2 are tested and the observed results are indicated by dots. Statistical analysis will indicate a significant difference. In Figure 2b the levels μ_3 and μ_4 are tested, but the method of analysis is much less precise and the observed difference is not significant. However, in reality the levels and differences were exactly the same! It is a question of *a priori* definition which difference between the levels is to be considered practically relevant. For any defined relevant difference standard statistical computation will then indicate sample sizes large enough to find such differences with a high probability!

7 Use Generally Accepted Rules (If Applicable to Your Aim)

A plethora of international standards, guidelines, protocols, *etc.* for testing and controlling the quality of measurement data in laboratories exist. Some of the more important are listed below.

A series of Harmonized Protocols have been and continue to be evolved[7-9] under the auspices of the International Union of Pure and Applied Chemistry, the International Organization for Standardization, and AOAC International.

A number of international standards and guides deal with the quality assurance in analytical chemical laboratories.

The most important and widely accepted international quality standard for testing laboratories is the ISO Guide 25 on General Requirements for the Competence of Calibration of Testing Laboratories.[10]

The International Standard series ISO 9000[11-15] concerns quality assurance in relation to a product, production process, or service and is particularly intended for industrial companies. It is focused on an efficient and reliable

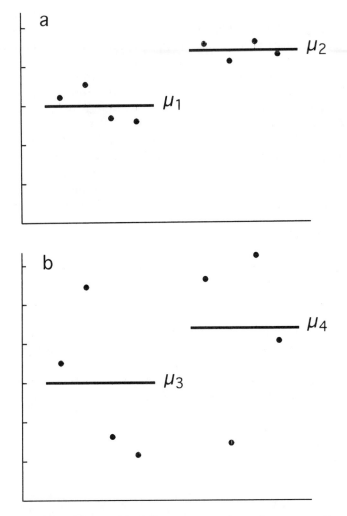

Figure 2 *Statistical significance. The difference $\mu_2 - \mu_1$ is equal to $\mu_4 - \mu_3$. By considering the precision, the difference between μ_2 and μ_1 is significant, while that between μ_4 and μ_3 is not*

management of affairs and is widely accepted in industry on a voluntary base. Laboratory evaluation is not a primary aim.

The European Standard series EN 45000,[16-18] partially based on ISO Guide 25, is aimed at quality assurance of information produced by test laboratories. As far as the requirements of a laboratory quality assurance system are concerned, EN 45000 and ISO 9000 have the same intention. In contrast with the latter, EN 45000 also considers the competence of the laboratory. Laboratory 'accreditation' according to EN 45000, guarantees that the test laboratory is able to fulfil the requirements for an adequate methodology of specified tests. Generally it is voluntary, in some EU countries necessary, when laboratories are designed as a 'notified body', obligatory, and in some countries legally required.

The target group for 'good laboratory practice' (GLP) are laboratories who perform (non-clinical) safety studies, aimed for the submission of dossiers for registration by national licensing bodies. In the seventies, the US Food and Drug Administration (FDA) evaluated a number of dossiers; it appeared that too much of the research had been performed and reported so carelessly that the reliability of the results was doubtful. Several instances of fraudulent research were also discovered. These incidents led the FDA to establish generally applicable rules[19] which passed into US law in 1979.[20] This meant that all safety research, *i.e.* technological studies and research into licensing of animal and human drugs, additives, and cosmetics had to be performed according to these rules of GLP. Likewise, the US Environmental Protection Agency (EPA) has formulated GLP regulations, intended for laboratories involved in environmental research.[21,22]

A Code of Good Laboratory Practice was also drawn up by the Organisation for Economic Cooperation and Development (OECD).[23,24] The OECD modified the FDA guidelines to match the administrative practices of the various member states. They are presently being used in a large number of industrialized countries.

The European Union implemented the OECD principles as a European standard.[25]

The International Standard ISO 5725-1986[26] is focused on measuring of the precision of test methods by repeatability, r, and reproducibility, R. However, it is now in a process to be replaced by a new standard ISO 5725,[27–31] that is split up into parts concerning also 'trueness' and 'intermediate measures of precision', as discussed already in Section 3.

Figure 3 *Photograph of Ransdorp with tower. The tower signal fulfils all criteria. Conclusion: The photo proves that a tower is present: true positive*

8 Use of Criteria Can Be Helpful

The underlying idea of precisely describing methods and validating them by collaborative tests is, that unequivocal description is a guarantee for similar results. However, this presumption is not true. Two cooks using the same recipe will make different cakes. It is well known, that even in the same laboratory, two identical instruments can give different results. Highly sophisticated methods like GC–MS cannot be described in a generally applicable way, because each laboratory arrangement will have specific peculiarities to be dealt with. Moreover, an exact description of method will inhibit improvements. The use of generally described criteria may be preferred to a detailed description of methods.[32]

In other fields the same approach is used. For example, a pass photograph has to fulfil certain criteria, but it is not described how such a photo has to be obtained (the choice of camera, film, exposure, and development), only the result counts. As a visual example, in Table 1 criteria are presented that prove

Table 1 *Criteria for proving the presence of a tower by a photo*

  (i)   Height object $> 2 \times$ surrounding objects (a tower shall be higher than the surrounding buildings)

 (ii) Height object $< 20 \times$ width object (to exclude chimney shafts)

(iii) Number of windows < 100 (to exlude offices or apartment buildings)

(iv) $10 \times$ height object $>$ height photo (to set a minimum signal requirement)

 (v) Object not obstructed by interfering signals

Figure 4 *Tower signal interferes with tree signal. Photo does not fulfil criterion (v), thus does not prove the presence of a tower: false negative*

Figure 5 *Very low signal. Photo does not fulfil criterion (iv) and thus does not prove the presence of a tower: false negative*

Figure 6 *A photo of Ransdorp without any tower signal. However, this photo does not prove that there is no tower in Randsorp*

Figure 7 *Fernsehturm in (East-)Berlin. This situation was not expected, when the criteria were stated. The TV tower does not fulfil criterion (ii). If we want to include such buildings in the concept 'tower', we have to adapt the criteria*

the presence of a tower in a picture. Let us assume that these criteria were laid down by experts in this field and are generally accepted. Then, Figure 3 proves the presence of a tower in the village of Ransdorp. But Figure 4 does not, because of interference of the tower signal with that of a tree: it is not fulfilling criterion (v). Similarly, Figure 5 is not fulfilling criterion (iv): the tower signal is too small. Both are false negative. A picture of a chimney will not fulfil criterion (ii): true negative. However, a picture without any tower (Figure 6) does *not* prove the absence of a tower in the village, may be a wrong part has been photographed. Thus, a negative result never proves the *absence* of the sought analyte! Even after having laid down the best criteria, later on new circumstances can arise, which force adaptation of the criteria. For example,

Figure 8 *Gedächtniskirche in (West-)Berlin. The building does not fulfil criterion (iii). The existence of towers comprised of windows only was also not expected beforehand*

do we want the Fernsehturm in (East-)Berlin (Figure 7) to be included in our idea about a tower? And in formulating criterion (iii), we have not realized that the tower at the Gedächtniskirche in (West-)Berlin (Figure 8) comprised of windows only. And what about 'indirect' pictures? A picture of a shadow of a tower, does that prove the presence of a tower? Or a drawing by Rembrandt (Figure 9)? In the EC decisions[33] it is stated, that 'reference methods of analysis must be based on at first molecular spectrometry providing direct information concerning the molecular structure of the substance under examination; or if not possible a combination of independent procedures providing indirect information concerning the molecular structure of the substance under examination'.

Figure 9 *'The church of Ransdorp in Waterland'. Pen drawing of the tower of Ransdorp, made by Rembrandt, about 1652/53, Oxford Ashmolean Museum. Is such a picture included in our criteria? Unexpected problems can arise when criteria are performed in practice*

Sharper criteria will lower the risk of false positives, but enhances the risk of false negatives. Conversely, by looser criteria less false negatives, but more false positives can be expected, as illustrated in Figure 10. By improving the method (at additional costs, *e.g.* by extra pre-purification), hopefully the number of false positives as well as that of the false negatives will be lowered.

Other criteria will result in other conclusions; in other costs as well.

What is the 'right criterion'? There is not one right one. Each protocol leads to a distinct probability of false negatives and a probability of false positives. Your client has to point out his requirements. But you have to make clear what is the quality of a given procedure, expressed in terms of false negative results with respect to the real positives ($\beta = FN/RP$) and the false positive results

SHARPER CRITERIA
This is expected to result in less false positive and more false negative results:

RESULT OF INSPECTION

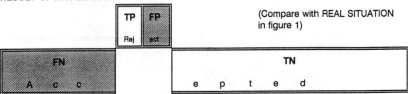

(Compare with REAL SITUATION in figure 1)

LOOSER CRITERIA
This is expected to give less false negative and more false positive results:

RESULT OF INSPECTION

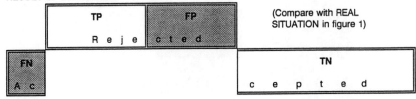

(Compare with REAL SITUATION in figure 1)

IMPROVING THE METHOD OF ANALYSIS
By improving the method of analysis, both the false positive and false negative results may be reduced:

RESULT OF INSPECTION

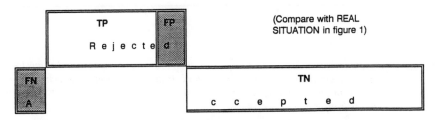

(Compare with REAL SITUATION in figure 1)

Figure 10 *The effects of sharper criteria, looser criteria, and of improving the method of analysis, on the result of inspection*

with respect to the real negatives ($\alpha = FP/RN$).

Are these concepts contradictory within the goal of harmonization? In our opinion this is not the case. Harmonization protocols for quality control and quality assurance are essential for guaranteeing the quality of the information about quality. However, the exact specification of quality criteria, should be a matter of consent between consumer and producer on the analytical market.

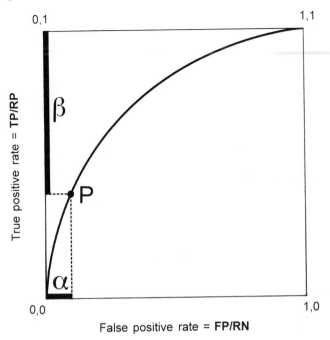

Figure 11 *Receiver operating characteristic (ROC) curve. For point P the probabilities of α and of β are heavy lined*

9 Use ROC Curves to Demonstrate the Quality of an Inspection Procedure

Insight into the probabilities on false positive and false negative results can be obtained from the so called receiver operating characteristic (ROC) curves, Figure 11.[34] Here the *true positive rate, i.e.* the probability of a true positive result (TP) under the real positives (RP) is plotted against the *false positive rate, i.e.* the probability of false positives (FP) under the real negatives (RN). Here $\alpha = FP/RN$. As β is defined as FN/RP and $FN + TP = RP$, $(1 - \beta) = TP/RP$.

Extremes are point $(0,0)$: all samples are denoted to be 'negative', that provides a 100% guarantee, that no 'good' sample will be disapproved: $\alpha = 0$, but also all 'bad' samples are approved: $\beta = 1$. In point $(1,1)$ all samples will be disapproved; thus providing a 100% guarantee that no 'bad' sample will be approved: $\beta = 0$, but also all 'good' samples are disapproved: $\alpha = 1$. Both cases are highly undesirable, however, the ideal case, point $(0,1)$: $\alpha = 0$ and $\beta = 0$, will never be reached. Generally, α has to be very small. But one has to be aware, whether the consequence will not be that β will become unacceptably high. Notice that all curves include the points $(0,0)$ and $(1,1)$, in other words

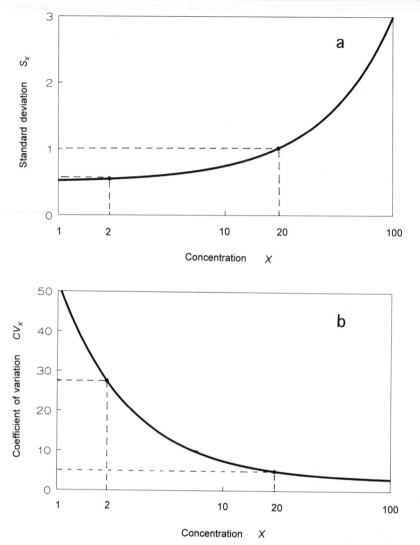

Figure 12 *Standard deviation and coefficient of variation as a function of the concentration*

by imposing an infinitely strict criterion on the ability to obtain a positive result, a positive result will never be found, and by an infinitely loose criterion, only positive results are found.

The precision of test results usually depends on the concentration of the analyte in the sample. As a general rule it can be stated that for many determinations for higher concentrations the standard deviation, S_x increases approximately linearly with the concentration. For very low concentrations a constant noise remains so that the standard deviation becomes constant (Figure

12a). Consequently, the coefficient of variance or relative standard deviation, $CV_x = (S_x/X) \times 100$, is constant at higher concentrations (as the ratio S_x/X is constant) and increases at lower concentrations (as although S_x is constant, X is decreasing) (Figure 12b).

The relationship can be described by a straight line with a positive intercept:[4]

$$S = a + b\mu \tag{2}$$

The closer the concentrations in the real positives and real negatives are to the stated limit, L, the higher the probability of false results.

The choice of a decision criterion, D, is not a matter for the public analyst or of statistics, but has to be judged by the authorities, just as, and together with, the choice of a maximum (or minimum) tolerance level, L. The analyst can provide an estimate of the precision at the level of interest, which can act as a basis for choosing an appropriate decision level.

As an example, the ROC curves will be calculated for a 'high' and for a 'low' maximum permissible level L for the case $a = 0.5$ and $b = 0.025$ in Equation (2); thus

$$S_x = 0.5 + 0.025X \quad \text{and} \tag{3}$$

$$CV_x = [(0.5 + 0.025X)/X] \times 100 \quad (\%) \tag{4}$$

'High' Maximum Permissible Level: $L = 20$

From the Equations (3) and (4) we see in Figure 12 that at this level $S_x = 1.0$ and $CV_x = 5.0\%$, which is quite good.

Let us assume that all 'good' samples have a real content equal to this maximum permissible level (MPL): $c_{RN} = 20$ (worst case, 'filling up to the tolerance') and that the real content in the 'bad' samples is $c_{RP} = 23$. In Figure 13 the real concentrations and the graphs for the distribution of the test results for $\mu = 20$ and 23 are presented. These are calculated from (3). The fraction false negatives, β, and false positives, α, depends on our choice of the decision limit, D. In Figure 13, β and α are indicated when for the limit of decision is chosen $D = 22$. By 'moving' D along the x-axis, other values for β and α are found. By plotting $(1 - \beta)$ against α the ROC curve for $c_{RN} = 20$ and $c_{RP} = 23$ is obtained, Figure 14, curve A. In this figure the ROC curves for $c_{RN} = 20$, $c_{RP} = 22$ (curve B) and for $c_{RN} = 20$, $c_{RP} = 21$ (curve C) are also given. From the figure, it is obvious, that the larger the difference between 'good' and 'bad' samples is, the better the results are obtained. This is felt intuitively, but by determining the standard deviations, and from that calculating the corresponding ROC curves, you can quantify the feeling.

'Low' Maximum Permissible Level: $L = 2$

At this level $S_x = 0.55$ and $CV_x = 27.7\%$ (see Figure 12), which is rather bad. Let us assume that all 'good' samples have a real concentration 2.0 and all

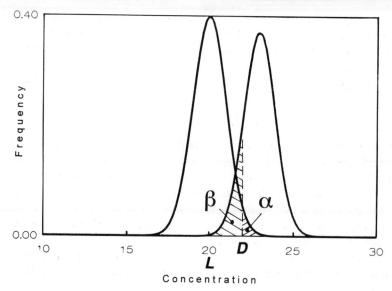

Figure 13 *Distributions of test results at C = 20 and C = 23. Maximum permissible level L = 20. Probability of false positive and false negative results for decision level D = 22*

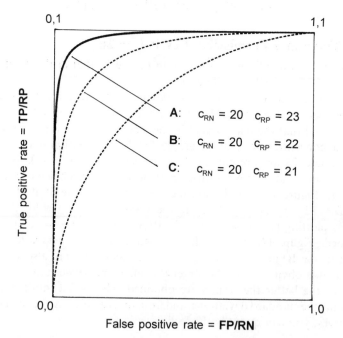

Figure 14 *ROC curves, where the maximum permissible level is 20 and the concentration of all real negative samples is 20 and the concentration of all real positive samples is 23 (curve A), 22 (curve B), and 21 (curve C), respectively. The greater the difference between the distribution of the real negatives and that of the real positives, the better the ROC curve*

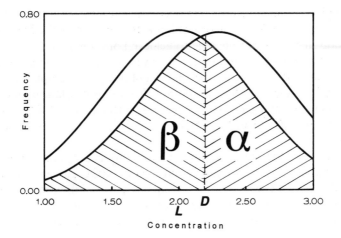

Figure 15 *Distributions of test results at C = 2.0 and C = 2.3. Maximum permissible level L = 2.0. Probability of false positive and false negative results for decision level D = 2.2*

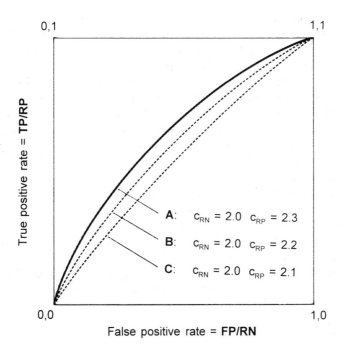

Figure 16 *ROC curves, where the maximum permissible level is 2.0 and the concentration of all real negative samples is 2.0 and the concentration of all real positive samples is 2.3 (curve A), 2.2 (curve B), and 2.1 (curve C), respectively.*
Here the ROC curve is very bad in all cases and hardly becomes better when the difference between the distribution of real negatives and that of real positives increases

'bad' samples 2.3, then the distributions of the test results are as given in Figure 15 and the resulting ROC curve in Figure 16. Because of the high coefficient of variation, here it is not possible to achieve both low β and low α.

These are hypothetical situations. However, for real situations the approach will be the same: determine the distribution of the 'good' samples. Do the same for the 'bad' samples. See how far both distributions overlap. By moving your decision point, D, along the x-axis, you can calculate α and β for each X. By plotting the $(1 - \beta)$ values against the α values, you are 'walking' along your ROC curve.

As already pointed out in Section 6, variability is a dynamic process rather than a static value. The variability, and thus the ROC curves and the risks of false results, β and α, which are based upon the variability, may vary in time. Therefore, ROC curves cannot be determined once for ever, but have to be followed in the time, in the framework of the normal routine inspection procedure.

10 Conclusions

It is the authors' view, that to make clear what the quality of an inspection procedure is, the following has to be done:

(i)   Make or collect samples in which the analyte content is:
      —at the maximum or minimum (as appropriate) permissible level;
      —at some non-permissible levels chosen such that a certain substantial fraction of false results is expected.
      This has to be done under realistic conditions. Repeatability conditions will give too optimistic results. If one ROC curve has to be established for the inspection procedure in all laboratories simultaneously, reproducibility conditions should be used. It will often be more sensible to establish ROC curves for each laboratory or apparatus separately; in that case appropriate 'intermediate' conditions should be used (see ISO 5725-3).
(ii) Analyse these samples in sufficient replicates, *e.g.* 20 times.
(iii) Calculate the standard deviations of these test results per level. Alternatively, a model relating standard deviation to real content can be fitted.
      From the standard deviations, construct the ROC curves of interest.
(iv) The ROC curves give the information, the client (*not* the analytical chemist!) needs to make a justified decision.
(v)   If the offered quality does not fulfil the requirements of the client, then it can be decided, to try to develop a better procedure, with a better precision.

Appendix

The Truncated Tower of Ransdorp

On the flat polder landscape, north of the Dutch capital Amsterdam, the heavy tower of the small village of Ransdorp is an imposing landmark that can be seen from afar. Against common usage in Holland, this tower does not have a spire, which sets one guessing: why?

It is said, that those of Ransdorp did not have enough money to finish their tower. Or, that the weak peat-bog could not carry such a heavy building. But that's too simple. Realize that Ransdorp is not far from the former Zuyderzee. Also that there are two other places with such flat-topped towers, Elburg and Muiderberg. Three truncated towers at the south rim of the Zuyderzee, that must have a common cause, must'nt it?

The truth is as follows.

We have to go back to olden times. Nowadays, a major part of the former sea is reclaimed land, and the remainder is separated from the North sea by a strong dike. However, formerly, the Zuyderzee could be a wild and dangerous water. Almost in the middle is Urk, a fishing village, in the reclaimed land now, but then a tiny island. In the olden times the Urk people were fishermen too—when there was nothing better to do. But they could be redoubtable pirates and many ships have fallen into their hands.

So, long, long ago, it happened that a ship enters the Zuyderzee. It was an awe-inspiring colossus: twenty men could not encompass its masts, and to get from port to starboard took half an hour. Surely, it was impudent of the captain, to come so close to the men of Urk.

As soon as they perceived the enormous prey, they smelt a rich booty and started their attack. But the captain of the giant ship kept his wits. He skilfully manipulated sails and helm. Yet the Urks, with their cockle-shell boats, surrounded him, trying to make the ship run aground on the sand banks to the south of the Zuyderzee, where they could make short work of their prey. They appeared close to success, but then, suddenly, the captain tacked, he put the ship about, and set all sails. Gathering momentum, he sailed into the midst of the astonished Urks—out of the Zuyderzee, into the expanse of the North sea and the ocean.

Where the ship came from nobody knows, nor why it entered into the Zuyderzee, nor where it went. Nobody has ever seen this mysterious ship again.

But, you shall ask, what has this ship to do with the truncated towers of Ransdorp, Elburg, and Muiderberg? Well, I told you before that the ship was close to the coast. Also that it was an enormous ship. It was so gigantic, that the bowsprit rose far above the land. And so it happened, that when they put the ship about, this bowsprit swung round over the landscape and in one enormous swing it cut off the spires on all three towers of the coastal villages.

This is the truth about the origin of the flat tower of Ransdorp, the true history, as my father told me when I was a little boy.

W. G. de Ruig

References

1. M. B. Carey, *Int. Stat. Rev.*, 1993, **61**, 27–40.
2. 'Quality control of analytical data produced in chemical laboratories'. Draft protocol, developed by the IUPAC/ISO/AOAC working party, 1993, ISO/REMCO N271.

 3. W. G. de Ruig, F. A. Huf, and A. A. M. Jansen, *Analyst*, 1992, **117**, 425.
 4. Analytical Methods Committee, Analytical Division, Royal Society of Chemistry, *Analyst (London)*, 1987, **112**, 199–204.
 5. L. A. Currie (ed.), 'Detection in analytical chemistry. Importance, theory and practice', American Chemical Society, Washington DC, 1988.
 6. 'Chemistry—Layouts for standards. Part 2: Methods of chemical analysis', ISO/DIS 78-2. International Organization for Standardization, Geneva, 1991.
 7. W. Horwitz, *Pure Appl. Chem.*, 1988, **60**, 855.
 8. W. D. Pocklington, *Pure Appl. Chem.*, 1990, **62**, 149.
 9. M. Thompson and R. Wood, *Pure Appl. Chem.*, 1993, **65**, 2123.
10. 'General requirements for the competence of calibration and testing laboratories'. ISO Guide 25, International Organization for Standardization, Geneva, 1990.
11. 'Quality management and quality assurance standards—Guidelines for selection and use'. ISO 9000: 1987 = EN 29000. International Organization for Standardization, Geneva, 1987—CEN/CENELEC Organisation Commune Européenne de Normalisation, Brussels, 1987.
12. 'Quality systems—Model for quality assurance in design/development, production, installation and servicing.' ISO 9001: 1987 = EN 29001. International Organization for Standardization, Geneva, 1987—CEN/CENELEC Organisation Commune Européenne de Normalisation, Brussels, 1987.
13. 'Quality systems—Model for quality assurance in production and installation.' ISO 9002: 1988 = EN 29002. International Organization for Standardization, Geneva, 1988—CEN/CENELEC Organisation Commune Européenne de Normalisation, Brussels, 1988.
14. 'Quality systems—Model for quality assurance in final inspection and test.' ISO 9003: 1987 = EN 29003. International Organization for Standardization, Geneva, 1987—CEN/CENELEC Organisation Commune Européenne de Normalisation, Brussels, 1987.
15. 'Quality management and quality system elements—Guidelines.' ISO 9004: 1985 = EN 29004. International Organization for Standardization, Geneva, 1985—CEN/CENELEC Organisation Commune Européenne de Normalisation, Brussels, 1985.
16. 'General criteria for the operation of testing laboratories'. EN 45001, 1989. CEN/CENELEC Organisation Commune Européenne de Normalisation, Brussels, 1989.
17. 'General criteria for the assessment of testing laboratories'. EN 45002, 1989. CEN/CENELEC Organisation Commune Européenne de Normalisation, Brussels, 1989.
18. 'General criteria for laboratory accreditation bodies'. EN 45003, 1989. CEN/CENELEC Organisation Commune Européenne de Normalisation, Brussels, 1989.
19. Good Laboratory Practice Regulations; Final Rule. US Food and Drug Administration. Federal Register 52, 33782, 4 September 1987 (update).
20. Good Laboratory Practice in Non-clinical Laboratory Studies. US Food and Drug Administration. Title 21, Code of Federal Regulations, Part 58, (21 CFR 58), 1979.
21. Good Laboratory Practice Standards. Federal Insecticide, Fungicide, and Rodenticide Act (FIFRA). US Environmental Protection Agency. Title 40, Code of Federal Regulations, Part 160, (40 CFR 160), 1989.
22. Good Laboratory Practice Standards. Toxic Substance Control Act (TSCA), US Environmental Protection Agency. Title 40, Code of Federal Regulations, Part 792, (40 CFR 792), 1987.

23. GLP in testing chemicals. Final report of the OECD Expert Group on Good Laboratory Practice (ISBN 92-64-112367-9), Paris, 1982.
24. OECD Principles of Good Laboratory Practice. Decision of the Council of 12 May 1981 concerning the mutual acceptance of data in the assessment of chemicals, C(81)30. Organization of Economic Cooperation and Development, Paris, 1981.
25. Council Decision of 28 July 1989 on the acceptance by the European Economic Community of an OECD decision/recommendation on compliance with principles of good laboratory practice (89/569/EEC). *Off. J. EEC*, 1989, **L 315**, 1.
26. 'Precision of test methods—Determination of repeatability and reproducibility for a standard test method by inter-laboratory tests'. ISO 5725: 1986. International Organization for Standardization, Geneva, 1986.
27. 'Accuracy (trueness and precision) of measurement methods and results—Part 1: General principles and definitions'. ISO 5725-1: 1994(?). International Organization for Standardization, Geneva (in press).
28. 'Accuracy (trueness and precision) of measurement methods and results—Part 2: A basic method for the determination of repeatability and reproducibility of a standard measurement method'. ISO 5725-2: 1994(?). International Organization for Standardization, Geneva (in press).
29. 'Accuracy (trueness and precision) of measurement methods and results—Part 3: Intermediate measures on the precision of a test method'. ISO 5725-3: 1994(?). International Organization for Standardization, Geneva (in press).
30. 'Accuracy (trueness and precision) of measurement methods and results—Part 4: Basic methods for estimating the trueness of a test method'. ISO 5725-4: 1994(?). International Organization for Standardization, Geneva (in press).
31. 'Accuracy (trueness and precision) of measurement methods and results—Part 6: Practical applications'. ISO 5725-6: 1994(?). International Organization for Standardization, Geneva (in press).
32. W. G. de Ruig, R. W. Stephany, and G. Dijkstra, *J. Assoc. Off. Anal. Chem.*, 1989, **72**, 487.
33. Commission Decision laying down the reference methods for detecting residues (93/257/EEC), *Off. J. EEC*, 1993, **L 118**, 75.
34. 'Criteria to limit the number of false positive and false negative results for analytes near the limit of determination', Codex Alimentarius Commission, CX/MAS 92/15, Rome, 1992.

Recent European Initiatives to Improve the Quality of Food Analysis

M. C. Walsh

STATE LABORATORY, DUBLIN 15, REPUBLIC OF IRELAND

1 Introduction

In recent years, a number of European organizations whose primary objective is 'quality of measurement', have been inaugurated. Many of these organizations are concerned with the quality of measurement in the food sector and this paper will describe some of the recent initiatives undertaken to achieve that objective.

In order to adequately cover this topic in the allocated space, it is necessary to adopt a pragmatic approach and deal with it through the medium of a few recent pieces of European Community (EC) food legislation. The legislative approaches to quality will be examined and it will be demonstrated how these approaches are being both augmented and complemented by the work of voluntary organizations such as EURACHEM (a focus for analytical chemistry in Europe), and the Measurement and Testing (M and T) strand of the European Communities Framework Research Programme (Figure 1).

Many of the current initiatives have been driven by a number of factors, *i.e.*:

- the Single Market;
- problems associated with the operation of technical legislation; and
- the need to demonstrate quality.

Single Market

In the White Paper on the Completion of the Internal Market,[1] the European Commission laid down a programme of measures that would be required in-order to make the Single Market work. For the food analytical sector, and indeed for the measurement and testing community in general, the White Paper dictated that certain measures would have to be taken (at this point in time many have) to achieve the twin objectives:

- the elimination of technical barriers to trade and
- the mutual acceptance of test data.

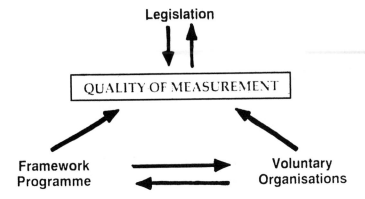

Figure 1

Laboratory personnel were faced with the implementation of these measures and they have had to put in place mechanisms which will achieve the above concepts.

Technical Legislation

Legislators are more concerned about the administrative aspects of legislation and less about the operation of the technical aspects. They usually assume that once the technical areas (*e.g.* maximum residue limits, methods of analysis) are prescribed in the legislation, the task is complete.

The laboratory manager however, has to:

- make the legislation work;
- demonstrate quality;
- evaluate data; and
- explain analytical data to non-analytical personnel and to the public at large.

Quality

It is no longer sufficient for a laboratory to generate quality data, it must now be in a position, to demonstrate to 'the world at large':

- that its analyses are under statistical control;
- that it is generating data of known and proven quality; and
- that its data is comparable to data generated in other laboratories.

In order to achieve all of the above objectives, the analyst requires a measurement and testing infrastructure. The building blocks of which are the components necessary to ensure comparability and, where appropriate, traceability of measurement.

The various initiatives that are currently taking place in Europe are geared towards the construction of such an infrastructure.

Due to limitations of time and space, remarks herein are confined to a few of the comparability components of measurement, *i.e.*:

- validated analytical methods;
- internal quality control;
- external quality control; and
- accreditation.

Hence, other comparability aspects:

- sampling;
- preparation of the sample in the laboratory, prior to the commencement of the analytical procedure;
- interpretation of analytical results;
- education, training;
- terminology, *etc.*

which are very important in ensuring that we are all operating from the same base-line and communicating on the same wavelength, have to be excluded.

For similar reasons the traceability issue is also excluded.

2 Food

Food embraces a diverse range of products and its analysis covers almost the complete range of analytical methods. It is necessary, for a variety of legislative purposes, for the food analyst to deal with:

- base products, such as, milk, meat, cereals, oilseeds, *etc.* and
- processed products, such as, baby food, meat pies, biscuits, chocolate, ice-cream, peanut butter, *etc.*

The legislative areas to be discussed are:

- Base Directives 64/443/EEC[2] and 86/469/EEC,[3] which concern residues in food of animal origin;
- Directive 92/16/EEC,[4] laying down the health rules for the production and placing on the market of raw milk, heat treated milk, and milk based products; and
- Directive 89/397/EEC[5] on the official control of foodstuffs.

Two approaches to quality have been adopted in the food sector:

- the first involves a network system [milk and residues in food of animal origin (base products)] and
- the second is an accreditation aproach [food control directive (processed products)].

In the network system, there is a hierarchy of laboratories:

- monitoring or approved laboratories;
- National Reference Laboratories; and
- Community Reference Laboratories.

This system allows for contact between the laboratory personnel and a bottom up and a top down approach is being developed. New methodologies, techniques, quality assurance, training, *etc.* come from the top down and problems, *etc.* from the bottom up.

Within this system an approved laboratory is defined as 'a laboratory approved by the competent authority of a Member State (MS) to make an examination of an official sample to detect the presence of residues', Council Directive 86/469/EEC.[3]

The responsibilities of national reference laboratories include:

'co-ordinate standards and methods of analysis for each residue or group of residues, including the arrangement of periodic comparative tests of split samples by the approved laboratories'.[3]

The approved laboratories and the national reference laboratories are appointed by the Member States. For residues in meat there are 36 National Reference Laboratories, Commission Decision 93/256/EEC.[6]

The Community Reference Laboratories are appointed by the Commission.
For meat residues the Community Reference Laboratories are located in Germany, France, Italy, and The Netherlands. Their terms of reference, which are laid down in Council Decision 89/187/EEC,[7] include the quality aspects of analysis and are summarized below:

(i) To provide the National Reference Laboratories with technical assistance and analytical methods.

(ii) To ensure the application of Good Laboratory Practice in the National Reference Laboratories and to organize comparative tests between those laboratories.

(iii) To promote and co-ordinate method research and development, to ensure National Reference Laboratories are kept informed of the 'state-of-the-art', and to engage in training.

(iv) To provide technical assistance to both the Commission services and the Measurement and Testing programme (previously known as BCR).

(v) To liaise with National Reference Laboratories designated by 'third countries' (non-EU countries).

In the meat residue area methods of analysis are not prescribed in legislation. The analyst is free to use the method of his choice, but in so doing, must prove that it is suitable for its intended purpose and that it conforms to criteria laid down in legislation [Commission Decision (93/256/EEC)].[6]

These criteria are very comprehensive and detailed as is illustrated by those summarized below for GC–MS:

(i) An internal standard should be used, preferably a stable, labelled isotope, form of the analyte.

(ii) The relative retention time of the analyte should be the same as that of the standard analyte in the appropriate matrix, within a margin of $\pm 0.5\%$.

(iii)   The designated peak should be separated from the nearest peak on the chromatogram by at least one full width at 10% of the maximum height.

(iv)   For confirmation, the intensities of preferably, at least, four diagnostic ions should be measured. If the compound does not yield four, then identification of the analyte should be based on the results of at least two independent methods with different derivatives and/or ionization techniques, each producing two or three diagnostic ions.

(v)   The relative abundance of the diagnostic ions of the analyte should match those of the standard analyte within a margin of $\pm 10\%$ for EI and $\pm 20\%$ for CI.

In the milk area,[4] reference methods of analysis are prescribed in legislation,[8] but monitoring methods are not. Monitoring methods may be used for routine screening purposes only, and they must be calibrated against the prescribed reference method.

The Community Reference Laboratory is located in France[4] and National Reference Laboratories have yet to be appointed.

Legislation considered in the previous examples has dealt with types of food which are termed base products (meat, milk, cereals, *etc.*). Before moving on to the other segments of the Measurement and Testing infrastructure, it is worthwhile considering what the likely scenario will be for processed foods (sausages, baby food, biscuits, *etc.*).

Council Directive on the Official Control of Foodstuffs (89/397/EEC),[5] is one of the 'new approach' Directives. Its implementation, is still under discussion, but it would appear that a probable scenario will be the following:

● a laboratory will be required to use validated methods;
● to participate in a proficiency testing scheme; and
● that accreditation will be mandatory.

Hence, in this Directive a laboratory's quality assurance programme, will be evaluated by the accreditation authorities.

Current indications are that analytical methods will not be prescribed in legislation and neither will criteria for method performance. A laboratory will either have to engage in method validation or to acquire validated methods from an external source.

Many validated methods exist for food and there are several initiatives currently in train which are addressing the deficit areas:

(i)   In Europe many National authorities already have validated methods and it is a question of making these methods available to the wider analytical community.

(ii)   The Measurement and Testing strand of the Third Framework Programme is funding several method development and validation projects,[9] *e.g.* vitamins in food, ochratoxin A in pig kidney and cereals, and triglycerides in fat.

(iii)   The *Journal of the Association of Public Analysts* has recently recommended publication and is providing laboratories with a forum for the publication of and access to validated methods. Examples of validated methods which were published recently are: dietary fibre (GLC);[10] biphenyl and 2-hydroxybiphenyl in citrus fruits;[11] moisture and ash in cocoa and chocolate products;[12] milk fat in cocoa and chocolate products;[13] and ice glaze on frozen fish fillets.[14]

(iv) The US laboratories have similar problems with the 'New Food Labelling' legislation. AOAC International are currently addressing this issue from a US perspective and a book on the topic[15] has just been published. It is available from AOAC International.

(v) CEN, the European standards organization, is also looking at methods of analysis.

Internal Quality Control (QC)

Once a laboratory has a validated method, it must be in a position to prove:

- that its methods are operating correctly and
- that its day-to-day operations are under statistical control.

This requires the use of certified reference materials (CRM), and secondary or in-house reference materials.

External QC

The next step in the production of quality data requires a laboratory to rate its performance against 'the outside world'. CRMs give a measure of this performance, but they are not always available and when they are, matrix matching can be a problem.

Proficiency testing (PT) is a very useful form of external QC. It helps a laboratory to assess its performance and to rate itself against other laboratories. It also pinpoints problems which may not be detected with internal QC.

There are many proficiency testing (PT) schemes in Europe, some are on a national basis others are transnational. Examples of some transnational schemes are:

(i) The Inspectorate for Health Protection, PO Box 465, 9700 AL Groningen, The Netherlands, organize the PT scheme, CHEK. Examples of areas tested in the 1993 CHEK programme are:

- sorbic acid and benzoic acid in salad;
- freezing point of milk;
- histamine in cheese;
- sulfite in dried fruits;
- glutamic acid in soup; and
- starch in meat products.

(ii) FAPAS, is the entitlement of a UK scheme organized by the Ministry of Agriculture, Fisheries, and Food, and is administered from The Food

Science Laboratory, Colney Lane, Norwich NR4 7UQ, UK. Some of the items that it addressed during the last 12 months were:

- nutrients and trace elements in food products;
- sulfadimidine residues in porcine tissues;
- aflatoxin in pistachio nuts; and
- organochlorines in beef fat.

(iii) CECA LAIT, is a French PT scheme and is devoted to the dairy sector. It is administered from, Poligny, BP 89129801, Cedex, France. Examples of its activities include:

- antibiotics in milk and
- coliform and aerobic plate count testing.

At the moment, it is very difficult for a laboratory to discover the existence of the various proficiency testing schemes which are organized by the different organizations, or indeed to objectively evaluate them.

With this in mind, the EURACHEM Proficiency Testing Working Group organized a Workshop on 8–9 November, 1993. The aim of the workshop was to bring European laboratories together who are involved in PT, or who intend to become involved in PT. Its objectives were:

- to integrate and disseminate expertise on the design and execution of proficiency tests;
- to improve the level of performance of proficiency testing in Europe;
- to promote co-operation between different organizations in the field of proficiency testing, through the creation of networks of laboratories, which conduct proficiency testing in specific fields of measurement; and
- to make information on existing schemes more widespread.

Proficiency Testing schemes may differ in their statistical approaches, *etc.* and this issue has been addressed by IUPAC, ISO, and AOAC International. They have completed work on 'The International Harmonized Protocol for the Proficiency Testing of (Chemical) Analytical Laboratories'. There was a high level of European input into this guide and it will be a useful tool for laboratories who would like to organize schemes. The protocol has recently been published by both IUPAC[16] and AOAC International.[17]

A Harmonized Protocol on 'Internal Quality Assurance for Chemical Laboratories' is currently under consideration and a UK draft proposal on this topic was discussed at a recent international meeting in Washington. An amended draft is being prepared and this was to be discussed in Delft in May 1994. Both guides are part of a series of International harmonization guides. Others deal with the collaborative study process: 'Protocol for the Design, Conduct and Interpretation of Collaborative Studies'[18] and 'Harmonized

Protocols for the Adoption of Standardized Analytical Methods and for the Presentation of their Performance Charactertistics'.[19]

External Quality Assurance (QA)

Accreditation is becoming increasingly important for laboratories engaged in chemical analysis and it is generally regarded as a visible outward sign that a laboratory is generating quality data. In Europe, the Accreditation authorities work to the EN 45000, series of standards.[20] EN are European Norms, produced by the European Standards Organization, (CEN). EN 45000, is similar to ISO Guide 25, which is produced by the International Standards Organization (ISO)[21].

Both documents lay down the criteria and the rules to which measurement and testing laboratories must conform, if they intend to seek accreditation. These are broad based documents, since they are intended to cover the complete spectrum of M and T laboratories, which cover a diverse range of topics, *e.g.*:

- metrology;
- uniformity of building material;
- microbiological;
- chemical, *etc.*

The interpretation of EN 45000 is particularly difficult for chemistry and there was a real danger that each accreditation authority could arrive at its own interpretation of the standards. Hence, instead of having a single set of interpretations for chemistry, 12–18 different systems could emerge, which would be neither comparable nor compatible with one another.

In order to prevent this from occurring, EURACHEM (a focus for analytical chemistry in Europe), in conjunction with WELAC (the umbrella group for European accreditation bodies) have compiled a guidance Document for the interpretation of EN 45000 series of standards and ISO Guide 25 for 'accreditation for chemical laboratories'. The guide is available from the EURACHEM, Secretariat, PO Box 46, Teddington, Middlesex TW11 0NH, UK, or from the Secretariat of National Accreditation organizations.

This Guide should be of considerable assistance to laboratories preparing for accreditation and also to the accreditation authorities who will be accrediting them. However, there is no point in the laboratories operating to harmonized guidelines, unless the accreditation authorities do likewise. Consequently, WELAC have issued the following Guidance documents for its members:

(i) Internal Quality Audits and Reviews;
(ii) Guidelines for Training Courses for Assessors Used by Laboratory Accreditation Schemes; and
(iii) Programme for Course for Tutors of Assessor Training.

A document for Criteria for Proficiency Testing in Accreditation is in the course of preparation. Further information on WELAC Documents is available from National Accreditation Organizations.

3 Future European Initiatives

Future initiatives include the following:

- A joint EURACHEM–WELAC working group is currently preparing a guide for the interpretation of EN 45000 for Microbiological laboratories;
- A workshop on measurement uncertainty is planned for 1994 (further information from, Professor W. Wegscheider, Technische Universitat Graz, Technikerstrasse 4, A-8010 Graz, Austria);
- AOAC Europe will hold a symposium in Switzerland in 1994 on Food and Feed analysis, and is entitled 'A Focus on Methods with a Minimum risk to Health and the Environment', (further information from Dr T. Rihs, Swiss Federal Research Station for Animal Production, CH-1725 Posieux, Switzerland);
- An inventory of methods affected by the Montreal Protocol is being prepared.

The Montreal Protocol 1/1/89 is effectively going to mean the unavailability of CCl_4 and $CH_3CH_2Cl_3$. Within the EU the protocols stipulations relating to the above solvents takes effect from the 1/1/95. Validated methods which use solvents effected by the protocol will either require revalidation with an alternative solvent system or complete redevelopment. EURACHEM, are currently endeavouring to assess what the analytical implications for Europe will be and would welcome positive input. AOAC International, following the European initiative, have identified 57 of their methods (mainly in the food sector)[22] which will be effected by the ban. A recent article[23] has examined the analytical implications from an environmental view point.

This review highlights some of the activities that are taking place in Europe regarding QA in food analysis. European food analysts are, of course, not working in isolation and many of the initiatives are of a transcontinental nature. In other initiatives the work has already been done and it is just a matter of collating and disseminating the information so that other food analytical laboratories can also avail of it to improve the quality of their analysis.

References

1. White paper from the European Commission to the European Council (Milan, 28–29 June 1985), 'Completing the Internal Market', COM(85) 310 final, Brussels, 14 June 1985.
2. *Off. J.*, No. L 121, 9.7.1964, p. 2012.
3. *Off. J.*, No. L 275, 26.9.1986, p. 36.
4. *Off. J.*, No. L 268, 14.9.1992, p. 1.
5. *Off. J.*, No. L 186, 30.6.1989, p. 23.
6. *Off. J.*, No. L 118, 14.5.1993, p. 64.
7. *Off. J.*, No. L 66, 10.3.1989, p. 37.
8. *Off. J.*, No. L 93, 13.4.1991, p. 1.
9. Personal Communication.
10. R. Wood, *J. Assoc. Public Anal.*, 1992, **28**, 25–35.
11. R. Wood, *J. Assoc. Public Anal.*, 1992, **28**, 43–49.

12. R. Wood, *J. Assoc. Public Anal.*, 1992, **28**, 63–72.
13. R. Wood, *J. Assoc. Public Anal.*, 1992, **28**, 89–94.
14. R. Wood, *J. Assoc. Public Anal.*, 1992, **28**, 89–102.
15. 'Methods of Analysis for Nutrition Labelling', ed. D. M. Sullivan and D. Carpenter, AOAC International (ISBN No. 0-935584-52-8).
16. M. Thompson and R. Wood, *Pure Appl. Chem.*, 1993, **65**, 2123–2144.
17. M. Thompson and R. Wood, *J. Assoc. Off. Anal. Chem.*, 1993, **76**, 926–940.
18. W. Horwitz, *Pure Appl. Chem.*, 1988, **60**, 856–864.
19. W. D. Pocklington, *Pure Appl. Chem.*, 1990, **62**, 150–162.
20. European Standards, The Joint European Standards Institution, CEN/CENELEC, Rue de Stassart 36, B-1050 Brussels, Belgium.
21. 'General Requirements for the Competence of Calibration and Testing Laboratories', ISO Guide 25, 3rd Edn., 1990, Geneva.
22. The Referee (AOAC International), September, 1993, p. 10.
23. D. Noble, *Anal. Chem.*, 1993, **65**, 693A–695A.

NB: References to European Communities publications, may also be obtained from, The Office for Official Publications of the European Communities, 2 Rue Mercier, L 2985 Luxembourg.

Chemical Contaminants in the Food Supply: Research in a Regulatory Setting

H. B. S. Conacher, P. Andrews, W. H. Newsome, and J. J. Ryan

FOOD RESEARCH DIVISION, BUREAU OF CHEMICAL SAFETY, FOOD DIRECTORATE, HEALTH PROTECTION BRANCH, HEALTH AND WELFARE CANADA, TUNNEY'S PASTURE, OTTAWA, ONTARIO, CANADA K1A OL2

Summary

An important component of any programme dealing with regulatory control of contaminants and other chemicals in food and drink is the availability of appropriate analytical methods. In this regard, the analytical estimation of the chlorinated hydrocarbon classes of contaminants represented by the polychlorinated dibenzo-dioxins and -furans (PCDDs/PCDFs), the polychlorinated biphenyls (PCBs), and toxaphene, represents a considerable challenge to regulatory agencies world-wide. The problem lies in the fact that each of these classes of contaminants contains more than one hundred individual closely related components (congeners) that can vary widely in toxicity. Thus, to obtain an indication of the toxicity of such congener mixtures found in foods and drinks, there is a requirement to analytically separate, identify, and measure the more toxic from the others. A considerable effort has been on-going within the Canadian Health Protection Branch laboratories over the past decade to resolve these difficult analytical problems. Some of the factors contributing to success in this area include the availability of analytical standards; the development of better clean-up techniques; the availability of better phases for capillary gas chromatography; and the development of more sophisticated instrumentation in terms of mass spectrometry and associated software. The application of these methods towards an assessment of the exposure of Canadians to these environmental contaminants has indicated that for the most part exposures are well below acceptable international standards.

1 Introduction

The assessment of risk to human health from potentially harmful chemical contaminants in the food supply requires two essential pieces of information—details of the inherent toxicity of the contaminant and details of the exposure of humans to the contaminant. An important component of the exposure assessment is the availability and application of sound analytical methods to determine the contaminants in question.

One of the more intriguing analytical challenges to regulatory agencies world-wide involves the analytical estimation of the chlorinated hydrocarbon classes of contaminants represented by the polychlorinated dibenzo-dioxins and -furans (PCDDs/PCDFs), the polychlorinated biphenyls (PCBs), and toxaphene.

The problem lies in the fact that each of these classes of contaminants contains more than one hundred individual closely related congeners that can vary widely in toxicity. Thus, to obtain an indication of the toxicity of such congener mixtures found in foods and drinks there is a requirement to analytically separate, identify, and measure the more toxic from the others.

A considerable effort has been devoted within the Canadian Health Protection Branch laboratories (and elsewhere) over the past decade or so to resolve these particular problems. A brief description of the current status with respect to the methodology, the toxicology, and the research remaining to be done for these three classes of chlorinated hydrocarbon contaminants is outlined in this report. A consistent theme throughout the work is the emphasis on the generation of valid data—an element critical to the conduct of any regulatory programme.

2 PCDDs/PCDFs

PCDDs/PCDFs are two closely related groups of environmental contaminants that can be found in a variety of industrial chemicals and can be produced through industrial processes such as incineration and chlorine bleaching.[1] There is a total of 210 discrete chemical compounds, called congeners, in which one to eight chlorine atoms are attached to the basic dibenzodioxin/furan skeleton (Figure 1). However, only 136 congeners, the tetra- to octa-chlorinated, comprising 49 dioxins and 87 furans, are of biological interest.[2]

The final measurements of PCDDs/PCDFs in the environment took place with industrial chemicals such as pesticides and wood preservatives. For these measurements at relatively high concentrations, packed gas chromatographic columns and electron capture detectors were sufficient. However, biological samples contain much lower levels of these contaminants, *e.g.* 2,3,7,8-TCDD in Great Lakes fish are in the low ng kg^{-1} range.[3] In this case, determinations have to be carried out with high resolution mass spectrometers with their low detection limits and quantitation is performed on the purified extract with stable isotopes as internal standards.

In addition to extensive internal laboratory quality control measures, an interlaboratory study that involved most of the laboratories in the United States and Canada generating data on fish in the Great Lakes was organized and conducted.[4] Levels of detection and agreement among the laboratories were surprisingly good considering the levels quantitated (p.p.t., 10^{-12} g) and the diverse methodology. This methodology was subsequently used to determine

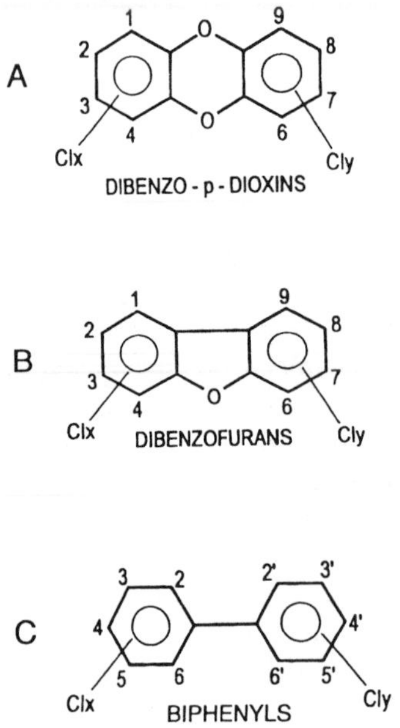

| Number of Chlorines | Number of | | |
x+y	Dioxins	Furans	PCBs
1	2	4	3
2	10	16	12
3	14	28	24
4	22	38	42
5	14	28	46
6	10	16	42
7	2	4	24
8	1	1	12
9	-	-	3
10	-	-	1
Total	75	135	209

Figure 1 *Structure of polychlorinated dibenzo-dioxins and -furans and of polychlorinated biphenyls*

the extent of the problem by determining the levels of 2,3,7,8-TCDD in Great Lakes fish.[5]

From this period, significant advances were made in both the toxicological and analytical fronts pertaining to these compounds. Synthesis of a wide array of standards[6] permitted the identification of many additional members of this class of compounds and simultaneously allowed their toxicological properties to be more readily defined.[7]

The widely different biological activity with structure led to the concept of '2,3,7,8-TCDD toxic equivalents' wherein 2,3,7,8-TCDD is a given factor of 1 while other dioxin and furan congeners are assigned lesser numbers.[8] These numbers function as multipliers for the corresponding concentration of residue.

Thus the methodology was expanded to include all of the 136 PCDD/PCDF congeners particularly the seventeen with 2,3,7,8-substitution. Some of the factors that permitted these advances included the commercial availability of better phases for capillary columns and the development of magnetic sector high resolution mass spectrometers with the ability to switch target ions more rapidly.[9] The latter allowed the determination of many more analytes in a single injection.

In this connection an assessment of the characteristics of the 136 PCDD/PCDF congeners on nine capillary phases has been conducted.[10] This study indicated that all 136 congeners, including the important 2,3,7,8-substituted variety, can be separated from each other mostly on two phases.

An outline of the methodology currently in place in our laboratory for human milk is shown in Figure 2. It involves an extraction with acetone–hexane (2:1), a defatting of the hexane by partitioning with concentrated sulfuric acid, and successive cleanups on florisil (to remove PCBs), on carbon dispersed on silica gel (to separate co-planar compounds), and on silica containing strong acid and base (to remove residual traces of fat). Dioxin and furan congeners in the final extract are identified and quantitated by capillary gas chromatography and high resolution mass spectrometry. The chromatographic procedure is used for the routine monitoring of samples for 49 dioxins and 87 furans,[11] although generally only the 2,3,7,8-substituted congeners are present in biological samples.

This methodology has been applied to selected samples from the Canadian Total Diet Program.[12,13] Calculation of the PCDD/PCDF intake, in terms of TCDD-equivalents, is illustrated in Figure 3. TDS refers to the data originally generated in the Total Diet Study—this included a relatively significant contribution from milk containing dioxins/furans leached from milk cartons. Since 1989 the introduction of new technologies into the manufacture of cardboard used for milk carton production has eliminated this source of contamination into milk. This 'new' milk data significantly reduces the overall dioxin intake. There is little variation in the intake of PCDD/PCDF across Canada and the overall mean intake of $0.6\,\mathrm{pg\,kg^{-1}\,day^{-1}}$ is considerably less than the WHO international guideline of $10\,\mathrm{pg\,kg^{-1}\,day^{-1}}$.[14]

The importance of sound quality control/quality assurance procedures in the development of such data cannot be over emphasized. For example, in the conduct of such a survey between 15 and 25% of the determinations are quality

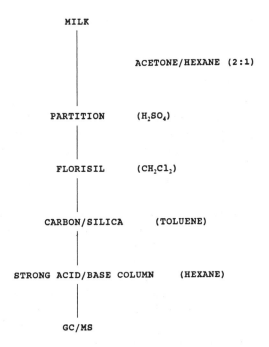

Figure 2 *Methodology for PCDDs/PCDFs in human milk*

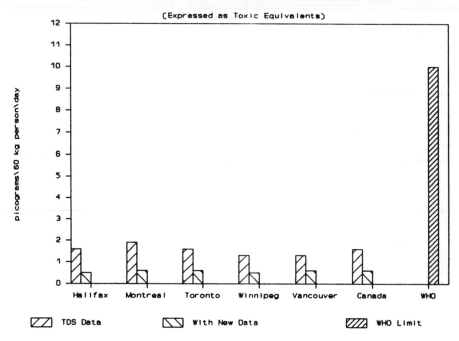

Figure 3 *PCDD/PCDF intake from Canadian foods*

control samples. Our laboratory is also an ongoing participant in the check sample programme conducted by WHO Europe.[15]

3 PCBs

Unlike the PCDDs/PCDFs that are essentially produced as by-products of industrial processes, the PCBs were produced commercially beginning in 1929 and subsequently used extensively in a number of industrial products.[16] Later, concerns regarding their toxicity and widespread presence in the environment led to their production being stopped in North America in the late 1970s.[17] The presence of PCBs in the food chain has been established for some 20 years.[18] There is a total of 209 congeners in which one to ten chlorine atoms are attached to biphenyl (Figure 1).

Until comparatively recently PCBs were determined by gas chromatography (GC) on packed columns using commercial mixtures of chlorinated biphenyls, such as the Aroclor series, as standards. It was, however, recognized that the GC elution patterns of PCBs observed in biological samples seldom exactly resembled that of any particular Aroclor on which the early toxicological studies had been conducted. This was likely due to the various metabolic changes occurring during transition through the various fish and animal species in the ecological food chain. This led to concerns as to the accuracy of PCB quantitation.[19]

The development of reliable capillary columns for gas chromatography provided us with an insight into the complexity of commercial PCB mixtures and enabled a finer definition of the nature of the contamination. In 1984, Mullin *et al.*[20] reported the synthesis and chromatographic properties of all 209 congeners, permitting the assignment of retention times and identification of those present in biological samples.

Questions then arose as to whether some congeners were more important toxicologically or whether a simple summation of the congeners present represented the toxic potential.

Subsequent toxicological studies[21] indicated that indeed there was considerable variation in toxicity among congeners, a situation similar to that observed with the PCDDs/PCDFs. Some congeners present in samples in only small quantities have a high potential for toxicity, while others present in greater amounts are less active. In this regard, it should be noted that the PCB congeners that can be considered approximate isostereoisomers of 2,3,7,8-TCDD, *i.e.* the 'co-planars', exhibited high biological activity.[22] The co-planars of most interest are the congeners #77, 126, and 169, numbered according to the system of Ballschmiter and Zell.[23] These can only be determined at present using the PCDD/PCDF methodology described earlier.

The information summarized in Table 1 indicates the current situation with regard to congener toxicity and relative amounts. Ideally a 'toxic-equivalents' concept similar to that developed for PCDDs/PCDFs is considered desirable but has not yet been completed.

Accordingly, although initially in surveys such as the National Survey of Human Milk, we determined total PCBs by packed column GC as Aroclor 1260,[24] we have recently moved towards individual congener determination on capillary columns as a measure of the PCBs present.[25] A flow scheme showing

Table 1 *Factors considered in development of toxic equivalents concept for PCBs*[a]

Congener[b]	2,3,7,8-TCDD type toxicity[c]	Neurotoxicity[c]	Quantity[c]
28		+++	
52		+++	
77	+++		
99	++		
105	+		+
118	++		++
126	+++		
128	+		
138	+		+++
153	+		+++
156	++		+
169	+++		
170	+		+
180	+		++
183			++
187			++

(a) E. Dewailly, personal communication; (b) numbered according to Ballschmiter and Zell;[23] (c) relative toxicities or amounts observed in biological samples; + lowest → + + + highest.

the methodology currently used for human milk samples is illustrated in Figure 4. The PCBs occur mainly in fraction 1 (hexane) with a small amount of the lower chlorinated congeners eluting in the second fraction (20% dichloromethane). The 39 congeners comprise approximately 85% of the mean of PCBs present in this substrate and represent approximately 80% of the value calculated by the previously described packed column GC procedure. The separation of a human milk extract on both packed and capillary columns is shown in Figure 5. The more effective separation of the PCB congeners (*i.e.* those eluting after *p*, *p*′-DDE) on the capillary column is readily apparent.

The methodology has also been applied to selected samples for the Canadian Total Diet Program.[12,13,26,27] Calculation of daily intake (in terms of a simple sum of congeners) is illustrated in Figure 6. The daily intakes were calculated in two ways, one in which non-detected congeners were assumed to be zero (ND = 0), and the other in which they were assumed to be present at their

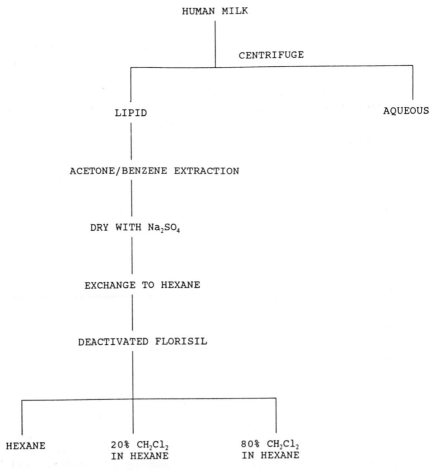

Figure 4 *Methodology for sample preparation in the determination of PCB in human milk*

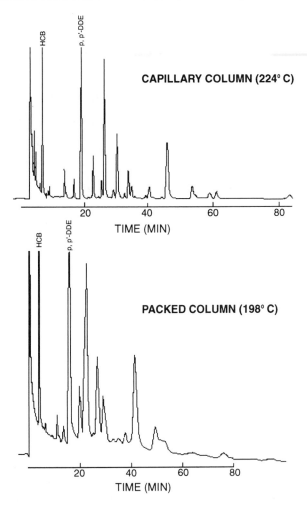

Figure 5 *GC–ECD chromatograms measuring PCBs in a human milk sample on packed and capillary columns*

respective minimum levels of detection (ND = MDL) indicated that no significant differences were observed across regions as was the case with the PCDDs/PCDFs. The mean intake for Canadians of approximately 7.5 ng kg^{-1} body weight day^{-1} is also considerably lower than the current Canadian guideline of 1000 ng kg^{-1} body weight day^{-1}.

As with PCDDs/PCDFs, extensive internal quality control measures were adopted to ensure the validity of the data. Such measures included the routine analysis of reagent blanks, fortified samples, confirmation of congeners by gas chromatography, and mass spectrometry.[26,27] The laboratory also routinely participated in the FAO/WHO Check Sample Program[28] and organized an international study which demonstrated the feasibility of estimating PCBs using specific polychlorinated biphenyl congeners.[29]

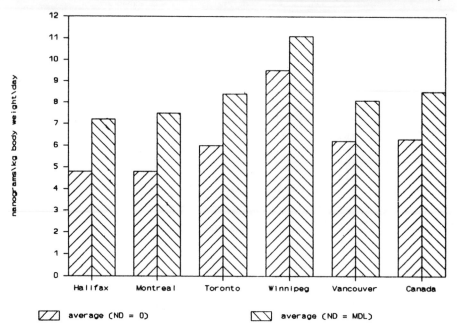

Figure 6 *PCB intake from Canadian foods*

4 Toxaphene

Toxaphene is a complex mixture of several hundred chlorinated bornanes formed by the photocatalytic chlorination of camphene (Figure 7).

Before it was banned in 1982, toxaphene was one of the most heavily used pesticides in the USA and in many other parts of the world.[30] It was recognized as widespread an environmental pollutant as were DDT and PCBs.[31]

Analytically, the situation with toxaphene is similar to that with PCBs in the late 1960s. The number of congeners present in the commercial material has been estimated at approximately 700. Capillary gas chromatograms are complex, and complete resolution of congeners is challenging. In addition, several other compounds commonly found as contaminants in fatty foods, such as PCBs, chlorinated pesticides, *etc.*, overlap the retention times of toxaphene (Figure 8). Although some biological testing has been conducted with toxaphene *per se*, congeners, with a few exceptions, have not been isolated nor their toxicity assessed.

Recently, however, eleven congeners have been prepared from the commercial mixture by preparative chromatographic techniques[32] and we have been conducting studies with them in an attempt to establish the optimum means of quantitation.

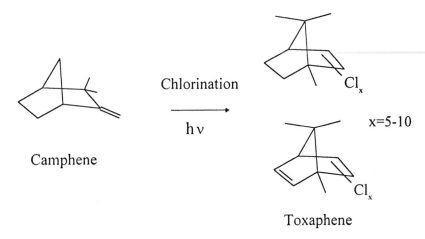

Figure 7 *Synthesis of toxaphene*

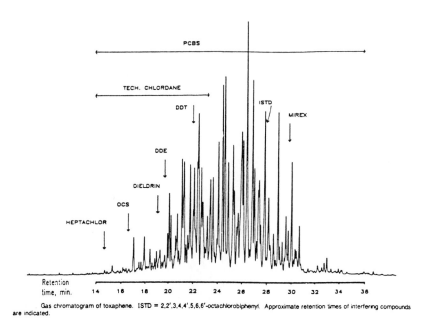

Gas chromatogram of toxaphene. ISTD = 2,2',3,4,4',5,6,6'-octachlorobiphenyl. Approximate retention times of interfering compounds are indicated.

Figure 8 *GC–ECD of technical toxaphene with approximate retention times of interfering compounds*
Abbreviations: OCS = Octachlorostyrene, DDE = Dichlorodiphenyldichloroethylene, DDT = Dichlorodiphenyltrichloroethane, ISTD = 2,2',3,4,4',5,6,6'-Octachlorobiphenyl
Reprinted with permission from Anal. Chem., *1987, **59**, 914. Copyright 1993 American Chemical Society*

Three analytical approaches currently being studied are:

(i)   gas chromatography with electron capture detection, (GC–ECD);
(ii) gas chromatography–high resolution mass spectrometry with electron impact and selective ion monitoring (GC–HRSIM–MS); and
(iii) gas chromatography–mass spectrometry with electron capture negative ionization (GC–ECNI–MS).

With regard to the GC–ECD technique,[33] the approach is similar to that utilized with the PCBs. A selected number of 'peaks' from standard toxaphene is compared quantitatively to the same peaks in the sample to obtain an estimate

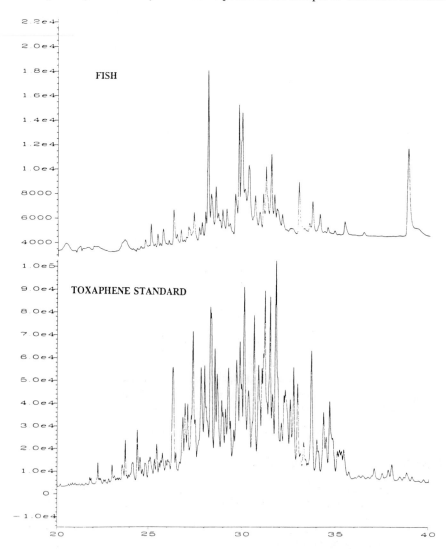

Figure 9 *GC–ECD chromatograms of toxaphene in fish and of toxaphene standard*

of the amount present. Although similar to that utilized with the PCBs, the comparison is much more difficult due to the more incomplete resolution of congeners (Figure 9). This approach also provides no structural information unless congener standards are employed.

The GC–MS–EI technique[34] using ions at m/z 159 and 161 is based on the formation of the dichlorotropylium ion by many of the congeners, resulting in a common fragment that appears unique to toxaphene (Figure 10). The sum of the area of the peaks generated by the 159 and 161 ions, relative to that of the toxaphene standard, results in a measure of the total toxaphene present (Figure 11). Again this technique provides no structural information unless congener standards are utilized (Figure 12). In addition, the technique is not too sensitive.

On the other hand the GC–MS–ECNI technique[35] is highly sensitive. It is based on the selective response of M–Cl chlorinated bornane ions and, consequently, even in the absence of congener standards, provides much more structural information than does either of the first two approaches. The main drawback to this approach is the varying response factors from congener to congener (Figure 13). For example, the differing response for peaks representing congeners 7, 10, and 11 should be noted.

A comparison of the results obtained by application of these three techniques to the same fish extract is presented in Table 2. Reasonable agreement is obtained

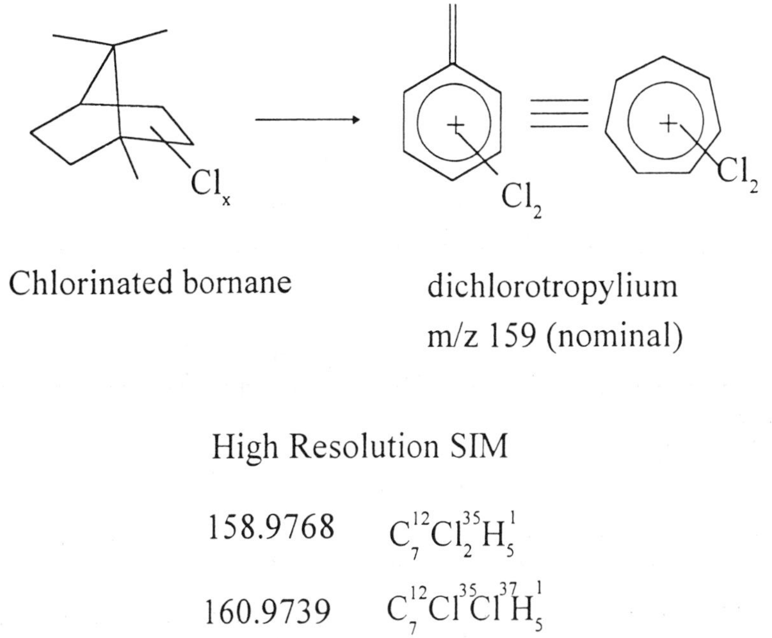

Chlorinated bornane   dichlorotropylium
                               m/z 159 (nominal)

High Resolution SIM

158.9768 $C_7^{12}Cl_2^{35}H_5^1$

160.9739 $C_7^{12}Cl^{35}Cl^{37}H_5^1$

Figure 10 *Formation of dichlorotropylium ion from toxaphene congeners*

Signal Monitored at m/z 158.9768

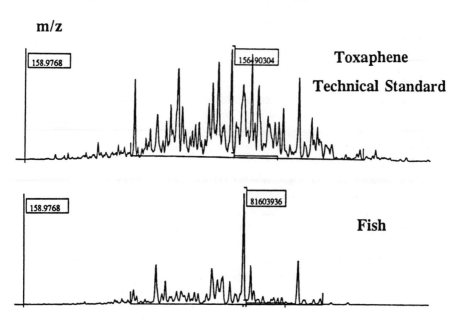

Figure 11 *HRSIM–GC–MS of technical toxaphene and of toxaphene in fish*

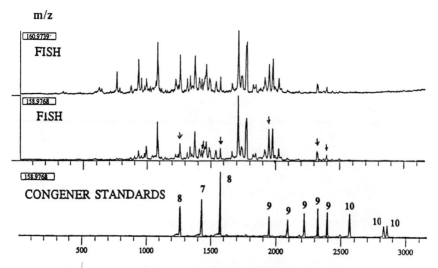

Figure 12 *HRSIM–GC–MS chromatograms of toxaphene in fish for eleven toxaphene congeners at 158.9768. The numbers 7,8,9,10 in the lower panel refer to the degree of chlorination in the toxaphene congeners*

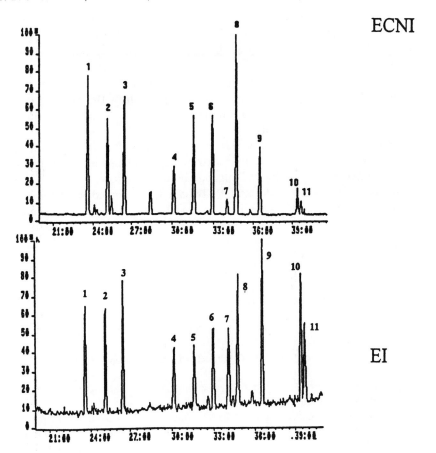

Figure 13 *Comparison of GC–MS–ECNI and GC–HRSIM–MS(EI) for eleven toxaphene congeners (Cl-7, -8, -9, and -10). Numbers on peaks given to facilitate comparison*

Table 2 *A comparison of analytical techniques for the analysis of toxaphene*

| Sample | Toxaphene in Fish ($\mu g\,g^{-1}$ wet weight $\pm$ std. dev.) | | |
	GC–ECD*	ECNI–MS#	HRSIM–MS#
EPA #3	90.8 ± 9.6	56.4	46.6
Fish #3	1.1 ± 0.6	0.7	0.6
Fish #4	9.5 ± 41.5	4.6	4.4

* Analysis performed by Environment Canada; $n = 3$.
# Analysis performed by Health and Welfare Canada; $n = 1$.

by the two MS techniques but the ECD produces a higher figure probably due to the more complicated chromatograms and associated interferences.

As may be evident from the preceding discussion, most groups involved in the analysis of toxaphene use a variety of approaches. Therefore, if quantitative results are to be utilized from various laboratories, it is imperative that some indication of the validity of data be obtained. To address this concern we are currently involved in the organization and conduct of a round robin study to assess the issue of compatibility.

In summary, the current state of the art techniques for estimation of these important classes of chlorinated hydrocarbon contaminants in foods and other substrates have been described. Utilizing these techniques, human exposure from foods to PCDDs/PCDFs and PCBs can be estimated with reasonable accuracy. Further research is required to develop more acceptable techniques for toxaphene.

References

1. H. B. S. Conacher, B. D. Page, and J. J. Ryan, *Food Add. Contam.*, 1993, **10(1)**, 129.
2. C. Rappe, *Environ. Sci. Technol.*, 1984, **18**, 78A.
3. R. L. Harless, E. O. Oswald, R. G. Lewis, A. E. Dupuy, Jr., D. D. McDaniel, and H. Tai, *Chemosphere*, 1982, **11(2)**, 1983.
4. J. J. Ryan, J. C. Pilon, H. B. S. Conacher and D. Firestone, *J. Assoc. Off. Anal. Chem.*, 1983, **66(3)**, 700.
5. J. J. Ryan, P. Y. Lau, J. C. Pilon, D. Lewis, H. A. McLeod, and A. Gervais, *Environ. Sci. Technol.*, 1984, **18**, 719.
6. R. A. Bell and A. Gara, 'Chlorinated Dioxins and Dibenzo-furans in the Total Environment II', Eds. L. H. Keith, C. Rappe, and G. Choudhary, Butterworths, Boston, 1985, Chapter 1, p. 3.
7. F. Iverson and D. L. Grant, 'Environmental Carcinogens: Methods of Analysis and Exposure, Vol. II, Polychlorinated Dioxins and Dibenzofurans', Eds. C. Rappe, H. R. Buser, B. Dodet, and I. K. O'Neill, IARC Publication 108, Lyon, 1991, Chapter 2, p. 5.
8. F. W. Kutz, D. G. Barnes, D. P. Bottimore, H. Greim, and E. W. Bretthauer, *Chemosphere*, 1990, **20**, 751.
9. J. J. Watson, 'Methods in Enzymology, Vol 193, Mass Spectrometry', Ed. J. A. McCloskey, Academic Press, San Diego, 1990, Chapter 4, p. 86.
10. J. J. Ryan, H. B. S. Conacher, L. G. Panopio, P. Y. Lau, J. A. Hardy, and Y. Masuda, *J. Chromatogr.*, 191, **541**, 131.
11. J. J. Ryan, C. Shewchuk, P. Y. Lau and W. F. Sun, *J. Agric. Food Chem.*, 1992, **40**, 919.
12. H. B. S. Conacher, R. A. Graham, W. H. Newsome, G. F. Graham, and P. Verdier, *Can. Inst. Food Sci. Technol. J.*, 1989, **22**, 322.
13. H. B. S. Conacher and J. Mes, *Food Add. Contam.*, 1993, **10(1)**, 5.
14. D. Kello and E. J. Yrjanheikki, *Chemosphere*, 1992, **25(7)**, 1067.
15. R. D. Stephens, C. Rappe, D. G. Hayward, M. Nygren, J. Startin, A. Esboll, J. Carlé and E J. Yrjanheikki, *Anal. Chem.*, 1992, **64**, 3109.
16. M. D. Erickson, 'Analytical Chemistry of PCBs', Butterworths, Boston, 1986, Chapter 1, p. 1.
17. M. D. Erickson, 'Analytical Chemistry of PCBs', Butterworths, Boston, 1986, Chapter 2, p. 5.

18. B. L. Sawhney and L. Hankin, *J. Food Prot.*, 1985, **48(5)**, 442.
19. M. D. Erickson, 'Analytical Chemistry of PCBs', Butterworths, Boston, 1986, Chapter 8, p. 267.
20. M. D. Mullin, C. M. Pochini, S. McCrindle, M. Romkes, S. H. Safe, and L. M. Safe, *Environ. Sci. Technol.*, 1984, **18**, 468.
21. S. Safe, L. Safe, and M. Mullin, *J. Agric. Food Chem.*, 1985, **33**, 24.
22. S. Safe, *Crit. Rev. Toxicol.*, 1990, **21(1)**, 51.
23. K. Ballschmiter and M. Zell, *Fresenius' Z. Anal. Chem.*, 1980, **320**, 20.
24. J. Mes, D. Davies, D. Turton, and W. F. Sun, *Food Add. Contam.*, 1986, **3**, 313.
25. J. Mes, D. Turton, D. Davies, W. F. Sun, P. Y. Lau, and D. Weber, *Int. J. Environ. Anal. Chem.*, 1987, **28**, 197.
26. J. Mes, W. H. Newsome, and H. B. S. Conacher, *Food Add. Contam.*, 1989, **6(3)**, 365.
27. J. Mes, W. H. Newsome, and H. B. S. Conacher, *Food Add. Contam.*, 1991, **8(3)**, 351.
28. Joint UNEP/FAO/WHO Food Contamination Monitoring Programme, Analytical Quality Assurance Studies, 1991, World Health Organization, Geneva.
29. J. Mes, H. B. S. Conacher, and S. Malcolm, *Int. J. Enviuron. Anal. Chem.*, 1993, **50**, 285.
30. M. A. Saleh, *Rev. Environ. Contam. Toxicol.*, 191, **118**, 1.
31. J. E. Casida, R. L. Holmstead, S. Khalifa, J. R. Knox, T. Ohsawa, K. J. Palmer, and R. Y. Wong, *Science*, 1974, **183**, 520.
32. H. Parlar, personal communication.
33. C. Lach and H. Parlar, *Toxicol., Environ. Chem.*, 1991, **31**, 209.
34. W. Newsome and P. Andrews, *J. Assoc. Off. Anal. Chem.*, 1993, **76**, 707.
35. D. Schackhammer, M. Charles, and R. Hites, *Anal. Chem.*, 1987, **59**, 913.

Water Quality Standards—Realistic Limits or Political Targets?

Gerry Best

PRINCIPAL FRESHWATER SCIENTIST, CLYDE RIVER PURIFICATION BOARD, RIVERS HOUSE, MURRAY ROAD, EAST KILBRIDE, UK

1 Introduction

The establishment of a water quality standard represents a target for which dischargers and enforcement authorities should aim. In the United Kingdom, water quality standards have been in existence for many decades. In 1912, the Royal Commission on Sewage Disposal established a target that, if a sewage effluent receives at least an eightfold dilution in a receiving stream, then it should not take up more than 20 p.p.m. of dissolved oxygen in five days at 65 °F and that it should not contain more than 30 p.p.m. of suspended solids. These standards for effluents were set so as to ensure that the level of dissolved oxygen in the river downstream of the discharge should not fall below 4 p.p.m. and so threaten the life of aquatic organisms. So the *objective* was to preserve fish stocks, the *water quality standard* was to maintain the dissolved oxygen above 4 p.p.m. and from this was derived the *effluent discharge consent condition* that the Biological Oxygen Demand (BOD) should not be greater than 20 p.p.m.

These consent conditions for sewage effluent have remained in force, with minor modifications, for over 90 years and despite efforts to replace the BOD test with a more rapid one, it is likely to remain as a standard for many more years.

The system of maintaining the quality of water by specifying the limits for particular parameters is still used and has become more refined as the knowledge of the effects of pollutants on aquatic life has improved and the range of polluting substances has increased. Essentially, the quality standards are proposed by experts after they have considered the wide range of evidence such as toxicity, bio-accumulation, persistence, and chemical speciation before proposing a limiting value. The number of parameters that have to be considered depends on the use to which the water is put. Thus, for example, for drinking water there are now 56 parameters that have to be tested to ensure the 'wholesomeness' of the supply.

When the UK entered the European Community, environmental pollution control was no longer solely a domestic issue. The UK was obliged to comply with EC Directives including those relating to water quality. The Directives are not laws so they have to become established into domestic UK legislation by passage through Parliament. For example, the EC Directive dealing with the discharge of dangerous substances into water (76/464/EEC) became the Surface Waters (Dangerous Substances) (Classification) Regulations 1989.

The pace of European legislation has been slow because it takes a long time for the Member States to agree on measures but the Directives now cover the main disposal routes of pollutants into the environment. The UK's interpretation of how to implement some of the Directives has led to some disputes with its European partners and often this country has had to 'go it alone' in the way a Directive has been applied.

2 Freshwater Fisheries Directive

One of the first Directives that was brought into force to protect the aquatic environment was the Freshwater Fisheries Directive (78/659/EEC).

The aim is to establish and maintain two types of fisheries—salmonid (trout, salmon) and cyprinid (coarse fish such as perch, pike, and carp). These classes of fish have different sensitivities to water quality and so two ranges of water quality parameters are quoted in the Directive. The selection of the criteria and the limits for them were discussed at length. For some parameters, mandatory (Imperative, or I) limits have been set:

- temperature increase;
- minimum dissolved oxygen;
- pH;
- ammonia, both total and un-ionized;
- total zinc;
- residual chlorine;
- taste of phenol in flesh; and
- no visible film of oil or taste in flesh.

For other parameters, discretionary (Guideline, or G) limits were specified for:

- BOD;
- nitrite;
- dissolved copper; and
- suspended solids.

The choice of the parameters is curious both for its inclusions (*e.g.* nitrite) and its omissions (other toxic metals such as chromium and lead). It is also not clear why total zinc is given an imperative value but soluble copper is only a guideline limit even though copper is the more toxic of the two:[1]

Water hardness	Maximum 95%ile for trout	
	Zn	Cu
$10\,mg\,CaCO_3\,l^{-1}$	$30\,\mu g\,l^{-1}$	$5\,\mu g\,l^{-1}$

When it came to applying the Directive the UK Government instructed the regulatory authorities (the Water Authorities in England and Wales and the River Purification Boards in Scotland) to designate only those rivers which complied with Directive's criteria even though fish may be present in other rivers. Furthermore, monitoring to check for compliance was to be confined to one sampling point on each river only. If the criteria were satisfied at that one point (usually the most downstream point) then it was assumed that the whole river is complying with the Directive. By this means the Government has managed to avoid problems with implementing the Directive by ensuring almost complete compliance at all times.

3 Dangerous Substances Directive

This Directive is the most important of those affecting water that has been implemented to date. Its aim is to **eliminate** pollution by a group of the most dangerous substances—the List I group—and **reduce** pollution by a less hazardous group—the List II substances. From the main enacting Directive, 'daughter' Directives covering specific List I substances have been developed. So far the substances controlled in the UK are (all from List I):

Aldrin, dieldrin, endrin, and isodrin;
Carbon tetrachloride and chloroform;
Hexachlorobenzene and hexachlorobutadiene;
Mercury and its compounds;
Cadmium and its compounds;
DDT and all its isomers;
Hexachlorocyclohexane HCH, (Lindane);
Pentachlorophenol;
Trichlorobenzene, trichloroethylene, and tetrachloroethylene; and
1,2 Dichloroethane.

The adoption of this Directive has highlighted the markedly different approach that the UK has taken compared with the rest of Europe and there was much acrimony in the discussions before the UK won its concession to use its own method. In mainland Europe, the List I substances are controlled by Limit Values (LVs) whilst the UK has adopted the technique of Environmental Quality Standards (EQSs).

The LV is the maximum permitted concentration for a substance (expressed as a monthly mean) in a discharge. This concentration (also known as the fixed emission standard) is applied irrespective of the size of the plant or the amount of dilution in the receiving stream. The LV is based on the best available technology to reduce the concentration in the discharge but also takes into account the cost of such treatment so as to avoid excessive expenditure. This approach embodies the spirit of the Treaty of Rome which set up the European Community. This required that no country would have any economic advantage over another when implementing the Directives. So, for example, a large factory in Germany situated in the upper reaches of the River Rhine and discharging cadmium in its effluent has to comply with the same standard as a small unit discharging to the same river in its lower reaches in Holland. Its applicability to preserving or improving the quality of waters in mainland Europe is doubtful because LVs take no account of the local situation. The small Dutch factory may have installed expensive effluent treatment facilities even though the wastewater has negligible effect on the river. Conversely, the large German factory may be complying with the LV but the concentration in the river may exceed the safe level for aquatic life.

The UK Government was opposed to the idea of using LVs to control pollution by dangerous substances and preferred to use EQSs. These are limits for particular substances which should not be exceeded in the receiving waters of a discharge. The limits are obtained by firstly deciding the use for the water (drinking water supply, salmon fishery, cooling water abstraction, *etc.*) and then setting limits for the quality parameters to ensure that the uses are not jeopardized. Amongst the arguments the UK put forward were that it was a self-contained island state and therefore didn't suffer from the problems of cross boundary pollution. It also firmly advocated that standards should reflect the local conditions so that advantage could be taken of the available dilution without adversely affecting the aquatic community. The concession was finally won and Table 1 shows the LVs and the EQSs that are now in use for the two List I metals cadmium and mercury. Despite the fact that the toxicity of both metals varies with water hardness and temperature, no allowance has been made for these factors in the derivation of the EQSs (*cf.* List II substances below).

List I Substances

Cadmium and mercury are the only metals included in the List I substances. The remainder are individual or groups of organic compounds. Of the 50 000 or so organic substances being used in the EC, 4500 are considered to be potential List I substances.[2] Many are used in small amounts only but 500 are estimated as being used in qualities of over 100 tonnes year^{-1}. Of these 500 a group of 132 was identified by a panel of experts as requiring priority control and of these just 18 have been officially classified. Five have been dropped from the list leaving 109 for future consideration. The intention is that each of these substances will have LVs or EQSs applied to them.

Table 1 *Limit Values and Environmental Standards for Mercury and Cadmium* ($\mu g\,l^{-1}$)

	Mercury			Cadmium		
	LV	Directive	Regulations	LV	Directive	Regulations
Freshwater	50	1 (total)	1 (total)	200	5 (total)	5 (total)
Estuaries	50	0.5 (diss)	0.3 (diss)	200	5 (diss)	2.5 (diss)
Coastal waters	50	0.3 (diss)	0.3 (diss)	200	2.5 (diss)	2.5 (diss)

In the UK, the next logical step after establishing the EQS for a particular substance, is to ensure that water courses or tidal waters are meeting the requirements. In those situations where an EQS may be breached then the effluent contributing the substance should be controlled. However, the Government have used a different approach from this and tackled the control of List I substances from the discharge point rather than the receiving water. Regulatory authorities have been asked to ensure that all discharges containing a List I substance have a consent limit applied to it and be monitored along with the receiving waters. 'When one or more of the substances is present in the discharge, information is required'. The difficulty with this request is to decide what is meant by 'being present'. Given sufficient sample and a good sensitive method of analysis most effluents could be shown to contain List I substances and should therefore be monitored. To counter this the concept of *de minimis* was introduced. In 1985, guidelines were issued as to what constituted a minimal amount below which it was unnecessary to consent discharges and monitor receiving waters. For mercury, for example, this was set at $3\,kg\,yr^{-1}$ loading. This numerical value has since been abandoned and the 1989 circular says that concentrations should be at their lowest possible level. Table 2 shows the results of monitoring a number of large discharges of treated and untreated sewage for Lindane in the West of Scotland and the substance is nearly always present because of its general spread around the environment. These data should be compared with the EQSs that have been set for Lindane:

Environmental Quality Standards
Annual mean concentration $(ng\,l^{-1})$

Inland surface waters	100
Estuary waters	20
Coastal waters	20

The monitoring has clearly demonstrated that Lindane is present in the effluents but are they at their lowest possible levels? All the effluents, except that from Dalmarnock STW, discharge into estuarine or coastal waters but after dilution will clearly not exceed the EQS. Do the values for Paisley STW represent the '*de minimis*' level or is it more likely to be about $50\,ng\,l^{-1}$? The same arguments apply to cadmium and mercury which are also found in trace quantities in discharges on most sampling occasions. What is required is a

Table 2 *Concentration of Lindane in Sewage Discharges* $(ng\,l^{-1})$

Discharge	$Conc^n$ in 1992
Dalmarnock sewage works effluent	44.0, 8.0, 26.2, 10.6
Dalmuir sewage works effluent	311.0, 32.0, 41.7, 43.4
Shieldhall sewage works effluent	28.7, 18.5, 72.3, 39.6
Paisley sewage works effluent	20.8, <1.0, <5.0, 4.9
Garnock Valley Sewer (crude sewage)	50.0, <1.0, 42.2, 31.3, 30.9, 38.8, 50.3
Irvine Valley Sewer (crude sewage)	62.7, 12.3, <10.0, 34.0, 16.3, 38.7

numerical value be given to '*de minimis*' instead of the present vague wording. The EQS should be applied in parallel with the limit value, so that where an effluent containing a list I substance, discharges into a river and the EQS is satisfactory, then the discharger still has to comply with an upper limit for the substance. This would be more workable than the present unspecified lower limit.

List II Substances

The EC Dangerous Substances Directive (76/464/EEC), obliges Member States to reduce pollution by List II substances using the EQS approach. The Department of the Environment contracted the Water Research Centre, WRC to recommend the appropriate EQSs for a selected list of List II substances for a range of possible uses of water. Their approach to the task has been outlined[3] and technical reports have been issued for:

	Technical Report Number
Chromium	207
Inorganic Lead	208
Zinc	209
Copper	210
Nickel	211
Arsenic	253
Inorganic Tin	254
Organo-tins	255
Boron	256
Sulfide	257
Iron	258
pH	259
Ammonia	260
Mothproofing Agents	261

EQSs have been proposed for the pesticides: endosulfan, atrazine, simazine, dichlorvos, trifluralin, azinphos-methyl, and malathion.

The approach taken by the WRC for each substance has been the same, namely:

- review available literature on the substance;
- evaluate laboratory toxicity data;
- determine the minimum adverse effect concentration;
- apply an appropriate safety factor;
- evaluate reported environmental levels;
- consider other factors, *e.g.* water hardness;
- publish proposed standard; and
- re-evaluate in the light of experience.

Once an EQS has been decided for a particular water use, the pollution control authorities then have to translate this figure into discharge consent limits using their knowledge of the use of the water, the flow rates of the discharge and the receiving water, and the quality of the water upstream.

It is interesting to examine the application of these EQSs for three substances, lead, permethrin, and atrazine, in different locations in the West of Scotland.

Lead. The recommended EQSs for inorganic lead are:

Water hardness (mg $CaCO_3 l^{-1}$)	Salmonid waters	Cyprinid waters
	(Annual average, µg soluble $Pb l^{-1}$)	
0–50	4	50
50–150	10	125
>150	20	250

In the upper reaches of the Glengonnar Water, a tributary of the River Clyde, there has been extensive mining for zinc and lead over many years. The last mine closed in 1930 but there are spoil heaps over a large area. The leachate from these and the inputs of drainage water from the shafts and tunnels contaminate the river. Elevated concentrations of lead are found in the upper reaches of the river and levels decline downstream with dilution. Table 3 shows the results of monitoring.[4]

Brown trout were found at all of the sampling sites even though the EQS was exceeded at each site. However, those that were caught at the topmost sites were found to have blackened tails, a symptom of lead poisoning. Those from the lowest sampling site appeared to be healthy.

This one survey suggests therefore that the EQS of $10 \, \mu g \, Pb \, l^{-1}$, for a water with a hardness of $50–150 \, mg \, CaCO_3 l^{-1}$, has been set at a correct level because no apparent effect was found at sample point 3 where the soluble Pb was $14.2 \, \mu g \, l^{-1}$ for a water hardness of $55 \, mg \, CaCO_3 l^{-1}$. However the margin from a safe level of lead to one where chronic effects can be found is narrow.

Permethrin. Permethrin is a synthetic pyrethroid insecticide which was first developed about 20 years ago. It has many uses including being used as a mothproofing chemical. The proposed EQS for the substance to protect freshwater life is at a very low concentration—the 95%ile concentration should not exceed $10 \, ng \, l^{-1}$.

In the village of Stewarton in Ayrshire, a company prepares wool for the garment and carpet manufacturing industries. The wastewater from the factory

Table 3 *Mean Concentration of Lead in Glengonnar Water* ($\mu g \, l^{-1}$)

Distance from Source (km)	Lead		Water Hardness (mg $CaCO_3 l^{-1}$)
	Total	Filterable	
3.3	148	65	50
3.8	61.4	24.2	73
9.0	27.6	14.2	55

Table 4 *Discharge of Permethrin from Stewarton STW into Annick Water Concentration $ng\,l^{-1}$ as Permethrin*

	River Annick at Stewarton	Stewarton Sewage Effluent	River Annick at Chapelton
22/7/92	< 50	< 50	< 50
17/9/92	63	2,140	40
14/10/92	< 50	135	81
17/11/92	1,700	912	1,460
15/12/92	< 50	120	< 50

is discharged into the public sewer for treatment at the sewage treatment works (STW) which discharges its effluent into the Annick Water. The results of the analysis of samples taken from the river and the STW effluent are shown in Table 4. The results are confusing because on occasions, mothproofing chemicals are discharged directly into the river from the wool processors because of spillages. The accidental discharges take place upstream of the sewage works effluent.

The analytical methods available at the moment for permethrin are such that it is difficult to detect concentrations as low as the EQS but clearly this value is exceeded on occasions.

The invertebrate fauna of the stream are sampled and examined twice yearly and the results expressed as a Biotic Index. The Average Score Per Taxa (ASPT) has a scale of 0–10 but in the West of Scotland the maximum value rarely exceeds 8. The results for the River Annick for 1992 are shown in Table 5. (The sewage works discharges its effluent about 250 m downstream of Lainshaw House.)

The results show that the invertebrate fauna are much depleted in the river below the STW outfall, however fish are seen throughout the river. There is obviously an insecticidal compound in the effluent because other analytical data for the effluent, such as BOD, show that it is of good quality. It is not possible to say whether the EQS for permethrin is appropriate for protecting the invertebrate community. The situation is complicated by evidence recently obtained that other pesticides are also present in the river. Samples taken from the lowest reaches of the river have been analysed for a range of pesticides and

Table 5 *Biotic Indices for Annick Water (ASPT)*

Sample Site	May 1992	December 1992
North Ayrshire Markets	6.57	6.37
B778 Bridge	5.18	5.36
Lainshaw House	4.25	5.21
Chapelton	3.50	3.40
Perceton	3.64	3.67
Broomlands Bridge	3.45	3.83

diazinon and propetamphos are frequently detected at the hundreds of $ng\,l^{-1}$ level. These substances may well originate from the sheep's wool as they are used as sheep-dip chemicals.

Atrazine. This compound is a triazine herbicide which has been used extensively in agriculture, forestry, and, in combination with simazine, as a total weed control chemical by local authorities, British Rail, *etc.*

The proposed EQS is that the annual average concentration of atrazine and simazine taken together should not exceed $2\,\mu g\,l^{-1}$, and that the maximum combined concentration (MAC_x) should not be greater than $10\,\mu g\,l^{-1}$ for any one occasion.

Recently the herbicide has been used in combination with another triazine compound, cyanazine, to combat weeds in a coniferous forestry planting scheme in Lanarkshire. As each tree is planted, the herbicide is applied around its base to eliminate competing undergrowth. The planting has taken place over a few months of 1993 and rainfall, sometimes heavy, has washed some of the herbicide into the drainage streams. Samples of the stream water were taken throughout the planting period to check for herbicide levels. The results for one of the most affected streams are shown in Table 6.

On two occasions the concentration of atrazine exceeded the maximum allowable concentration of $10\,\mu g\,l^{-1}$. However the total concentration of triazines is likely to be about double the figures shown because the herbicide mixture contains equal amounts of the two active ingredients.

Laboratory test data has shown that the highest no-effect concentration in benthic invertebrates is $10\,\mu g$ atrazine l^{-1}. One would anticipate therefore that there would be some adverse effect on the insect life in the Drumalbin Burn from the planting activity. However samples of the stream invertebrates taken in early August showed that there was a good variety of healthy organisms. There also didn't appear to be any adverse effect on the plant life.

This limited investigation suggests that the proposed EQS for the triazine herbicides is too stringent. There are of course other reasons for restricting their use and imposing a strict limit, including their regular appearance in trace quantities in drinking water samples.[5]

In general, therefore, it would appear that the use of EQSs to control the effect of polluting substances is an effective technique that preserves the aquatic

Table 6 *Lochlyock Afforestation Scheme. Concentration of Atrazine in Dumalbin Burn at Stonehill Farm*

	$\mu g\,l^{-1}$
27/4/93	24.5
6/5/93	3.0
14/5/93	16.9
15/6/93	2.3

life whilst not imposing too harsh a financial burden on dischargers for treating wastes containing the substances.

However, although a proposed EQS may seem sensible, and based on good scientific principles, they can be changed when discussed by the Member States. Bargains can be struck to allow, for example, a relaxation of a particular standard for water in exchange for the tightening of a completely unrelated environmental standard such as lawnmower noise!

4 Red List Substances

The Red List of substances comprises those materials that are regarded as the most hazardous based on the criteria of toxicity, persistence, and bioaccumulation. Their production rates and physicochemical properties have also been taken into account before inclusion on the list.

The derivation of the Red List of substances and the action plan to reduce their input into the North Sea by the littoral States can be largely attributed to the success of the Green Movement in former West Germany.

Using emotive and alarmist language, they persuaded the German government to host an international conference to discuss ways of halting the deterioration in the quality of the North Sea. The first conference was held in Bremen in 1984, followed by a second in London in 1987, and the third in the Hague in 1990.

It was at the second conference that the participating countries decided on a target of reducing by about 50% by 1995, from a base year of 1985, the loading of hazardous substances into the North Sea. A further target was made that by the year 2000 the inputs should be reduced to a point where hazardous substances 'no longer represented a threat to man or nature'.

At the third conference the substances were identified and these became known as the Red List as shown in Table 7. The target reduction was modified at the third conference; it was decided that the inputs of cadmium, mercury, lead, and dioxins should be reduced by 70% or more by 1995 and that PCBs should be eliminated completely.

These fine-sounding political statements have produced many problems for the organizations that have to implement them. In the UK there was a distinct lack of data available on the loadings of all the Red List substances for 1985 because they had not been tested for in the major discharges and rivers at that time. In 1989 the Clyde River Purification Board (CRPB), in common with other pollution control authorities, instigated a cash programme to purchase the necessary equipment to measure the Red List substances, to modify laboratories, recruit new staff, and liaise with dischargers about the installation of the effluent sampling devices. By 1991, a comprehensive programme was in place for the sampling and analysis of major rivers at their tidal limits and the direct discharges to tidal waters of sewage and industrial wastewater, and to measure their flow rates. The intention was to use these data as a substitute for the 1985 baseline and to effect the reductions in a shorter timescale.

It has now become apparent that it will not be possible to demonstrate that a 50% reduction in loading has taken place in 1995 even if substantial

Table 7 *The UK Red List*

Aldrin
Atrazine
Arsenic
Azinphos-methyl
Azinphos-ethyl
Cadmium and its compounds
Carbon tetrachloride
Chloroform
Chromium
Copper
DDT (including metabolites DDD and DDE)
1,2-Dichloroethane
Dichlorvos
Dieldrin
Dioxin
Endosulfan
Endrin
Fenitrothion
Fenthion
Hexachlorobenzene
Hexachlorobutadiene
γ-Hexachlorocyclohexane
Lead
Malathion
Mercury and its compounds
Nickel
Parathion
Parathion methyl
PCBs (polychlorinated biphenyls)
Pentachlorophenol
Simazine
Trichlorobenzene (all isomers)
Trichloroethylene
Trichloroethane
Tetrachloroethane
Trifluralin
Triorgano-tin compounds
Zinc

improvements are made in treatment processes. This is because there is such a large variability in both flow rates and the concentration of those few Red List substances that have been found to be present, that it would require an enormous sampling effort to calculate the actual loading figures with any degree of confidence. The data in Table 8 shows the variation in loadings for two discharges in the West of Scotland.

A statistical analysis of these data shows there is a wide variation in the flow and concentrations figures that go to make up the loadings calculations. If the loadings were to be obtained with, say, a 20% confidence in the results, then about 650 samples would have to be taken in the year for obtaining the loading

Table 8 *Irvine Valley Sewer–Cadmium*

	1991			*1992*		
	Flow $(m^3 d^{-1})$	*Concentration* $(\mu g l^{-1})$	*Loading* $(kg\,yr^{-1})$	*Flow* $(m^3 d^{-1})$	*Concentration* $(\mu g l^{-1})$	*Loading* $(kg\,yr^{-1})$
Mean	82 350	3.56	104	96 100	3.0	105
Min	50 600	0.75	17.7	65 130	0.6	14.2
Max	157 100	7.6	204	134 000	6.0	208
SD		2.0	59		1.9	65

Dalmuir Sewage Works–Mercury

	1991			*1992*		
	Flow $(m^3 d^{-1})$	*Concentration* $(ng\,l^{-1})$	*Loading* $(kg\,yr^{-1})$	*Flow* $(m^3 d^{-1})$	*Concentration* $(ng\,l^{-1})$	*Loading* $(kg\,yr^{-1})$
Mean	223 300	66	5.1	223 400	72	6.0
Min	161 800	28	2.0	175 000	29	2.4
Max	444 000	117	7.6	280 000	149	15.0
SD		28	2.2		35	3.6

of cadmium in the Irvine Valley sewer and at least 30 samples per year for calculating the loading of mercury from Dalmuir STW.

An evaluation has been carried out[6] on the relative contributions of some of the Red List substances that are made by rivers, direct discharges, and from the dumping of sewage sludge and dredgings into the North Sea. This found that, even if the contributions from the direct discharges were eliminated, between 73–94% of metals entering the North Sea originates from rivers and from dredgings. The dredgings, of course, result from the deposition of particulate matter into the estuaries from the rivers so the loadings from this source have been largely accounted for in the monitoring of rivers at their tidal limits. So the conclusion was that the main emphasis of the effort at reducing loadings should be directed at the many inland discharges if the target reductions were to be achieved. Furthermore it was pointed out that, even if the targets were achieved, there will not necessarily be any direct benefit to the North Sea. Contamination of shellfish and fish flesh by metals and organic pollutants are not considered to be hazardous for the major part of this body of water.[7] Indeed, it is considered to be highly productive and supplies fish for consumption to many of the surrounding countries. There are problems though in some of the estuaries which receive industrial waste.

It would appear then, that when 1995 arrives it will not be possible to say whether the 50% or 70% reductions have been achieved (though politicians, with their skill at statistics will probably say otherwise!).

The North Sea Conference proposal for a reduction in the loading of a list of dangerous substances has been of value though to the aquatic environment.

It has ensured that the pollution control authorities examined more critically the concentration of a range of dangerous chemicals that enter the tidal waters in their areas. Where elevated levels have been found then pressure can be applied on the users of the substances to either install additional treatment facilities or to look for alternative, less hazardous materials. This has happened in the West of Scotland; significant contributions of copper, cadmium, pentachlorophenol, and mercury have been identified and discussions are currently taking place with the dischargers on how to effect a reduction in their release into the environment.

Thus, although the political target cannot be attained, the environment of the North Sea and its estuaries will be cleaner because of the effort that has been put in to reduce the loading of a range of polluting substances.

The views expressed in this paper are those of the author and are not necessarily those of the Clyde River Purification Board.

References

1. J. S. Alabaster and R. Lloyd, 'Water Quality Criteria for Freshwater Fish', Butterworth, London, 1980, Chapter 8, p. 160; Chapter 9, p. 190.
2. ENDS, 'Dangerous Substances in Water', Environmental Data Services Ltd., Bowling Green Road, London, 1992, Chapter 2, p. 4.
3. J. Gardiner and G. Mance, 'Proposed Environmental Quality Standards for List II Substances in Water: Introduction', Water Research Centre Technical Report TR 206, Water Research Centre, Medmenham, Berks, UK, 1984.
4. Clyde River Purification Board, 'Concentration of trace metals in water and brown trout from the Glengonnar Water, Leadhills', Clyde River Purification Board, East Kilbride, Scotland, 1983.
5. Department of the Environment, 'Drinking Water 1990. A report by the Chief Inspector', *Drinking Water Inspection*, Her Majesty's Stationery Office, London, 1991.
6. G. Mance, 'Pollution Threat of Heavy Metals in Aquatic Environments', Elsevier Applied Science, Barking, Essex, 1987, Chapter 11, pp. 324–329.
7. T. F. Zabel and D. G. Miller, 'Water Quality of the North Sea: Concerns and Control Measures', *J. Inst. Water Environ. Manage.*, 1992, **6**, 31–49.

Effects of Redox Variations on Metal Speciation—Implications on Sediment Quality Criteria Assessment

U. Förstner, W. Calmano, J. Hong, and M. Kersten

SECTION OF ENVIRONMENTAL ENGINEERING, UNIVERSITY OF TECHNOLOGY OF HAMBURG-HARBURG, EIßENDORFERSTRASSE 40, 21073 HAMBURG, GERMANY

Summary

New objectives regarding the improvement of water quality, as well as problems with the resuspension and land deposition of dredged materials, require a standardized assessment of sediment quality. Numerical criteria approaches, which are based on pore water concentrations, and the accumulation, solid/liquid partition, and elution properties of contaminants, usually do not imply variations of chemical interactions with solid matrices. Regarding the potential release of metals from sediments, changes in pH and redox conditions are of prime importance. To incorporate new experience with non-linear and time-delayed processes, special emphasis should be put on the characteristics of the mineral and organic solid matrices, *e.g.* to capacity controlling properties, and in particular, the buffer capacity against pH-depression.

1 Introduction

Sediment quality criteria were developed in the mid-eighties for the following reasons.

(i) In contrast to the strong temporal variability in the water phase, sediments integrate contaminant concentrations over time. One can, therefore, reduce the *number of samples* in monitoring, surveillance, and survey activities.

(ii) Long-term perspectives in water management need *'integrated strategies'*, in which sediment-associated pollutants have to be considered as well; generally, the contaminant level in the sediment may have greater impact on the survival of benthic organisms than do aqueous concentrations.

(iii)   Management plans have increasingly been based on the *assimilative capacity* of a certain receiving system, and this requires knowledge of the properties of sedimentary components as the major sink.

(iv)   Finally, there is the wide spectrum of problems with dredged sediments: permission for *dredging activities* and *deposition of dredged material* have to be based on standardized sampling protocols and test procedures. In this context, the conventions for the marine environment—Oslo and London—should be mentioned.

Predictions of the short-, middle-, and long-term effects of metal contaminated particulates, in the framework of sediment quality objectives and criteria development, have to consider the environmental 'speciation' of critical metals, which in this context means 'describing the distribution and transformation of metal species and various media'.[1] Long-term prognosis, in particular, of the behaviour of metals at critical sites requires both the knowledge of interactions of element species in solid matter and solution, and an estimation of the future borderline (particularly 'worst case') conditions in a dynamically evolving medium.[2] In sediments, typical driving forces for intensified matrix–element interactions are strong chemical gradients of redox conditions, pH-values, and organic ligand concentrations, all three factors being mainly induced by the degradation of organic matter.

2 Redox Variations: Implications for Sediment Quality Criteria

Numerical approaches for the assessment of the environmental impact of sediment-associated pollutants are based on: (i) accumulation; (ii) pore water concentrations; (iii) solid–liquid equilibrium partition (both sediment–water and organism–water); and (iv) elution properties of contaminants.[3]

Background Data Comparisons: This procedure compares data from the test area with that of natural or insignificant pollutant concentrations. Particularly useful are samples from deeper layers of the sediment sequence at a given site, for example, from drill holes, since this material is derived from the same catchment area and usually is similar in its substrate composition. Nonetheless, standardization with respect to grain size distribution is indispensable.

Porewater Analysis: This is based on the experience, that the composition of interstitial waters is the most sensitive indicator of the reactions that take place between pollutants on particles and the aqueous phase which contacts them. There is the advantage of a direct recovery and analysis of water-borne constituents. However, there are several disadvantages, mainly arising from the sampling and sample preparation, which need considerable precautions, such as for exclusion of oxygen, which could alter redox states during storage.

Equilibrium Partition Studies: These approaches are related to the broad toxicological basis of food and water quality data and as such, offer a very important advantage. On the other hand, there are the effects of sample preparation, *e.g.* the drying procedures; the separation techniques, *e.g.* filtration or centrifugation; there are strong effects of grain size composition and the influence of suspended matter concentration in the aquatic system, which is even more important, if the kinetics of sorption and desorption are too slow for equilibria to be achieved in a given time of interaction. Unlike non-ionic organic chemicals, the K_D-values of *metals* are not only correlated to organic substances but also with other sorption-active surfaces. Hence, the equilibrium partition approach exhibits strong limitations for metallic elements.

Remobilization: Short-term effects may be studied from water–sediment suspensions, medium-term effects from experiments using tanks, and long-term effects by applying chemical extractants, either singly or in sequence. Field observations often do not show clear effects, as has been demonstrated for the release of metals from anoxic sediments during oxidation. Such implications for future criteria development, particularly important for dredging and the management of dredged material, will be discussed on the basis of experience from metal speciation studies on soils and sediments.[4]

Effects of Redox and pH Variations on the Mobility of Metals

Acidity is perhaps the most important long-term factor in the mobilization of metals from metal-bearing wastes and, in some instances, sediments. The threat is especially great in waters with little buffer capacity, *i.e.* in carbonate-deficient areas where dissolved-metal pollution can be spread over great distances. The acidity production can develop many years after disposal, *e.g.* when the neutralizing or buffering capacity in a pyrite-containing waste is exceeded.

The major process affecting the lowering of pH-values (down from pH 2 to 3) is the exposure of pyrite (FeS_2) and other sulfide minerals to atmospheric oxygen and moisture, whereby the sulfidic component is oxidized to sulfate and acidity (H^+-ions) is generated. Bacterial action can assist the oxidation of Fe^{2+} (aq) and precipitation as $Fe(OH)_3$ in the presence of dissolved oxygen.

The acidification problem in a sediment–water system arises after hydrogen ions are generated during oxidation, *e.g.* during dredging or resuspension of mainly fine grained material containing less carbonate than needed for long-term neutralization.[5,6] Primary emissions of high metal concentration occur from waste rocks and tailings and secondary effects on ground-water take place from ponds. An important and long-term source of metals are the sediments reworked from the floodplain, mainly by repeated oxidation and reduction processes.[7]

If anoxic sediments are exposed to the atmosphere, redox conditions will change and a new distribution and transformation of heavy metal species bound to the sediments takes place. Periodic changes, for example, are typical for the sediments of older fluvial or marine terraces in monsoon regions, which have a seasonally changing water level caused by submergence with rain water. Under these conditions, hydrology favours either lateral or vertical drainage and leads to a variation in redox potential.[8]

Tidal currents lead to daily periodic redox changes in coastal and estuarine sediments. In some coastal plains the sediments and soils are seasonally flooded.

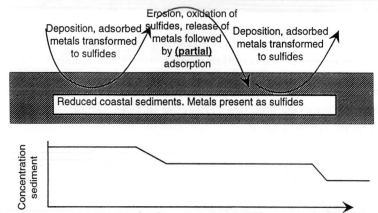

Figure 1 *Decrease in metal concentrations in coastal sediments through cyclic processes of erosion and deposition (Dr. W. Salomons, Haren/Netherlands, unpublished results)*

By these processes, the chemical properties of sediment compounds are changed. For example, iron(II) sulfide (or pyrite) is formed by reduction of sulfate from sea water, while dissolved hydrogen carbonate is partly removed by diffusion or convection. Over several years or decades appreciable amounts of sulfide can be accumulated in tidal sediments.[9]

For coastal marine environments, such as the Wadden Sea coast, the possible implications of cyclic processes on metal concentrations in sediments have been modeled schematically by Salomons.[10] Figure 1 suggests, that after deposition and during storage in reduced coastal sediments, adsorbed metals are transferred to sulfidic bonding forms. During erosion processes, oxidation can take place involving remobilization of a certain proportion of the sediment-associated metals, which may subsequently be readsorbed in part by particulate matter. Under marine conditions, the extent of adsorption will mainly depend on the stability of complexes formed by chloride and competition for adsorption sites by cations like Mg^{2+}, Ca^{2+}, and Na^+. Such oxidative events are relatively short compared to the longer periods of reduced conditions during deposition. However, it can be expected that through repeated, though irregular (in contrast to the tidal interactions mentioned before) cyclic processes of erosion and deposition, a significant release of metals from coastal sediments will take place.

During the last two decades, transformation of heavy metal binding forms under changing redox conditions has received much attention. Examples have been given for seasonally changing cadmium mobilities in Corpus Christi Bay Harbour,[11] for releases of Hg, Pb, and Zn from estuarine sediments of Mobile Bay,[12] for the transformation of Cd compounds in Mississippi River sediments under controlled redox and pH conditions,[13] for the removal of sulfide from pore waters and subsequent Cd mobilization via ventilation of the upper sediment layers,[14,15] and for the release of Pb and Cd from contaminated dredged material after dumping in a harbor environment.[16] Typical early diagenetic geochemical changes and subsequent element mobilization via the porewater, as a result of effects of oxidation during dredging activities, were studied by Darby *et al.*[17] in a man-made estuarine marsh.

From enclosure experiments in Narragansett Bay it has been estimated by Hunt and Smith[18] that, by mechanisms such as oxidation of organic and sulfidic material, the anthropogenic proportion of cadmium in marine sediments is released to the water within approximately three years; where as for remobilization of copper and lead, approximately 40 and 400 years, respectively, is needed, according to extrapolations.

Chemical extraction experiments, for estimating characteristic particulate binding forms of metals in anoxic marine and freshwater sediments, were carried out by Kersten and Förstner,[19] both in the presence and absence of atmospheric oxygen during the experiments. By drying the sediments under oxygen, the proportion of sulfidic bound metals was significantly decreased.

In Table 1, an example is given for the possibilities of standardizing the data from elution experiments with respect to numerical evaluation. An 'elution index' for sediment samples from various rivers in West Germany is based on the metal concentrations exchangeable with 1 M ammonium acetate at pH 7. Higher values imply greater mobility. These metal fractions are considered to be remobilizable from polluted sediments, in the relative short term, under more saline conditions, for example, in the estuarine mixing zone. Comparison of the release rates from oxic and anoxic sediments clearly indicates that the oxidation of samples gives rise to a very significant increase in the mobilization of the metals studied.

Limitations of Available Test Procedures

It is clear that when considering the mobilization of heavy metals from sediments, there are two dominant variables: redox potential and pH, with the latter directly influenced by the former. For the practice of criteria development and application, available test procedures should be checked regarding their relevance for predicting the middle- and long-term behaviour of metal-bearing sediment matrices under changing redox and pH conditions.

Table 1 *Elution Index for Selected River Sediments, as Determined from Exchangeable Properties (Extracted with 1 M Ammonium Acetate)*

	Neckar	*Main*	*Elbe*	*Weser*
Copper	< 1	—	1	—
Lead	1	1	1	1
Zinc	7	10	40	10
Cadmium	22	22	25	—
Toxic oxic	*30*	*33*	*67*	*11*
(Anoxic	0.5	0.3	> 4	4)

Thermodynamic Models. With respect to the possible implications of sediment-bound metal species, *e.g.* in the framework of sediment quality assessment studies, it seems that thermodynamic models are still of limited use for various reasons: (i) adsorption characteristics are related not only to the system conditions (*i.e.* solid types, concentrations, and adsorbing species), but also to changes in the net system surface properties resulting from particle–particle interactions, such as coagulation and ageing processes; (ii) influences of organic ligands in the aqueous phase can rarely be predicted as yet; (iii) the effects of competition between various sorption sites should be considered; and (iv) the reaction kinetics of the individual constituents cannot be evaluated in a mixture of sedimentary components.

Pore Water Chemistry. As mentioned previously, the composition of pore water is a highly sensitive indicator for reactions between chemicals on solid substrates and the aqueous phase which contacts them. For fine-grained material, in particular, the large surface area compared with the small volume of its entrapped interstitial water, ensures that minor reactions with the solid phase will be indicated by major changes on the composition of the aqueous phase. While the direct recovery and analysis of water-borne constituents can be seen as a major advantage of this approach, there are several disadvantages, particularly arising from sampling and sample preparation, which are not yet routine procedures, and usually involve considerable precautionary measures, such as exclusion of oxygen.

Elutriate Tests. To estimate short-term chemical transformations, the inter-relations between solid phases and water have been increasingly subjected to laboratory studies. The advantage of such experiments is that especially important parameters can be directly observed and particularly unfavourable conditions simulated. The US Army Corps of Engineers and the US Environmental Protection Agency have developed an elutriate test that is designed to detect any significant release of chemical contaminants in dredged material. This test involves the mixing of one volume of the dredged sediment with four volumes of the disposal site water for a 30 min shaking period. If the soluble chemical constituent in water exceeds 1.5 times the ambient concentration in the disposal site water, special conditions will be imposed and will govern the disposal of the dredged material.

Sequential Extraction. In connection with the problems arising from the disposal of solid wastes, particulary of dredged materials, chemical extraction sequences have been applied, which are designed to differentiate between the exchangeable, carbonatic, reducible (hydrous Fe/Mn oxides), oxidizable (sulfides and organic phases), and residual fractions. The undisputed advantage of this approach with respect to the estimation of long-term effects on metal mobilities lies in the fact that rearrangements of specific solid 'phases' can be evaluated prior to the actual remobilization of certain proportions of an element into the dissolved phase.

3 Prognostic Tools for Long-term Behaviour of Metals in Sediments

Redox and pH variations and their effects on metals in natural systems are typical non-linear and delayed—'chemical time bomb'—processes, *e.g.* toxic metals can 'break through' once the specific buffer capacity of a sediment system has been surpassed. Within a scientific perspective of the chemical time bomb concept, the aspect of the storage capacity controlling properties (CCP's) of solid substrates is playing a key role.[20] There are two mechanisms for potential time bomb evaluation. The first is direct saturation, by which the capacity of a sediment for toxic chemicals becomes exhausted; the second way to 'trigger' a time bomb is through a fundamental change in a chemical property of the substrate, that reduces its capacity to adsorb (or keep adsorbed) toxic materials. In the aquatic environment, such effects are mainly induced by redox variations and it is, in particular, the carbonate buffer capacity of the sediments, which controls the pH values and the most significant interactions between solid and dissolved metal phases, as well as transfer processes of trace metals between inorganic and organic solid substrates.

A typical example, dredged material from Hamburg Harbour, indicates typical differences in the kinetics of proton release from organic and sulfidic sources.[21] Recent deposits are characterized by low concentrations of nitrate, cadmium, and zinc; when these low-buffered sediments are oxidized during a time period of a fews months to years, the concentrations of ammonia and iron in the pore water typically decrease, whereas those of cadmium and zinc increase (with the result that these metals are easily transferred into agricultural crops!). The different steps are schematically given in Figure 2. Oxidation of sulfides during stage B strongly increases the concentrations of cadmium and zinc in a relative short time. When acidity is consumed by buffer reactions (phase C), cadmium and zinc concentrations drop, but are still higher than in the original sulfidic system. In phase D, oxidation of organic matter again lowers pH values and can induce a long-term mobilization of Zn and Cd. The latter development, in particular, can rarely be predicted with available procedures.

Criteria for the prognosis of the middle- and long-term behaviour of metals should include the abilities of sediment matrices for producing acidity and for neutralizing such acid constituents.

Acid Producing and Consuming Capacity of Sediments

The concept of the acid-producing capacity (APC) was initially developed as part of the prediction and calculation of acid mine drainage and waste tailings management.[22] Periodic redox processes can cause an increase or decrease in APC or pH in a sediment–water system.[23] In a *closed* system, periodic redox processes can lead to a change or transfer between APC(s) and APC(aq), but the total APC of the system does not change. The processes are reversible. The hydrogen ions produced in the oxidation cycle will be consumed by the following reduction cycle. However, in an *open* system, the total APC of the system will

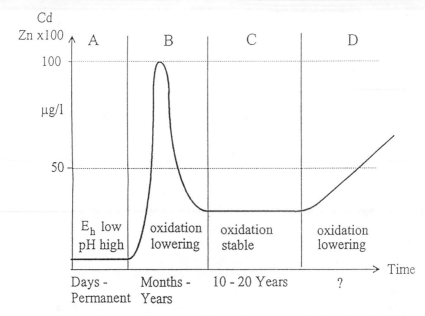

Figure 2 *Schematic diagram illustrating different phases of metal release from land-disposed (low-carbonate) dredged material (Maaß and Miehlich,[21] modified)*

change depending on the properties of the system and the reaction processes. Under certain conditions, the total APC in the system increases, where as under other conditions, the total APC in the system decreases. Some processes are irreversible. The components producing or consuming H^+ leave the system and cause the changes in APC(s), APC(aq), and the permanent acid consuming capacity (ACC). Important acid-producing oxidation reactions, leading to a decrease of pH in low buffered sediments, are listed in Table 2.

Table 2 *Important Acid-Producing Oxidation Reactions in Aquatic Systems*

Elements	Reactions	Acid formation coefficient, f
inorganic S	$H_2S + 2O_2 = SO_4^{2-} + 2H^+$	2
S	$S^0 + 3/2O_2 + H_2O = SO_4^{2-} + 2H^+$	2
S, Fe	$FeS + 9/4O_2 + 3/2H_2O = FeOOH + SO_4^{2-} + 2H^+$	2
S, Fe	$FeS_2 + 15/4O_2 + 5/2H_2O = FeOOH + 2SO_4^{2-} + 4H^+$	4
Fe	$Fe^{2+} + 1/4O_2 + 5/2H_2O = Fe(OH)_3 + 2H^+$	2
N	$NH_4^+ + 2O_2 = NO_3^- + H_2O + 2H^+$	2
N	$NO_x + 1/4(5 - 2x)O_2 + 1/2H_2O = NO_3^- + H^+$	1
organic N	$R{-}NH_2 + 2O_2 = R{-}OH + NO_3^- + H^+$	1
S	$R{-}SH + H_2O + 2O_2 = R{-}OH + SO_4^{2-} + 2H^+$	2

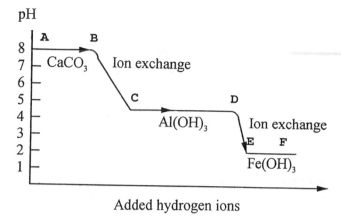

Figure 3 *Acid buffering processes in soils.[25] This curve summarizes important aspects of acidification in a soil–water system*

The acid-producing capacity is not only related to the oxidation of sulfides, but oxidation of organic matter must be considered as well. For example, the contribution of protons from organic nitrogen and organic sulfur in a sample containing approximately 5% organic carbon is equivalent to the acid-producing capacity of 1% FeS_2.[24] In total, APC can be defined as the largest amount of hydrogen ions (H^+) produced by the oxidation of the components per unit weight of sediment, each unit volume of water, or in the whole sediment–water system.

Acid consuming processes in terrestrial and aquatic ecosystems include:[23] (i) mineralization of cations; (ii) assimilation of anions; (iii) protonation of anions; (iv) reductions; (v) weathering of metal oxide components; (vi) reverse weathering of anions; and, in particular, (vii) dissolution of carbonate minerals. A schematic curve of different buffering levels in soil–water systems (Figure 3 after Prenzel[25]) combines two factors: the horizontal axis, which usually refers to the 'capacity' factor, indicating the amount of strong acid added, while the vertical axis indicates the 'intensity' factor, *i.e.* the pH or the degree of base saturation of the exchange complex. As long as $CaCO_3$ is present, added protons will be consumed by dissolution reactions, and the pH of the system will remain constant (A → B in Figure 3). Between B and C the exchangeable sites of clay minerals are displaced by hydrogen ions. Ultimately, this process leads to the dissolution of Al, *e.g.* clay minerals (C → D); due to the abundance of Al in clay minerals, the buffering capacity in this process is high. When the accessible reservoir of Al is exhausted, the dissolution of ferric oxide will consume hydrogen ions and resist the pH decrease (E → F). With regard to long-term effects, major processes consuming hydrogen ions in ecosystems include the release of basic cations by decomposition of organic matter, specific anion adsorption, mineral weathering, and unbalanced reductions of oxidized compounds.[26] The process of specific anion adsorption in sediments mainly refers to interactions of sulfate and phosphate with aluminium and iron sesquioxides.[27]

Estimating Acid Producing and Consuming Capacities

Experimental approaches for calculating APC and ACC for sulfidic mining residues have been summarized by Ferguson and Erickson.[28] A test described by Sobek *et al.*[29] involves the analysis of total pyritic sulfur; potential acidity is then subtracted from the neutralizing potential, which can be obtained by adding a known amount of HCl, heating the sample, and titrating with standardized NaOH to pH 7. Bruynesteyn and Hackl[30] calculated APC from total sulfur analysis; here the acid-producing capacity was subtracted from the acid-consuming capacity, obtained by titration with standardized sulfuric acid to pH 3.5. The APC-relationships of sediments are more complex than those in sulfidic ores because the APC of organic matter must be considered; also, the time scale plays a major role in sediments (Figure 2).

The acid neutralization capacity of a natural aquatic system is composed of the acid consuming capacity of the solids ACC_s and that of the dissolved phase ACC_{aq}:

$$ACC = ACC_s + ACC_{aq}$$

The ACC_{aq} of the dissolved phase (in aquatic systems this is always much smaller than ACC_s) may be described by the following equation:

$$ACC_{aq} = [HCO_3^-] + 2[CO_3^{2-}] + 2[S^{2-}] + [HS^-] + [NH_3] - [H^+]$$

A similar rule applies to the ACC_s of the solids at pH > 5:

$$ACC_s = 2[CaO] + 2[MgO] + 2[Na_2O] + 2[K_2O] + 2[MnO] + 2[FeO] \\ - 2[SO_3] - 2[P_2O_5] - [HCl]$$

From the two parameters, ACC and APC, an effective acid producing capacity APC_{eff} can be calculated. This may be helpful in practice for the specification of sediment quality criteria. The APC_{eff} of a sediment suspension may be defined as:

$$APC_{eff} = V/W([H^+]_e - [H^+]_o)$$

where APC_{eff} is effective acid producing capacity; V is suspension volume; W is solid mass; and $[H^+]_{o,e}$ is hydrogen ion concentration before and after oxidation of the suspension.

For example, for an anoxic Elbe River sediment with an original pH_o of 7.03 and a pH_e of 3.29, an APC_{eff} of 5.12 mmol kg^{-1} was calculated after an oxidation time of 35 days. Such simple test procedures allow better statements of the sediment quality than simple considerations of limiting values.

4 Implications for Metal-related Sediment Criteria

While the actual focus in sediment criteria is on individual pollutant concentrations and their biological effects, inclusion of data on potential matrix variations by redox and pH changes has not been considered, as yet, to any significant extent. The only exception is the initiative of the United States Environmental Protection Agency to introduce the acid volatile sulfide (AVS) procedure as a measure for the non-availability of critical metals at a surplus of reactive sulfide ions.[31] Such effects have been investigated experimentally with a multi-chamber device, where the individual components are separated by membranes.[32] Metal transfer processes observed in the experimental device, usually cannot be measured under field conditions, due to the complexity of natural systems and short residence time of released metals in the water phase. Extrapolation of such findings, therefore, into quantitative models seems to be nearly impossible, and this may also restrict their incorporation into criteria. Three approaches are given below for providing a basis for the decision-making processes.

Mobilization of Metals Following Resuspension of Anoxic Sediments

As has been shown in previous sections, metals can be released into the dissolved phase, but may subsequently be readsorbed or precipitated, in part, to solid phases. Resuspension of sediments from the Elbe River, which were characterized by a high acid producing potential and low neutralizing capacity, was studied from a regional and long-run perspective. Experiments were undertaken at constantly neutral pH values, and both total release and scavenging rates were extrapolated from time series of measured net release values over 630 h.[33]

The example in Figure 4 shows that the initial lead release was high, but then decreased continuously due to readsorption. After about 100 h the relative equilibrium between release and readsorption was reached. Curves for copper indicate continuous release until about 350 h, when the scavenging rate became faster than the release rate. With respect to cadmium and zinc, release was

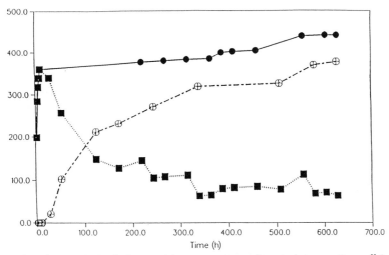

Figure 4 *Laboratory simulation experiments on wet sediments in a reaction cell indicate typical time evolution of interactive release and scavenging processes: lead release from Elbe sediments.[33] Full quadrangles: net release; full circles: total release; open circles: scavenging (adsorption/coprecipitation). y-axis denotes Pb concentration released from sediment ($\mu g\,kg^{-1}$)*

found to be higher than the scavenging rate during the whole experiment.

Based on these experimental results and relevant literature data, a four stage interaction model has been developed for metals in anoxic sediments subsequent to oxidation:

- *Release Stage*: Includes amorphous sulfide oxidation, some pyrite oxidation, and some organic detritus decomposition. At the beginning, amorphous sulfide oxidation is a dominant chemical process; the metals, which are bound to acid volatile sulfides will be released at the early stage. The second contribution is characterized by mineralization of ammonium and pyrite oxidation. (At the end of the experiment, about 50% of the pyrite content was oxidized.)
- *Transition Stage of Release and Scavenging*: Acid-volatile sulfide compounds have been exhausted; pyrite oxidation and organic detritus decomposition continued. On the other hand, readsorption becomes stronger. Phosphate, derived from the decomposed organic detritus, accumulates and then precipitates with some metals.
- *Scavenging Stage*: At this stage all processes including pyrite oxidation, organic detritus decomposition, and release from solid organic and inorganic compounds decline. Iron and manganese oxide formation continues and scavenging of trace metals becomes a dominant process.
- *Equilibrium Stage*: Release and scavenging processes have basically reached a new equilibrium, *i.e.* the release rate is equated to the scavenging rate. The metal concentration in solution will remain constant. The formation of new solid metal species and their redistribution still proceed during this period of 'early diagenesis'.

The order of total release from the sediment has been found to be Cd (5%) > Zn (1.5%) > Cu (1%) > Pb (0.3%). The percentage scavenging of released metals was in the order of Pb (85%) > Cu (53%) > Zn (35%) > Cd (30%). The order of net release percentage of total metals from the sediment is Cd > Zn > Cu > Pb. For the released amount, however, the order is Zn > Cu > Pb > Cd. At the end of the experiment this order and the magnitude of released metals, were similar to that observed in the natural waters of the river Elbe at Hamburg.

Geochemical Characterization of the Potential Metal Mobility

The most efficient fixation process within anoxic sediments for trace metals is production of free sulfide during anaerobic degradation of organic matter and reduction of sulfate (study of heavy metal associations with sulfide and carbonate formations in anoxic sediments, therefore, provides insight into early diagenetic processes[34,35]).

Whereas the ability of the sediment to produce free sulfide is determined by the sulfate reduction rates, the ability to remove all the free HS^- produced is given by the reactive metal concentrations (predominantly reducible Fe^{3+}), available to form sulfide minerals ('Available Sulfide Capacity' [ASC][36]). Simultaneous application of standard sequential leaching techniques[37]

(BCR-Version: Ure et al.[38]) on critical trace elements and matrix components, can be used for geochemical characterization of anoxic, sulfide-bearing sediments in relation to the potential mobility of critical trace metals[39]:

Step 1	Exchangeable ions Carbonates	Ca^{2+} and Fe^{2+} from carbonates and phosphates	Shake 0.5 g sediment for 5 h with 20 ml of acetic acid $(0.11 \text{ mol} l^{-1})$
Step 2	Reducible metals	Fe^{3+} from reactive oxides	then, shake for 16 h with 20 ml $NH_2OH \cdot HCl$ $(0.1 \text{ mol} l^{-1})$, HNO_3 pH 2
Step 3	Sulfidic and organic bound metals	S from easily oxidizable sulfides (FeS)	then, add 10 ml hydrogen peroxide 8.8 mol l^{-1} (twice)—followed by extraction with 50 ml $NH_4Ac/HOAc$ pH 5

For determination of the acid producing capacity ('maximum APC') in anoxic sediments, both the FeS pool ('actual APC') and the maximum iron (II) sulfide-producing capacity (worst case: pyrite) upon disposal has to be taken into consideration. The latter is given by the 'sulfide binding capacity' (ASC, see above), which as determined in our example of Hamburg harbour mud (Table 3) using the Fe-concentrations in Step 2 of the sequential leaching results. The stoichiometry of the oxidizable S- and Fe-fractions of Step 3 indicate that the iron (II) sulfide extracted in this step was in the FeS form. The sum of both the ASC and the 'actual APC' gives the 'maximum "APC"' for the samples as shown in Table 3. The acid consuming capacity values for these samples was more simply determined from the Ca-concentration released from reactive carbonates by the Na-acetate solution of Step 1, which has to be multiplied by a factor of 0.5 to account for the stoichiometric ratio between pyrite and calcite within the redox reaction:

$$FeS_2 + 2CaCO_3 + 15/4 O_2 + 1/2 H_2O = FeOOH + 2SO_4^{2-} + 2Ca^{2+} + 2CO_2$$

The negative balance between the ACC and APC indicates that the mud sample from Hamburg harbour has a significant acidification potential.

Table 3 *Balance Between Acid Producing Capacity (APC) and Acid Consuming Capacity (ACC) Values for an Anoxic Sediment Sample from Hamburg Harbour*

Compound	Function	Value	Parameter	(mmol kg^{-1})
Ca in Step 1*	base potential	$(\times 0.5 =)$	ACC	90
Fe in Step 2*	ASC†	(335)		
S in Step 3*	sulfide sulfur	$(+ 85 =)$	max. APC	420
Balance			ACC − APC =	−330

*Steps in sequential extraction procedure.[38]
†Available Sulfide Capacity.

Assessing Long-term Mobility of Metals in Sediments

With regard to prediction of long-term effects of sediment-bound metals, chemical extraction procedures are of limited value because they usually involve neither reaction-mechanistic nor kinetic considerations. Those limitations can be avoided, by an experimental approach originally used by Patrick *et al.*[40] and Herms and Brümmer,[41] where sediments can be treated in a circulation system under controlled changes of significant release parameters such as pH, redox-potential, and temperature. Our experimental design[42] includes an ion exchange system for extracting and analysing the released metals at an adequate frequency, and compares sequential extraction results before and after treatment of the sample in the circulation apparatus. Individual metal species are released at different time intervals. By taking into account the level of elements released during the ten week experiments (equivalent to several thousand years of solid–water interaction) and those obtained by extrapolation from extraction 'pools', concentrations can be calculated for different scenarios.

These extrapolations have been made from pH 5 conditions, but titration curves for investigations on a wide spectrum of metal-bearing waste materials[43] suggest that pH 4 may be more appropriate for long-term predictions of potential metal release from contaminated sediments. In this pragmatic approach, which is outlined in Figure 5, the pH is automatically adjusted to 4 over a time period of

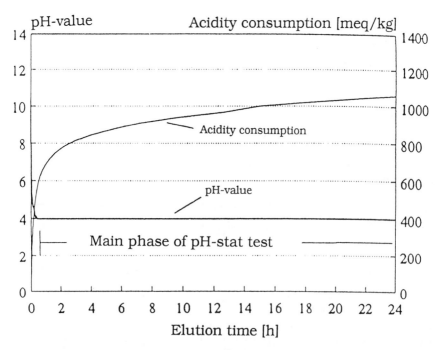

Figure 5 *Patterns of a suspension titration.*[43] *The initial phase, during which the target pH value is not attained, is followed by a 50-fold longer main phase with highly stable pH conditions*

24 hours; in addition to the release rates of metals, which can be determined from samples taken at different time intervals, the sum curve of acid consumption provides information on (a) the potential changes of the matrix composition during acidification and (b) the availability of buffer capacity at different time scales. In this respect a diagram of discrete acid additions would be valuable!

The acid consumption curve in Figure 5 reflects slow long-term metal release from sediments. Because calcite dissolution was fast, the acid consumption in the first stage increased drastically within a short period of time. Cation release and alumosilicate dissolution were dominant factors consuming acid in the later stage. The reactions can be treated as:

$$SO\equiv Me + 2H^+ = SO\equiv H_2 + Me^{2+}, \text{ and}$$

$$Al_2O_3 + 6H^+ = 2Al^{3+} + 3H_2O$$

where $SO\equiv$ and Me are the reaction groups on the solid phase, and the metal ion, respectively. The reactions of the hydrogen ions with metals and alumosilicate are delayed due to the complex sediment structure and the composition, *e.g.* clay minerals coated with organic matter or biofilms. Rates of reactions can be estimated by determination of the metal concentrations in solution.

5 Final Discussion and Outlook

Regarding the potential release of metals from contaminated sediments, changes of redox and pH conditions are of prime importance. In practice, therefore, characterization of sediment substrates with respect to their buffer capacity is an indispensable step for the prognosis of middle- and long-term processes of metal mobilization.

Description of Basic Processes

Experimental studies in the multi-chamber device and reactor cell provided information on metal transfer between different sedimentary matrices and on inter-relations between mobilization and readsorption mechanisms (Section 4). Further studies should focus on the role of iron sulfides in large-scale changes in metal species, *i.e.* solid–solution interactions and the bioavailability of trace metals in aquatic systems. The combination of matrix and trace element data in sediments, extracted simultaneously with different reagents, can provide additional information, which can be used to establish models on chemical interactions occurring under characteristic borderline conditions, *e.g.* with respect to sulfide, chloride, and organic complexing agents.[44]

Reliability of long-term prognoses on the impact of metal-bearing sediments is restricted due to the dominance of non-linear and delayed processes in redox-controlled systems. While measurements of actual acid producing and consuming capacities can be undertaken with available methods (Section 3), a possible 'split' or translocation of their inventories between solid and dissolved phases renders such parameters difficult to predict in a long-term perspective. It seems that calculation of the maximum acid producing capacity (APC), involving 'available sulfide capacity', *i.e.* the reducible Fe(III) concentration (Section 4), gives a relatively safe basis for such considerations, because iron generally is less mobile than some sulfur species. From data on the long-term acid producing potential of anoxic sediments, appropriate management techniques for dredged materials can be derived (see below).

Definition of Parameters

The pragmatic pH_{stat} approach (presented in Section 4) for evaluating the behaviour of solid waste subsequent to landfilling involves the 'spot test' of lowering the pH value and, in addition, provides data on the buffer capacity of the solid matrix. Direct assessment of the acid producing capacity can be performed by ventilation with air or oxygen; our data, however, have shown that these experiments should be designed for relatively long time periods, *i.e.* for several weeks duration.[33]

Although the three parameters forming the framework of future criteria for assessing long-term effects from metal-bearing sediments, *viz.* the acid producing potential, the acid consuming capacity, and the metal concentration, are closely inter-related, regulatory activities will focus on the concentration term, mainly because its priority function as an effect parameter in biological systems. In this respect, however, the long-term prognostic approach is different from the assessment of the short-term impact by metal-contaminated sediments, which is preferentially done by applying bioassays on sediment porewater and elutriates (recent developments are directed towards solid phase bioassays). Long-term predictions, *i.e.* extrapolations of experimental or field data to a hypothetical situation, necessarily implies some type of numerical method. From the available approaches (Section 2), the group of elution techniques generally seems to be best suited for this purpose; within this group, the long-term predictive capacity is predominantly reflected by the more rigid extractants, in particular, acid solutions. Guideline values based on acid extraction data have been proposed for soil and solid waste materials. The most common pH level of about 4 was primarily aimed to simulate conditions of acid precipitation; however, as has been shown from sediment and solid waste data, internal processes such as oxidation of sulfides, can, by far, be more effective in the lowering of pH.

Usually, criteria will focus on direct effects by ionic trace metals released from polluted sediments. However, it should be mentioned that even if it is not a metal-polluted sediment, Al dissolution from clay materials should also be a matter of concern, as Al^{3+} can be classified as a toxic metal ion in aquatic and terrestrial systems.

Criteria for Dredging and Up-land Deposition of Sediments

In dredged material management, two different target areas for combined matrix/metal criteria can be distinguished: (1) sediment resuspension and (2) dredged material disposal. With respect to 'resuspension of aquatic sediments', which involves more short-term effects than the disposal of dredged material, special emphasis should be placed on the factor, 'available metal species'. Within certain categories of acid producing and consuming capacities, guideline values for individual metals should be based on elutriate data, preferentially at pH 4, for better comparison with other solid matrices (e.g. Swiss Ordinance for Waste Materials,[45] which includes two categories of limit values: 'Inert-Construction Materials' and 'Residue Material Deposits', the latter mainly comprising pretreated products of municipal solid waste incineration).

Environmental impact of sediment deposits is influenced by the internal chemical conditions rather than by the concentration and extractability of metals. Priority, therefore, should be given to the optimization of long-term chemical stability ('geochemical engineering').

Consequences for Technical Measures

Regarding the various containment strategies, it has been argued that up-land containment, e.g. on heap-like deposits, could provide a more controlled management than containment in the marine environment. However, contaminents released either gradually from an imperfect impermeable barrier (including to groundwater) or catastrophically from failure of the barrier, could produce substantial damage. Up-land disposal of contaminated sediments at least requires provision of a long-term buffer capacity against acidity, arising from the oxidation of sulfides and organic substances. Laboratory studies on the efficiency of stabilization processes, undertaken by Calmano,[46] suggest that best results are attained with calcium carbonate, since pH conditions are not changed significantly upon addition of $CaCO_3$.

On the other hand, near-shore marine containment, e.g. in capped mound deposits, offers several advantages, particularly with respect to the protection of groundwater resources, since the underlying water is saline and chemical processes are favourable for the immobilization or degradation of priority pollutants. In a review of various marine disposal options Kester et al.[47] suggested that the best strategy for disposing of contaminated sediments is to isolate them in a permanently reducing environment. Under sub-sediment conditions there is a particular low solubility of metal sulfides, compared to the respective carbonate, phosphate, and oxide compounds. Marine sulfidic conditions, in addition, seem to repress the formation of mono-methyl mercury, one of the most toxic substances in the aquatic environment, by a process of disproportionation into volatile dimethyl mercury and insoluble mercury sulfide.[48]

With regard to the immobilization of contaminants in municipal and industrial waste materials, the term 'final storage quality' has been brought into discussion.[49] Solid residues with final storage quality should have properties very similar to the earth's crust (natural sediments, rocks, ores, soil). Landfills with solids of final storage quality need no further treatment of emissions into air and water. It seems that for large-volume waste materials, such as dredged sediments, deposition under permanent anoxic conditions both provides immediate economic advantages and could fulfil the requirement of 'final storage quality'.

Outlook

The preceding discussion has shown that the potential mobility, bioavailability, and toxicity of metals in sediments strongly depend on redox conditions. From earlier chemical extraction procedures, as proposed by Tessier *et al.*[37] to more recent SEM/AVS ratio determinations,[31,50] metal speciation is a major characteristic describing its potential mobility in sedimentary environments. The inherent link between the parameters to describe 'metal speciation' is the characterization of chemical stability and bioavailability of metal compounds in sediments under certain redox conditions.

In a sulfidic anoxic environment, even if the sediment is heavily polluted by metals, organisms are considered to be safe due to the strong fixation of metal ions by S^{2-} or HS^- as a source of acid volatile sulfide. Such heavily polluted sediment, however, can behave as a time bomb which is triggered by only one factor, a redox increase to a critical point, *e.g.* by exposure to oxygen-rich overlying water or directly to the air (a possible pathway also is oxygen transfer via plant roots). Once this situation occurs, toxic metals in the sediments will be released to the water phase or transformed into more bioavailable species.[51]

At the moment, research on long-term effects of redox variations on metal behaviour in sediments is mostly based on thermodynamic considerations. Future research should emphasize studies on the kinetics of metal species transformations, hydrogen ion production, and metal release, as affected by changing redox conditions. Additional important aspects involve the bridging of the gaps between numerical criteria approaches, as reflected by matrix composition and metal mobility, and biological approaches. Assessment for bioavailability of metal species is still based on the results of water phase and elutriate analysis, mostly using acute toxicity data.[52,53,50] However, initial data on solid phase bioassays suggest good prospects for future applications in sediment quality assessment. It may well be that for such systems, which are much less disturbed than artifical sediment elutriates, relationships between matrix conditions, as reflected, *e.g.* by redox indices, and metal species bioavailability may be found, which may serve as a more solid basis for the interpretation of results from bioassays, eventually with respect to chronic toxicity.

References

1. 'The Importance of Chemical "Speciation" in Environmental Processes', eds. M. Bernhard, F. E. Brinckman, and P. S. Sadler, Springer, Berlin, 1986, p. 1.
2. U. Förstner, *Int. J. Environ. Anal. Chem.*, 1993, **51**, 5.
3. D. Shea, *Environ. Sci. Technol.*, 1988, **22**, 1256.
4. G. Rauret and Ph. Quevauviller (eds.), *Int. J. Environ. Anal. Chem.*, 1993, **51**, 1.
5. J. I. Drever, 'The Geochemistry of Natural Waters', Prentice-Hall, Englewood Cliffs, NY, 1982.
6. N. van Breemen, C. T. Driscoll, and J. Mulder, *Nature (London)*, 1984, **307**, 599.
7. J. N. Moore and S. N. Luoma, *Environ. Sci. Technol.*, 1990, **23**, 1278.
8. N. van Breeman, *Neth. J. Agric. Sci.*, 1987, **35**, 271.

9. J. A. Kittrick, D. S. Fanning, and L. R. Hossner, 'Acid Sulfate Weathering', SSSA Special Publication 10, Soil Science Society of America, Madison, 1982, p. 247.
10. W. L. Salomons, unpublished data.
11. C. W. Holmes, E. A. Slade, and C. I. McLerran, *Environ. Sci. Technol.*, 1974, **8**, 255.
12. R. P. Gambrell, R. A. Khalid, and W. H. Patrick, Jr., *Environ. Sci. Technol.*, 1980, **14**, 431.
13. R. A. Khalid, R. P. Garnbrell, and W. H. Patrick, Jr., *J. Environ. Qual.*, 1981, **10**, 523.
14. S. Emerson, R. Jahnke, and D. Heggie, *J. Mar. Res.*, 1984, **42**, 709.
15. M. E. Hines, W. B. Lyons, P. B. Armstrong, W. H. Orem, and G. E. Jones, *Mar. Chem.*, 1984, **15**, 173.
16. B. Prause, E. Rehm, and M. Schulz-Baldes, *Environ. Technol. Letts.*, 1985, **6**, 261.
17. D. A. Darby, D. D. Adams, and W. T. Nivens in 'Sediment and Water Interactions', ed. P. G. Sly, Springer, New York, 1986, p. 343.
18. C. D. Hunt and D. L. Smith, *Can. J. Fish. Aquat. Sci.*, 1983, **40**, 132.
19. M. Kersten and U. Förstner, *Water Sci. Technol.*, 1986, **18**, 121.
20. W. M. Stigliani in 'Chemical Time Bombs, European State-of-the-Art Conference on Delayed Effects of Chemicals in Soils and Sediments', Veldhoven, The Netherlands, 2–5 September 1992, p. 12.
21. B. Maaß and G. Miehlich, *Mitt. Dtsch. Bodenkundl. Ges.*, 1988, **56**, 289.
22. Anonymous, 'Suggested Guidelines for Method of Operation in Surface Mining of Areas with Potentially Acid-Producing Materials', West Virginia Surface Mine Drainage Task Force, WV. Dep. Nat. Resour., Charleston, WV, 1979, p. 18.
23. J. Hong, U. Förstner, and W. Calmano in 'Bioavailability: Physical, Chemical and Biological Interactions,' eds. J. L. Hamelik, P. F. Landrum, H. L. Bergman, and W. H. Benson, CRC Press, Boca Raton, Fl., 1994, p. 129.
24. R. S. Swift in 'Soil Organic Matter Studies', IAEA Vienna, 1977, p. 275.
25. L. Prenzel, *Z. Dtsch. Geol. Ges.*, 1985, **136**, 293.
26. D. Binkley, C. T. Driscoll, H. L. Allen, P. Schoenberger, and D. McAvoy, 'Acidic Deposition and Forest Soils', Springer, New York, 1989, p. 146.
27. M. A. Tabatabei, *J. Am. Phys. Chem. Assoc.*, 1987, **37**, 34.
28. K. D. Ferguson and P. M. Erickson in 'Environmental Management of Solid Waste—Mine Tailings and Dredged Material', eds. W. Salomons and U. Förstner, Springer, Berlin, 1988, p. 24.
29. A. A. Sobek, W. A. Schuller, J. R. Freeman, and R. M. Smith, 'Field and Laboratory Methods Applicable to Overburden and Mine Soils', US EPA Rep., EPA-600/2-78-054, Cincinnati, Ohio, 1978.
30. A. Bruynesteyn and R. P. Hackl, *Miner. Environ.*, 1984, **4**, 5.
31. D. M. DiToro, J. D. Mahony, D. J. Hansen, K. J. Scott, M. B. Hicks, S. M. Mayr, and M. S. Redmond, *Environ. Toxicol. Chem.*, 1990, **8**, 1487.
32. W. Calmano, W. Ahlf, and U. Förstner, *Environ. Geol. Water Sci.*, 1988, **11**, 77.
33. W. Calmano, U. Förstner, and J. Hong in 'Environmental Chemistry of Sulfide Oxidation', eds. C. N. Alpers and D. W. Blowes, ACS Symposium Series, American Chemical Society, Washington DC, 1993, in press.
34. R. A. Berner, *J. Sediment. Petrol.*, 1981, **51**, 359.
35. J. W. Morse and F. T. Mackenzie, 'Geochemistry of Sedimentary Carbonates', Elsevier, New York, 1990.
36. K. J. Williamson and D. A. Bella, *J. Environ. Eng. Div. ASCE*, 1980, **106**, 695.
37. A. Tessier, P. G. C. Campbell, and M. Bisson, *Anal. Chem.*, 1979, **51**, 844.
38. A. M. Ure, Ph. Quevauviller, H. Muntau, and B. Griepink, *Int. J. Environ. Anal. Chem.*, 1993, **51**, 135.
39. M. Kersten and U. Förstner, *Geo-Marine Letts.*, 1991, **11**, 184.

40. W. H. Patrick, Jr., B. G. Williams, and J. T. Moraghan, *Soi. Sci. Soc. Am. Proc.*, 1973, **37**, 331.
41. U. Herms and G. Brümmer, *Mitt. Dtsch. Bodenkundl. Ges.*, 1978, **27**, 23.
42. J. Schoer and U. Förstner, *Vom Wasser*, 1987, **69**, 23.
43. P. Obermann and S. Cremer, 'Mobilisierung von Schwermetallen in Porenwässern von belasteten Böden und Deponien: Entwicklung eines aussagekräftigen Elutionsverfahrens', Landesamt für Wasser und Abfall Nordrhein-Westfalen Düsseldorf/ Germany, 1992, Vol. 6, p. 127.
44. K. Wallmann, M. Kersten, J. Gruber, and U. Förstner, *Int J. Environ. Anal. Chem.*, 1993, **51**, 187.
45. Anonymous, 'Technische Verordnung über Abfälle (TVA)', Der Schweizerische Bundesrat (Swiss Federal Parliament), SR 814.015, Bern/Switzerland, 10 Dec. 1990.
46. W. Calmano in 'Environmental Management of Solid Waste—Mine Tailings and Dredged Material', eds. W. Salomons and U. Förstner, Springer, Berlin, 1988, p. 80.
47. 'Wastes in the Ocean. Vol. 2: Dredged-Material Disposal in the Ocean', eds. B. H. Ketchurn, I. W. Duedall, and P. K. Park, Wiley, New York, 1983, p. 299.
48. P. J. Craig and P. A. Moreton, *Mar. Pollut. Bull.*, 1984, **15**, 406.
49. 'The Landfill—Reactor and Final Storage', ed. P. Baccini, Springer, Berlin, 1989.
50. D. M. DiToro, J. D. Mahony, D. J. Hansen, K. J. Scott, A. R. Carlson, and G. T. Ankley, *Environ. Sci. Technol.*, 1992, **26**, 96.
51. W. Calmano, J. Hong, and U. Förstner, *Vom Wasser*, 1992, **78**, 245.
52. G. T. Ankley, G. L. Phipps, P. Kosian, A. Cotter, V. R. Mattson, and J. D. Mahony, *Environ. Toxicol. Chem.*, 1991, **10**, 1299.
53. W. Ahlf, J. Gunkel, W. Liß, H. Neumann-Hensel, K. Rönnpagel, and U. Förstner, *Schriftenr. Ver. Wasser, Boden, Lufthyg.*, *Berlin Dahlem*, 1992, **89**, 427.

Speciation of Organic Sulfur Forms in Heavy Oils, Petroleum Source Rocks, and Coals

C. E. Snape, S. C. Mitchell, K. Ismail, and R. Garcia

DEPARTMENT OF PURE AND APPLIED CHEMISTRY, UNIVERSITY OF STRATHCLYDE, THOMAS GRAHAM BUILDING, 295 CATHEDRAL ST, GLASGOW G1 1XL, UK

Summary

To illustrate the key role of advanced analytical techniques in elucidating the complex structure of heavy oils, petroleum source rocks, coals, and other sedimentary organic matter, the recent advances that have been achieved with X-ray techniques, temperature programmed reduction (TPR) and oxidation (TPO), and selective chemical modification for the speciation of the organic sulfur forms present are reviewed. Although each of these approaches has its limitations, a consistent overall picture is beginning to emerge on the nature of organic sulfur groups in solid fuels and heavy oil fractions. As might be anticipated, the more stable sulfur groups are most abundant in the more mature coals and petroleum source rocks. Further, the use of these advanced techniques has enabled the changes to be monitored that occur in the refining of heavy oils and chemical treatments being developed to reduce the sulfur levels of coals.

1 Introduction

Coal and petroleum are derived from sedimentary organic matter of vastly different composition (woody plants for coals and algae plus non-woody higher plants for petroleum) but both contain varying amounts of sulfur. This was incorporated during the early stages of diagenesis via the products obtained from the bacterial reduction of sulfate ions (S^0, H_2S, HS_x^-).[1-3] In sediments where iron was abundant, pyrite was the predominant form of FeS_2[1] produced with the remainder of the sulfur being bound into the organic structure of the sediments which are at the peat stage in the case of coal. Oil is generated from source rocks (an oil shale is merely a source rock containing a sufficiently high concentration of organic matter to make direct utilization for either combustion or retorting to oil feasible) by the effect of temperature over long periods of geological time (catagenesis).[4] The sulfur contents of both petroleum and coal vary over the wide range of 1–10% w/w (Orimulsion comprises high sulfur heavy oil). During maturation of both coal and petroleum source rocks, the distribution of organic sulfur groups are thought to alter markedly with sulfides being converted into thiophenes and the degree of condensation of thiophenic moieties increasing.[2,3,5] Further, the ease of sulfur removal from oil fractions during petroleum refining processes is known to be heavily dependent on the nature of the organic sulfur groups present.[6] Thus, their speciation is important from the standpoints of being able to fully understand coal and petroleum formation and also to effectively model refinery processes, particularly catalytic cracking and hydrotreatment. Further, as EC legislation pertaining to sulfur emissions from fossil fuel combustion becomes more stringent (*e.g.* Directive 88/609 for large combustion sources), there is an increasing need to have analytical techniques to monitor the different forms of sulfur present in fuels.

Standard methods are available for the determination of total sulfur contents in all fuels, together with pyritic sulfur in coals (ATSM D2492).[1,7] Many studies have addressed the molecular characterization of individual sulfur compounds in (i) light oils and (ii) extracts and pyrolysis products of petroleum source rocks and coals using gas chromatography–mass spectrometry (GC–MS) after appropriate pre-fractionation via liquid chromatography. The reviews by Sinninghe Damste and de Leeuw provide a comprehensive coverage of this topic[2,3] but, in general, only small proportions of the organic sulfur actually present in coals and petroleum source rocks are actually observed. This is because of their macromolecular character which confers low solubilities in common organic extracting solvents. Indeed, until recently, virtually no information was available on the overall distributions of the thiophenic and non-thiophenic (sulfuric) forms in fossil fuels and the need for suitable techniques has provided a major challenge to the analytical chemist. This review covers the significant advances achieved primarily over the past five years using *X*-ray techniques,[8-18] temperature programmed reduction (TPR)[19-27] and pyrolysis,[28-30] temperature programmed oxidation (TPO),[31-33] and selective chemical modification.[34-40] The latter approach comprises oxidation,[34] reduction,[35-39] and quaternization reactions[40] and invariably involves the use of appropriate techniques, such as nuclear magnetic resonance (NMR) to monitor the resultant changes. Space precludes a comprehensive coverage of all the research on organic sulfur conducted using these techniques but it is hoped that the review illustrates the benefits that can arise from using a variety of analytical approaches to address a complex structural challenge and provides a platform for further reading. For a more detailed account of the application of these techniques to coal, the reader is referred to the recent article by Davidson.[7]

2 General Aspects

As well as pyritic and organic sulfur, sedimentary organic matter (particularly coal) is likely to contain significant amounts of both sulfatic and elemental sulfur. However, there is now overwhelming evidence to indicate that these are both oxidation products derived from pyrite.[7] In such weathered samples, it is likely that some of the labile non-thiophenic organic sulfur forms have also been oxidized which gives rise to sulfoxides and sulfones from sulfides and sulfonic acids from di/polysulfides and thiols (Section 3). Hence, it is extremely important to ensure that appropriate handling precautions are taken to avoid oxidation.

The presence of pyrite in sedimentary organic matter poses a problem for most of the techniques described here and its removal prior to analysis is thus desirable. Pyrite can be removed via oxidation with dilute nitric acid as in the standard procedure (ASTM D 2492) for its determination and via reduction using lithium aluminium hydride.[41,42] However, care is required with both procedures as some of the organic sulfur forms can be affected; nitric acid can oxidize non-thiophenes whilst LiAlH₄ cleaves poly/disulfides to yield thiols. As a result, researchers have preferred to work with samples low in pyrite, for example, the two-high sulfur lignites, Mequienza and Rasa which both contain *ca.* 10% organic sulfur but less than 0.5% w/w pyrite. Further, for coals, it has proved relatively easy to compare the results from the different analytical approaches as most researchers have also used the samples across the rank range supplied through the Argonne Premium Coal Sample (APCS) programme. For crude oils and petroleum source rocks, high-sulfur samples have again attracted the most attention, particularly for reductive studies[35-37] where a seep oil from Utah (Rozel Point, 14% w/w sulfur) and a Middle East shale (Jurf Ed Darawish, 16% w/w sulfur) have received the most attention.

3 *X*-Ray Techniques

X-Ray Photoelectron Spectroscopy

Sulfur at or near the surface or in particles (electron escape depth is *ca.* 3 nm) can be observed via *X*-ray photoelectron spectroscopy (XPS) using the S 2*p* peak.[8-10,34] The binding energies of all sulfur compounds fall within a narrow range of less than 10 eV (Table 1) and the differences in binding energy between thiophenes/aromatic sulfides and aliphatic sulfides (sulfur bound to at least 1 *sp*³ carbon) are only *ca.* 1 eV (Table 1). Further, the S 2*p* peak comprises a 2*p*₃/₂ and a 2*p*₁/₂ component which have a 2:1 intensity ratio and are separated by 1.2 eV. In addition, there does not appear to be one unique binding energy for thiophenic/aromatic and aliphatic sulfur (Table 1). Not surprisingly, the first investigators to apply XPS to fuels[8,34] both concluded that it was not feasible to resolve thiophenic and different sulfidic environments directly by XPS and both proposed the use of selective oxidation to first convert sulfides into sulfones (Section 5). This gives rise to visually observable differences of 1–5 eV between the resultant sulfones and non-oxidized forms.

Figure 1 shows the XPS spectrum of Rasa coal obtained by Kelemen *et al.*[9] that has been fitted to two signals with binding energies of 164.1 and 163.3 eV corresponding to the binding energies of thiophenes and alkyl sulfides, respectively (Table 1). From the peak fitting procedure, they concluded that thiophenic sulfur accounts for 70% of the organic sulfur present in this coal.

Table 1 *XPS Sulfur 2p Binding Energies for Some Model Compounds*

Compound	Binding energy, eV
Dibenzothiophenesulfone	168.2
Polyphenylene ether sulfone	168.2
DL-Methionine sulfoxide	165.8
6-Ethoxy-2-mercaptobenzothiazole (thiazole)	164.2
Dibenzothiophene polymer	164.1
1,2-Benzodiphenylene sulfide	164.0
Sulfur	163.7
Polyphenylene sulfide	163.7
S-Methyl-L-cysteine	163.3
DL-Cysteine	162.7
6-Ethoxy-2-mercaptobenzothiazole (mercapto)	162.1

Taken from Kelemen *et al.* (ref. 9).

Diaryl (aromatic) sulfides were not explicitly accounted for with polyphenylene sulfide having a binding energy of 163.7 p.p.m. midway between those above for thiophenes and alkyl sulfides. Nonetheless, after air oxidation (5 days at 125 °C) to convert the non-thiophenes into a mixture of sulfoxides, sulfones, and sulfonic acids, thiophenic component still accounted for 69% of the sulfur with most of the sulfides (23% out of the original 30% of the sulfur) having been oxidized.

For coals containing significant quantities of pyrite, the peak-fitting procedure becomes more complex with a third component having to be included although the surface concentration of pyrite appears to be less than in the bulk.[10] Despite the obvious uncertainties concerning the absolute accuracy of the peak-fitting

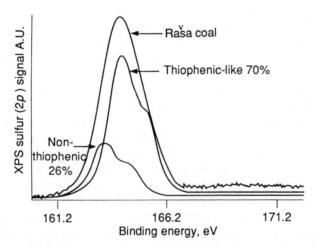

Figure 1 *XPS spectrum of Rasa coal showing fit into thiophenic and alkylsulfide components (Kelemen et al., ref. 9). Reproduced by permission of the publishers, Butterworth Heinemann Ltd.*

Table 2 *Comparison of Sulfidic (Non-thiophenic) Sulfur Contents Derived by Different Techniques for Some Coals*

			% of Organic Sulfur		
	Rasa	Mequinenza	Illinois No. 6	Wyodak	Upper Freeport
C, % dmmf basis	80.2	68.4	80.7	76.0	88.0
Total sulfur	11.8	9.0	4.8	0.63	2.3
Pyritic sulfur	0.4	0.4	2.8	0.17	0.5
Organic sulfur (%w/w dry basis)	11.4	8.0	2.0	0.43	1.8
High pressure TPR	35	25	25	20	<10
XPS[a]	30	66	31	37	19
XANES	26	48	33–37	33–44	5–15
Fluid-bed pyrolysis[a]	47	67	—	36	—
TPO[a]	12	36	26	43	37

[a] Refers to aliphatic sulfides with the implication that the total sulfidic content may be significantly higher.

procedure, the proportion of the thiophenic sulfur component has been found to increase with rank (maturity) for the Argonne coals (Table 2).[10]

X-Ray Absorption Near Edge Structure Spectroscopy

X-Ray absorption near edge structure (XANES) spectroscopy has now been used by a number of research groups[10-18] to distinguish between the different organic sulfur forms in coals and heavy oil fractions. However, the requirement of a synchrotron source to provide the intense *X*-ray beam limits the availability of the technique. Until the very recent report of the sulfur *L*-edge,[17-18] all the studies on fuels have involved observation of the sulfur *K*-edge where electrons are ejected from the 1*s* shell.[10-16] The positions of the first inflection points for a number of sulfur compounds are summarized in Table 3 and, as with XPS, there is good resolution between sulfones, sulfoxides, and non-oxidized sulfur forms. XANES differs from XPS in that data can be obtained in modes to monitor either the surface or the bulk of a material. Further, the data in Table 3 indicates that the edge energies of thiophenes and diarylsulfides coincide but are resolved from those of aliphatic sulfides (sulfur bound to one or two sp^3 carbons) and disulfides. However, as with XPS, the energy differences between the different thiophenic and sulfidic environments are small (0.5–1.0 eV, Table 3) meaning that there is virtually no visible resolution observed in the region 2470 eV for fuels which is demonstrated in Figures 2 and 3 for heavy petroleum and coal samples, respectively. Thus, curve fitting procedures once again come into play for quantification. The two approaches that have been used to estimate aromatic- and aliphatic-bound sulfur involve (i) taking the third derivatives of

Table 3 *Sulfur K-Edge Energies in XANES Spectra for Some Model Compounds*

| Compound | Energy at first inflection point, eV | | Formal Oxidation state |
	Actual value (ref. 11)	Relative to elemental S (refs. 11 and 15)	
Dibutylsulfone	—	— (7.5)	4
Diphenylsulfone	2474.8	4.7	4
Dibenzothiophenesulfone	2474.7	4.6	4
Dibenzylsulfoxide	—	— (3.4)	2
Dimethylsulfoxide	2472.8	2.7	2
Benzothiophene	2470.4	1.3	0
Dibenzothiophene	2470.4	1.3 (1.3)	0
Diphenylsulfide	2470.4	1.3	0
Thiaanthrene	2470.2	1.1	0
Dioctylsulfide	2470.1	1.0	0
Benzylphenylsulfide	2469.9	0.8	0
Dibenzylsulfide	2469.8	0.7 (0.7)	0
Cysteine	2469.2	0.1	0
Diphenyldisulfide	2469.2	0.1 (0.1)	0
Sulfur	2469.1	0	0
Pyrite	2468.4	−0.7 (−0.5)	−1

Values taken from Gorbaty *et al.* (ref. 11) and, in brackets, from Hufman *et al.* (ref. 15).

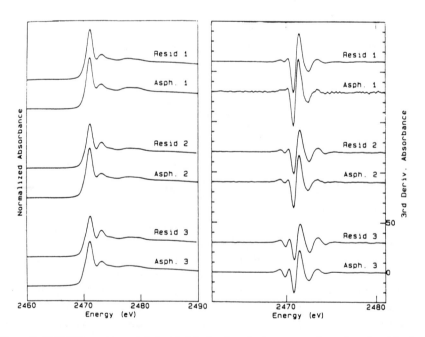

Figure 2 *XANES spectra and their third derivatives for heavy petroleum fractions (Gorbaty et al., ref. 11). Reproduced by permission of the publishers, Butterworth Heinemann Ltd.*

the spectra (Figure 2) and using the heights of the features at 2469.8 and 2470.4 eV[10-13] and (ii) least-squares analysis of the near-edge structure into a series of components (Figure 3) for pyrite, the aromatic- and aliphatic-bound sulfur, sulfoxide, sulfone, and sulfate.[15,16] If oxidized sulfur forms are present, calibrations need to be carried out as the observed intensities increase with oxidation state making peaks from sulfones/sulfates look pronounced even when their concentrations are fairly small (Figure 3).

Within the estimated accuracy of both fitting procedures (*ca.* ±10% of the organic sulfur), remarkably close agreement has been achieved for the Argonne coals (Table 2) with the fraction of aromatic-bound sulfur increasing from *ca.* 60% in sub-bituminous coals to over 85% in low-volatile bituminous coal.[7,10,12,15] Although, for the reasons outlined above, the results from XPS and XANES are probably not strictly comparable, good agreement has generally been

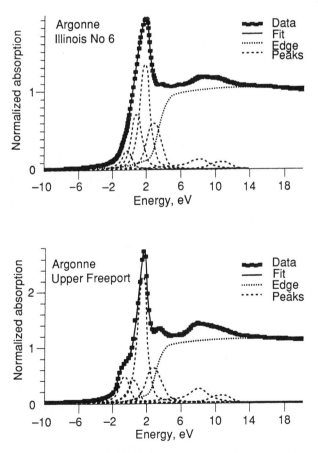

Figure 3 *XANES spectra of Illinois No. 6 and Upper Freeport coals showing deconvolution into components for pyrite, aliphatic- and aromatic-bound sulfur, sulfoxide, sulfone, and sulfate. Reprinted from Huffman et al., ref. 15. Copyright 1991, American Chemical Society*

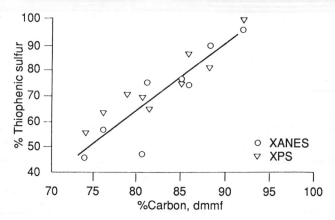

Figure 4 *Proportions of thiophenic or, probably more strictly, aromatic-bound sulfur, derived from XPS and XANES for the Argonne coals. Reprinted from Gorbaty et al., ref. 10. Copyright 1991, American Chemical Society*

achieved for heavy petroleum fractions and coals investigated by both techniques (Figure 4).[7,11] However, there is evidence to suggest that XANES might underestimate aromatic-bound sulfur. For example, it appears unrealistic that lignites and sub-bituminous coals[7,11] contain similar proportions of aliphatic sulfur to heavy petroleum fractions despite being significantly more aromatic in character. Further, although the high temperature *in-situ* measurements conducted by Huffman *et al.*[16] have shown the anticipated decrease in the proportion of aliphatic sulfur with increasing temperature, significant amounts remain above 500 °C which is certainly not expected from their thermal behaviour. As with XPS, the use of selective oxidation in which non-thiophenic/aromatic sulfur forms only are converted into sulfoxides and sulfones (S^{II} and S^{IV} species) gives rise to vastly improved resolution in XANES. Where air oxidation has been used to achieve this, the agreement between XPS and XANES has again been fairly good.[13]

Sulfur forms can also be examined by *L*-edge XANES spectra in which transitions from the 2*p* orbitals are observed and the first spectrum of a coal was published recently.[17] Although distinct peaks are visible because the linewidths are narrower and photon resolution is greater, considerable overlap occurs between thiophenes and the other forms and curve-fitting is still required. In fact, the distribution of 70% thiophenic and 30% arylsulfide which gave the best fit[17] was considerably at variance with that derived from the *K*-edge spectra[18] and was considered as unrealistic by Calkins, Gorbaty, and their co-workers.[44] However, this apparent discrepancy probably arises from the fact that not enough standards have yet been analysed under the high-vacuum conditions required for *L*-edge XANES.[18]

4 Thermal Techniques

Reductive Techniques

Low Pressure TPR. TPR is based on the principle that different organic sulfur forms present in solid fuels have different characteristic reduction temperatures at which hydrogen sulfide (H_2S) evolves. Calibration with model compounds has indicated that the ease of reduction is in the order of thiols > aliphatic sulfides > aromatic sulfides > thiophenes[19] (Table 4). The method for coals was pioneered by Attar[19,20] and has been used by others[21-23] with few modifications to the original design of the reactor in which coal is refluxed in a mixture of low-boiling solvents and H_2S is swept from the reactor by a stream of carrier gas; a condenser prevents the escape of tar. Attar originally used lead acetate paper to detect H_2S but later workers have used potentiometry[21,22] and flame photometric detection.[23] However, only limited success has been achieved thus far primarily because only the labile non-thiophenic forms have actually been observed with sulfur balances being poor. Virtually all the thiophenic sulfur remains in the char due to the use of low hydrogen partial pressures (max. 1 atm.) and inappropriate low boiling reducing agents, such as tetralin. Further, no account has been taken of the reduction of pyrite to pyrrhotite and retrogressive reactions including the conversion of sulfides into thiophenes which are extremely likely due to the long residence time of tar in the reactor. Nevertheless, Attar[19,20] was confident enough to estimate the concentration of thiolic and aliphatic and aromatic sulfidic sulfur directly from the TPR traces. However, the reported thiol concentrations (*ca.* 20% of the organic sulfur in bituminous coals) are now considered to be unrealistically high,[78] whilst those for aromatic sulfides of below 10% are probably too low.

High Pressure TPR. The extent of desulfurization in the pyrolysis of sedimentary organic matter generally increases with hydrogen pressure[44] and typically, in a fixed-bed reactor at a pressure of 150 bar with a dispersed sulfided molybdenum catalyst, over 95% of the organic sulfur in both lignites and bituminous coals can be released. Moreover, only about 20% is released as thiophenic compounds in the tars with the remaining 75–80% appearing in the gas phase as H_2S.

Table 4 T_{max} *for Solid Calibrants Used in High Pressure TPR*

Calibrant	T_{max}, °C
Silica-immobilized dibenzothiophene	470
Silica-immobilized diphenylsulfide	350
Silica-immobilized phenylbenzylsulfide	300
Silica-immobilized thioanisole (S—CH$_3$)	250–350
Di/polysulfides in vulcanized coal tar pitch	150–250
Thiol, cysteine	180

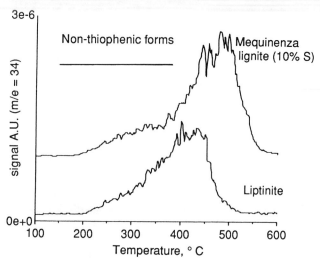

Figure 5 *High pressure TPR traces for Mequinenza lignite and its liptinite concentrate (Mitchell et al., ref. 25)*

These findings prompted the authors to overcome the inherent drawbacks of TPR by using a high pressure technique, in which the evolved H_2S is measured with a quadrupole mass spectrometer.[24,25] Pyrite contributes significantly to the H_2S profiles and, where present in high concentrations, it has been removed with $LiAlH_4$.

Figure 5 compares the high pressure TPR traces for Mequinenza lignite and a liptinite concentrate (H rich) separated by density gradient centrifugation.[45] The H_2S evolution profiles comprise a broad shoulder between 200 and 400 °C from non-thiophenic sulfur forms followed by the dominant peak above 400 °C attributable to thiophenes. These reduction temperatures have been confirmed using novel silica-immobilized sulfur compounds[26] and other appropriate solid calibrants (Table 4). Indeed, the reduction temperature at which H_2S evolution reaches a maximum (T_{max}) of 480 °C for immobilized dibenzothiophene is close to those of low-rank coals. Given that the T_{max} for diphenylsulfide is *ca.* 350 °C, it can be concluded that much of the non-thiophenic sulfur in both Mequinenza and Rasa lignites (*ca.* 30% of the total, Rasa gives a similar trace to Mequinenza[26]) occurs as aliphatic sulfides and disulfides with approximately half of the H_2S below 400 °C evolving between 200 and 320 °C (Figure 5). That below 250 °C is attributed to disulfides but the lack of any intensity much below 200 °C suggests that thiols are not present in significant concentrations. The resolution is thus considerably superior to that achieved for *X*-ray techniques, and as might be anticipated from its lower aromaticity, the liptinite contains a greater proportion of non-thiophenic sulfur forms than the whole lignite. Further, T_{max} for the thiophenic sulfur in the liptinite occurs at a much lower temperature strongly suggesting that single ring thiophenes are the dominant form present (Figure 5) compared to probably a mixture of 1–3 ring structures in the whole lignite. These, together with the parallel trends for the distributions of thiophenic

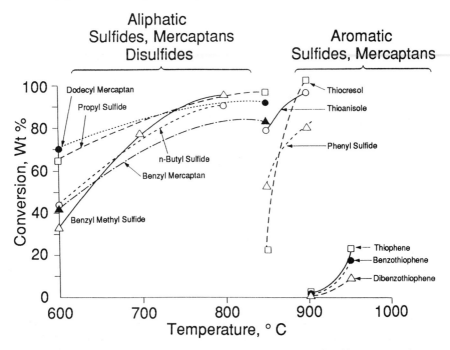

Figure 6 *Hydrogen sulfide evolution from reduction of model sulfur compounds in a fluidized-bed reactor. Reprinted from Calkins, ref. 28. Copyright 1991, American Chemical Society*

compounds in the TPR tars[27] indicate the potential of high pressure TPR to give a good indication of the average ring size of the thiophenic structures present.

The high pressure technique has been applied to Rasa and Mequinenza and indicated that thiophenic forms account for *ca.* 70% of the total[24,25] (Table 2). In common with XPS and XANES and other pyrolysis methods, high pressure TPR indicates that thiophenic sulfur increases with rank. However, the proportions of thiophenic sulfur derived from high pressure TPR are consistently higher than by the *X*-ray techniques with bituminous coals containing no more than 10% of the organic sulfur in non-thiophenic forms (Table 2).

Flash Pyrolysis. Calkins and co-workers[28-30] studied the reduction behaviour of a number of sulfur compounds in a fluidized-bed reactor as a function of temperature. The results summarized in Figure 6 indicate that the reduction of aliphatic sulfides, thiols (mercaptans), and disulfides is nearly completed by 800 °C whilst aromatic sulfides are stable until higher temperatures are reached and thiophenes are only partially reduced above 900 °C. These cut-off points were used for coals to estimate the proportions of aliphatic and aromatic sulfides/thiols. The proportions of aliphatic sulfides derived by this method are generally higher than those from both *X*-ray and TPR techniques (Table 2) with surprisingly low values for aromatic sulphides.[7] These differences might arise from the fact that the desulfurization behaviour of solids is probably different to that of volatile species in fluidized-beds in that their residence times are much longer implying that some aromatic-bound sulfur may evolve as H_2S below 800 °C.

Temperature Programmed Oxidation

The principal of TPO is the same as that of TPR except the sulfur forms are being oxidized to yield primarily sulfur dioxide (small amounts of COS are also released) at characteristic temperatures, the evolved SO_2 being monitored by Fourier transform infra-red spectrometry. The technique has been developed by LaCount and co-workers[31–33] over a number of years and is now referred to as 'controlled-atmosphere programmed temperature oxidation' (CAPTO). To limit heating effects, coals are diluted in an excess of tungsten oxide (WO_3) which also acts as a combustion catalyst. Early work was conducted using a mixture of 10% v/v of oxygen in helium,[31] but it was found that pyrite produced SO_2 in the temperature range 250–380 °C, the same as that for aliphatic sulfide polymers. This problem was solved using pure oxygen which curtails pyrite oxidation until above 470 °C[32,33] which is beyond the oxidation temperature found for aryl sulfides and thiophenes of *ca.* 420 °C. Typical CAPTO traces for samples of untreated and biotreated Illinois coal are shown in Figure 7 which shows the distinct peaks obtained for aliphatic- and aromatic-bound sulfur (290 and 420 °C) and pyrite (480 and 580 °C). The particular biotreatment used has removed all the pyrite, together with some of the aliphatic- and aromatic-bound sulfur (Figure 7).

The data in Table 2 and the information presented by Davidson[7] indicate

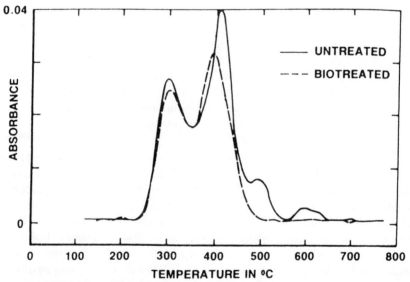

Figure 7 *CAPTO sulfur dioxide evolution profiles for samples of untreated and biotreated Illinois No. 6 coal (LaCount et al., ref. 33). Reproduced by permission of the publishers, Butterworth Heinemann Ltd.*

that, compared to the other techniques described, the estimates of aliphatic sulfide contents from CAPTO occur across a relatively narrow range (10–35%) and there is no clear rank trend. It seems surprising that a low-volatile bituminous coal, such as Upper Freeport (Table 2), contains similar or greater proportions of aliphatic sulfides than lower rank coals, such as Rasa and Mequinenza lignites. These apparent anomalies might be a consequence of the limited number of standards that have been used to calibrate the technique. As with high pressure TPR, only solid calibrants which will not soften upon heating can be used in the well-swept fixed-bed reactor.

5 Chemical Modification

Oxidation

The selective oxidation and reduction of particular non-thiophenic sulfur forms prior to analysis offers an attractive approach for speciation. It has already been described how simple air oxidation appears to be fairly effective for non-thiophenic forms (Section 3). Of the mainly available oxidizing agents, t-butylhydroperoxide oxidizes aliphatic sulfides selectively to sulfoxides.[34] This approach was used in conjunction with XPS and LiAlH₄ reduction (see following) to show that asphalt contains approximately 50% thiophenic sulfur which appears to be reasonably consistent with direct XPS[9] and XANES results,[11] albeit for different samples.

For heavy oils, selective oxidation of the sulfides present to sulfoxides has also been achieved in solution using tetrabutylammonium periodate.[46] Chromatographic separation of the resultant sulfoxides and subsequent reduction with LiAlH₄ to regenerate the sulfides provided concentrates for GC–MS analysis. After separation of the sulfides from the n-heptane soluble fractions, the remaining thiophenes were then treated in a similar manner but being selectively oxidized using the stronger reagent, m-chlorobenzoic acid. For the suite of samples investigated, the proportions of thiophenes and sulfides varied significantly and the compositions of the sulfur fractions were complex although a good correlation was found between the depth of burial and a maturity index based on the pattern of alkyl substitution for the dibenzothiophenes. Stepwise perchloric acid oxidation has been used to determine pyritic and total organic sulfur contents of coals[47] and this approach shows promise in attempting to resolve sulfidic and thiophenic sulfur.[48] However, as for the reduction of insoluble macromolecular organic matter, there is always going to be the uncertainty over accessibility of reagents.

Reduction/Hydrodesulfurization

The points of sulfur incorporation into sedimentary organic matter during diagenesis have been a topic of great interest to organic geochemists (see, for example, references 35–37). For high-sulfur oils and source rocks, the hydrocarbons released using relatively specific reducing agents or desulfurization catalysts indicate the actual moieties involved in sulfur linkages as well as helping to identify the depositional environment. For kerogens (insoluble material in petroleum source rocks), soluble reductants are clearly going to be more effective than solid desulfurization catalysts, such as Raney nickel, which is nonetheless effective for sulfides and simple thiophenes in heavy oily fractions. Indeed, the alkane products obtained using nickelocene/LiAlH$_4$ and lithium in ethylamine (Li/EtNH$_2$) have been found to be similar to those with Raney Ni for heavy oil fractions.[36,37] Moreover, the use of deuterated reducing agents enables the points of attachment of the released alkanes to be assigned with some certainty. As an example of this elegant approach, Figure 8 shows the reconstructed ion chromatogram, together with the partial m/z 218 mass chromatogram which is indicative of the sterane distribution for the alkane fraction from Li/EtND$_2$ treatment of the Jurf EdDarawish (JED) kerogen.[37] The deuterium labelled products identified were dominated by the steranes of algal origin with the positions of deuterium incorporation providing confirmatory evidence of sulfide linkages. In contrast, the labelled products from a bituminous kerogen (Serpiano) were dominated by hopanes of bacterial origin which were linked via the side chain.[37]

For coals, Lochmann's base, an equimolar mixture on n-butyllithium and potassium t-butoxide has been used to reduce sulfides in what, from XANES evidence, appears to be in a fairly selective manner;[38] a 30% reduction in organic sulfur was achieved for the Illinois No. 6 coal from the Argonne suite. The Lochmann's base treatment was followed by potassium naphthelenide in tetrahydrofuran which converts dibenzothiophene to sulfur free products.[39] This subsequent treatment reduced the organic sulfur content of Illinois No. 6 coal by a further 0.55% w/w leaving 0.7% w/w in the final product which suggests that, possibly due to a combination of mass transfer and chemical factors, it is only partially effective for the thiophenic sulfur in coals.

Quaternization

The low receptivity of ^{33}S and the fact that it has a quadrupole has made it extremely difficult to observe organic sulfur in fuels directly by NMR. However, for oils and extracts, quaternary iodide salts can readily be formed in solution from the sulfur species present and if ^{13}C-enriched iodomethane is used, then ^{13}C NMR can be used to estimate the concentrations of non-thiophenic forms present.[40] We have applied this approach to the chloroform extract of Mequinenza lignite and Figure 9 shows the ^{13}C NMR spectra of the extract before and after quaternization with ^{13}C-enriched iodomethane. The aliphatic region in the quaternized fraction is dominated by methyl peaks from the S$^+$—CH$_3$ groups formed and most of the intensity occurs in the region 20–30 p.p.m., where peaks from non-thiophenes occur.[40] Indeed, from the intensity of this region, it is estimated that non-thiophenes account for over 60% of the total sulfur which is not surprising in light of the low carbon aromaticity of this particular fraction (*ca.* 1% w/w) in relation to the whole lignite (0.25 *cf.* 0.75).

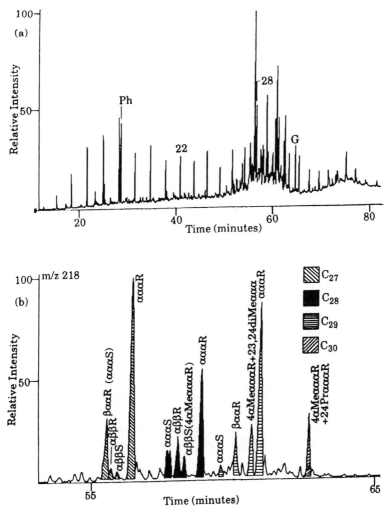

Figure 8 (a) Reconstructed ion chromatogram and (b) partial m/z 218 mass chromatogram (showing the sterane distribution) from GC–MS analysis of the alkane fraction obtained via Li/EtND₂ treatment of JED kerogen. The sterane peaks are marked with stereochemistry at positions 5, 14, and 17 and configuration at position 20. The components in brackets are the minor contributors to peaks for coeluting species. Reprinted from Hofmann et al., ref. 37. Copyright 1991, with kind permission from Elsevier Science Ltd., The Boulevard, Langford Lane, Kidlington OX5 1GG, UK

6 Conclusions

Although each of the approaches described here has its limitations, a consistent overall picture is beginning to emerge on the nature of organic sulfur groups in sedimentary organic matter and fuels. As might be anticipated, the more stable sulfur groups are most abundant in the more mature coals and petroleum

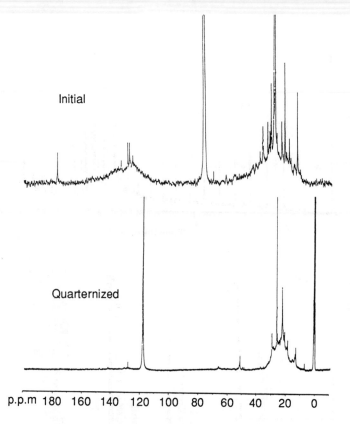

Initial

Quarternized

p.p.m 180 160 140 120 100 80 60 40 20 0

Figure 9 100 MHz ^{13}C *NMR spectra of the chloroform extract of Mequinenza lignite before and after quaternization with ^{13}C-enriched iodomethane*

source rocks. Further, these *X*-ray and pyrolytic techniques are now at the stage that the changes in organic sulfur distributions which occur during the formation and refining of heavy oils and the chemical desulfurization treatments for coals can be monitored effectively.

Acknowledgements

The authors thank the Science and Engineering Research Council for financial support (Grant No. GR/G/26600). The sample of Rasa lignite for high pressure TPR was supplied courtesy of Dr C. White (Pittsburgh Energy Technology Centre) and the liptinite concentrate was prepared by Dr D. Taulbee at the University of Kentucky, Centre for Applied Energy Research. RG thanks FICYT for supporting his studies at the University of Strathclyde.

References

1. C.-L. Chou, 'Geochemistry of Sulfur in Coal', in 'Geochemistry of Sulfur in Fossil Fuels', American Chemical Society Symposium Series No. 429, eds. W. L. Orr and C. M. White, American Chemical Society, Washington, 1990, p. 30.
2. J. S. Sinninghe Damste and J. W. de Leeuw, *Fuel Process Technol.*, 1992, **30**, 109 and references therein.
3. J. S. Sinninghe Damste and J. W. de Leeuw, *Org. Geochem.*, 1990, **16**, 1077 and references therein.
4. B. P. Tissot and D. H. Welte, 'Petroleum Formation and Occurrence', 2nd edn., Springer, Heidelberg, 1984.
5. T. I. Eglinton, J. S. Sinninghe Damste, M. E. L. Kohnen, J. W. de Leeuw, S. R. Larter, and R. L. Patience, 'Analysis of maturity-related changes in the organic sulfur composition of kerogens by flash pyrolysis-gas chromatography', in 'Geochemistry of Sulfur in Fossil Fuels', American Chemical Society Symposium Series No. 429, eds. W. L. Orr and C. M. White, American Chemical Society, Washington, 1990, p. 529.
6. J.-F. LePage, 'Applied Heterogeneous Catalysis', IFP EditionsTechnip, Paris, 1987.
7. R. M. Davidson, 'Organic Sulfur in Coal', IEA Coal Research CR/60, August 1993.
8. R. B. Jones, C. B. McCourt, and P. Swift, Proceedings of the 1981 International Coal Science Conference, Verlag Gluckauf, Essen, 1981, p. 657.
9. S. R. Kelemen, G. N. George, and M. L. Gorbaty, *Fuel*, 1990, **69**, 939.
10. M. L. Gorbaty, G. N. George, and S. R. Kelemen, 'Direct determination and quantification of sulfur forms in heavy petroleum and coal: sulfur K-edge X-ray absorption and X-ray photoelectron spectroscopic approaches', in 'Coal Science II', American Chemical Society Symposium Series No. 461, eds. H. H. Schobert, K. D. Bartle, and L. J. Lynch, American Chemical Society, Washington, 1991, Chapter 2, p. 127.
11. M. L. Gorbaty, G. N. George, and S. R. Kelemen, *Fuel*, 1990, **69**, 945.
12. G. N. George, M. L. Gorbaty, S. R. Kelemen, and M. Sansome, *Energy Fuels*, 1991, **5**, 93.
13. M. L. Gorbaty, S. R. Kelemen, G. N. George, and P. J. Kwiatek, *Fuel*, 1992, **71**, 1255.
14. R. Fiedler and D. Bendler, *Fuel*, 1992, **71**, 751.
15. G. P. Huffman, S. Mitra, F. E. Huggins, N. Shah, S. Vaidya, and F. Lu, *Energy Fuels*, 1991, **5**, 574.
16. M. M. Taghiei, F. E. Huggins, N. Shah, and G. P. Huffman, *Energy Fuels*, 1992, **6**, 293.
17. J. R. Brown, M. Kasrai, G. M. Bancroft, K. H. Tan, and J. M. Chen, *Fuel*, 1992, **71**, 649.
18. J. R. Brown, M. Kasrai, G. M. Bancroft, K. H. Tan, and J. M. Chen, *Fuel*, 1993, **72**, 900.
19. A. Attar, 'Sulphur groups in coal and their determinations', in 'Analytical Methods for Coal and Coal Products', Vol. III, Academic, 1979, Ch. 56 and DOE/PC/30145TI Technical Report.
20. A. Attar and F. Dupois, 'Data on the distribution of organic sulphur functional groups in coals', in 'Coal Structure', Advances in Chemistry Series No. 192, eds. M. L. Gorbaty and K. Ouchi, American Chemical Society, Washington, 1981, p. 239.
21. B. B. Majchrowicz, J. Yperman, G. Reggers, J. M. Gelan, H. J. Martens, J. Mullens, and L. C. van Poucke, *Fuel Process. Technol.*, 1987, **15**, 363.
22. B. B. Majchrowicz, J. Yperman, H. J. Martens, J. M. Gelan, S. Wallace, C. J. Jones, M. Baxby, N. Taylor, and K. D. Bartle, *Fuel Process. Technol.*, 1990, **24**, 195.
23. B. T. Dunstan and L. V. Walker, Final Report to Australian Nat. Energy Res. Dev. and Dem. Council, 1988.
24. C. J. Lafferty, S. C. Mitchell, R. Garcia, and C. E. Snape, *Fuel*, 1993, **72**, 367.
25. S. C. Mitchell, C. E. Snape, and R. Garcia, *Fuel*, 1994, **73**, 1159.

26. S. Mitchell, C. J. Lafferty, R. Garcia, K. Ismail, C. E. Snape, A. C. Buchanan III, P. E. Britt, and E. Klavetter, *Prep. Am. Chem. Soc. Div. Fuel Chem.*, 1992, **37** (4), 1691 and *Proc. 1993 Int. Conf on Coal Sci.*, in press.

27. R. Garcia, S. C. Mitchell, C. E. Snape, S. Moinelo, and K. D. Bartle, *Proc. 1993 Int Conf. on Coal Sci.*, Vol. 1, 445.

28. W. H. Calkins, *Energy Fuels*, 1987, **1**, 59.

29. R. J. Torres-Ordonez, W. H. Calkins, and M. T. Klein, 'Distribution of organic-sulfur containing structures in high organic sulfur coals', in 'Geochemistry of Sulfur in Fossil Fuels', American Chemical Society Symposium Series No. 429, eds. W. L. Orr and C. M. White, American Chemical Society, Washington, 1990, p. 287.

30. W. H. Calkins, R. J. Torres-Ordonez, B. Jung, M L. Gorbaty, G. N. George, and S. R. Kelemen, *Proc. 1991 Int. Conf. on Coal Sci.*, Butterworth-Heinemann, 1991, p. 985 and *Energy Fuels*, 1992, **6**, 411.

31. R. B. LaCount, R. R. Anderson, S. Friedman, and S. Blaustein, *Fuel*, 1987, **66**, 909.

32. R. B. LaCount, D. G. Kern, W. P. King, T. K. Trulli, and D. K. Walker, *Prep. Am. Chem. Soc. Div. Fuel Chem.*, 1992, **37** (3), 1083.

33. R. B. LaCount, D. G. Kern, W. P. King, R. B. LaCount Jr., D. J. Miltz Jr., A. L. Stewart, T. K. Trulli, D. K. Walker, and R. K. Wicker, *Fuel*, 1993, **72**, 1203.

34. J.-M. Ruiz, B. M. Carden, I. J. Lena, and E.-J. Vincent, *Anal. Chem.*, 1982, **54**, 688.

35. M. E. L. Kohnen, J. S. Sinninghe Damste, A. C. Kock-van-Dalen, and J. W. de Leeuw, *Geochim. Cosmochim. Acta*, 1991, **55**, 1375.

36. H. H. Richnow, A. Jenisch, and W. Michaelis, *Geochim. Cosmochim. Acta*, 1992, and in 'Advances in Organic Geochemistry', eds. C. B. Eckard *et al.*, Pergamon, Oxford, 1991, p. 351.

37. I. C. Hofmann, J. Hutchinson, J. N. Robson, M. I. Chicarelli, and J. R. Maxwell, *Org. Geochem.*, 1992, **19**, 371.

38. K. Chatterjee and L. M. Stock, *Energy Fuels*, 1991, **5**, 704.

39. K. Chatterjee, R. Wolny, and L. M. Stock, *Energy Fuels*, 1990, **4**, 402.

40. T. Green, W. G. Lloyd, L. Gan, P. Whitely, and K. Wu, *Prep. Am. Chem. Soc. Div. Fuel Chem.*, 1992, **37** (2), 664.

41. D. L. Lawlor, J. I. Fester, and W. E. Robinson, *Fuel*, 1963, **42**, 239.

42. J. W. Smith, N. B. Young, and D. L. Lawlor, *Anal. Chem.*, 1964, **36**, 618.

43. W. H. Calkins, M. L. Gorbaty, and S. R. Kelemen, *Fuel*, 1993, **72**, 900.

44. R. Garcia, S. R. Moinelo, C. J. Lafferty, and C. E. Snape, *Energy Fuels*, 1991, **5**, 582.

45. D. N. Taulbee, S. H. Poe, T. L. Robl, and B. Keogh, *Energy Fuels*, 1989, **3**, 662.

46. J. D. Payzant, T. W. Mojelsky, and O. P. Strausz, *Energy Fuels*, 1989, **3**, 449.

47. C. W. McGowan and R. Markuszewski, *Fuel*, 1988, **67**, 1091.

48. R. Markuszewski, X. Zhou, and C. D. Chriswell, *Prep. Am. Chem. Soc. Div. Fuel Chem.*, 1992, **37** (1), 417.

Recent Progress in Plasma Mass Spectrometry

N. Jakubowski and D. Stuewer

INSTITUT FÜR SPEKTROCHEMIE UND ANGEWANDTE SPEKTROSKOPIE, POSTFACH 10 13 52, D-44 013 DORTMUND, GERMANY

1 Introduction

Current industrial developments and increasing social awareness of the inherent problems have forced analytical chemistry into the limelight, with the role of securing high quality living standards. This task requires the development of techniques for the analysis of substances to previously inaccessible levels of sensitivity, not only in the field of modern material sciences, but also in geochemistry, the bio-sciences in general, and medicine in particular.[1] It is especially relevant to environmental research in all its different spheres.

This analytical task looks very simple when the use of atom spectroscopy is considered for elemental analysis, because all matter in our universe is made from atoms. Analysing these atoms with an adequate instrument will result in information about the composition of the sample under investigation. However, for application of atom spectroscopy, the sample must necessarily be atomized, excited, or ionized. Only if this is done well can the accurate composition of a sample be determined, which highlights the essential role of the spectrochemical source.

The application of plasma sources for atomization, excitation, and ionization in elemental spectrochemistry has a long and distinguished history, in particular in atomic emission spectrometry (AES), where such sources have proved as reliable, precise, and fast in research as well as in manifold industrial applications. Most frequently, they are used as excitation sources in the form of arcs, sparks, glow discharges, inductively coupled plasmas (ICP), microwave induced plasmas (MIP), and so on.[2,3] Some of them have additionally been used as ion sources in mass spectrometry (MS), which in comparison to AES displays several basic advantages. In particular MS shows excellent detection limits, which are lower by several orders of magnitude and additionally more uniform across the periodic table than in AES. With MS, true multi-element analysis can be realized easily, because (nearly) all elements of the periodic table can be analysed with an extraordinary large dynamic range, which for some instruments may amount to 10 orders of magnitude. The result of an analysis is recorded as a simple mass spectrum which represents the isotopes of the elements present in the sample. This isotope specific information can additionally be useful for analytical purposes, in particular for application of the isotope dilution technique which is unsurpassed for accurate quantification. When a quadrupole is used as the mass analyser, a spectrum can be obtained for all elements in a sample in less than one second.

Despite these advantages, it took a long time for suitable and powerful ion sources to be developed for elemental mass spectrometry on the basis of stationary discharges. Some 15 years ago, commercial sources for elemental mass spectrometry[4,5] were nearly exclusively realized by beam techniques as in secondary ion mass spectrometry (SIMS),[6] plasma desorption mass spectrometry (PDMS),[7] or by non-stationary sources as in spark source mass spectrometry (SSMS)[8] and laser induced mass spectrometry (LIMS).[9] Of these techniques, SIMS and SSMS were rather successful in spite of several disadvantages (matrix effects, tedious evaluation, poor precision). Nowadays, the apparent victorious advance of elemental mass spectrometry is mainly based on the recent development of stationary plasma ion sources to which this paper will be restricted.

2 Fundamentals

From a basic physical point of view, a plasma is nothing else but a partially or fully ionized gas. For operation as a spectrochemical source, any energy carrier (particle or photon) can be used to obtain relevant information on the composition by interaction with the sample. In most techniques the sample has to be atomized first for subsequent exposition to the physical interaction. In this way plasmas are applied as atomization sources in atomic absorption[10] and fluorescence spectroscopy[11] or are used as sources for excitation or ionization in emission and mass spectrometry. Whenever the steps of atomization and ionization are separated in space and time, the technique is less likely to suffer from matrix effects.

Different concepts of applying plasma sources in combination with MS have been realized commercially. For direct bulk analysis of conducting and semiconducting materials, glow discharge mass spectrometry (GDMS) with direct current (d.c.) sources is without competition. A special radio frequency (RF) source has been applied for surface analysis in sputtered neutrals mass spectrometry (SNMS). Inductively coupled plasma mass spectrometry (ICP-MS) has gained wide-spread acceptance mainly for the analysis of liquid samples. In this paper, these techniques shall be reviewed briefly, and progress will be demonstrated by several examples selected from the authors' work.

In general, all instruments in plasma source mass spectrometry (PSMS) have similar components comprising an ion source and equipment for ion transfer, mass separation, and detection. The most important component, though not considered as such in many cases, is the sample itself, the way it is taken and prepared for analysis, and the system for its introduction to the source. In the case of a plasma ion source, a working gas for the discharge is always needed. The source pressure may range from 10^{-3} mbar in special RF sources, up to

10 mbar in GDMS, and up to atmospheric pressure in ICP-MS. For reduction of the source pressure down to a pressure of about 10^{-6} mbar, which is necessary in the mass analyser in order to enable separation by electrical or magnetic fields, an interface system with differential pumping is required. An electrostatic lens system is used for ion transfer from the source to the analyser.

Up to now, quadrupole filters have been applied as mass analysers in most cases, because they enable fast and simple operation at only moderate expense. In order to overcome limitations from spectral interferences, there is an increasing interest in the use of double focusing high resolution mass spectrometers. As a detector system, secondary electron multipliers (SEM) are always used, providing an extraordinary high dynamic range of up to 10 orders of magnitude.

Having the advantages of PSMS in mind, some limitations should also be mentioned here. Spectral interferences due to isobaric atoms or multiple charged ions and molecules formed from atoms of the discharge gas, its contaminants or the main constituents of the sample, may effect the mass spectra. Non-spectral interferences, well known as matrix effects, may influence the sensitivity for some elements, complicating quantitative analysis. The expenses of the instrumentation and the operating staff are further limitations of increasing significance. Mass discrimination effects and noise effects may complicate the determination of isotope ratios. In comparison to optical emission spectroscopy, the increase in analysis time should also be mentioned, because real simultaneous mass spectrometers are not commercially available. Some plasma sources, in particular the ICP, have only a limited tolerance to the dissolved salt load of a solution and too high load may result in matrix and drift effects or the clogging of orifices in the interface. Concerning quantitation, mass spectrometry is only a relative technique so that standards are always needed for calibration. Also, for some applications, mainly the analysis of solid samples, there is a lack of standard reference materials certified for low level analyte concentrations.

3 Instrumentation

The practical limitation is, of course, always determined by the performance of the instrument. A short overview of commercially available instruments, which are equipped with a plasma ion source, is given in the following sections.

SNMS

The SNMS instrument ('INA 3') with a RF plasma ion source was developed by Leybold–Heraeus and is now available from Specs. About 25 instruments have been sold world-wide so far.

GDMS

The first instrument in this field was the 'VG 9000', developed by VG, now Fisons Instruments, combining at first a GD source to a thermionic instrument with a magnetic sector MS only. Later the instrument was improved to a double focusing sector field instrument, and more than 50 of these instruments are now working world-wide in routine analysis. Although successful use of quadrupoles with a GD source had been reported earlier,[12,13] it was not till 1989 that VG (Fisons Instruments) introduced the first low resolution instrument ('GloQuad'). Up to now about 15 of these instruments are in routine use world-wide. A combined ICP-GD-MS instrument ('TS Sola') was introduced in 1989 by Turner Scientific, now available from Finnigan MAT. About 10 instruments equipped with a GD source have been sold so far. Very recently, Finnigan MAT has announced a GD source for their new high resolution ICP-MS instrument ('Element'). In comparison to GD-AES, which is applied in about 315 systems, it can be concluded that GDMS is rapidly becoming an established technique with promising aspects for the future.

ICP-MS

The first quadrupole based ICP-MS systems were introduced to the market in 1983 by Sciex, now the Perkin Elmer Corporation, and by VG, now Fisons Instruments. For a long time the market was dominated by these instruments. Many improvements have been made over the years so that a new generation of reliable and powerful instruments is now working successfully in many routine applications. Some years ago, a special low cost version for environmental analysis ('PQe') was introduced to the market by VG (Fisons Instruments). As mentioned before, with the 'TS Sola', Turner Scientific introduced a combined ICP-GD-MS system. Very recently some other companies established in the ICP-AES market, like Varian Associates and Spectro Analytical Instruments, have also announced low resolution ICP-MS systems.

The first high resolution instrument ('Plasma Trace') was introduced to the market in 1988 by VG, based on the '7070' a double focusing sector field instrument for organic analysis. Two further high resolution instruments have been announced by Jeol Ltd. and Finnigan MAT. The latter, the above mentioned 'Element', seems to compete with quadrupole based insruments concerning price and scan speed, but is clearly superior through its high mass resolution and the typical sensitivity provided by a high resolution instrument.

Summarizing, up to now about 800 ICP-MS systems are installed world-wide. Obviously there is an increasing trend to develop, not a dedicated instrument, but a family of instruments with different plasma sources for different analytical applications, as suggested in our research over many years. For the new generation of instruments, automation is obviously indispensable, and increasing competition will no doubt benefit the users of the technique.

Laboratory Instruments

Besides the commercial instruments, some laboratory instruments were developed in research groups, be it in anticipation of later commercial development, or for special investigation purposes. Concerning GDMS, the group of Harrison should be mentioned here;[12] and the ICP-MS work of the groups of Gray,[14] Hieftje,[15] and Houk[16] should also be mentioned. In our own case, we developed first a low resolution GDMS[13] instrument and later also an ICP-MS[17] instrument, following the concept of an instrument family, by using mainly identical MS equipment for both instruments.

4 Plasma Sources for Elemental Mass Spectrometry

Radio Frequency Sources

About 20 years ago RF discharges were introduced by Coburn and Kay[18-20] for analysis of technical surface layers and for the analysis of non-conducting samples. Simultaneously, a new inductively coupled plasma ion source for post-ionization of sputtered neutrals under high vacuum conditions was developed by Oechsner.[21] This pioneering work culminated in the development of a commercial instrument. It should be mentioned that different concepts of post-ionization techniques have been realized using electron beams, different plasma sources, or laser systems, but in this paper only commercially available plasma source systems will be considered.

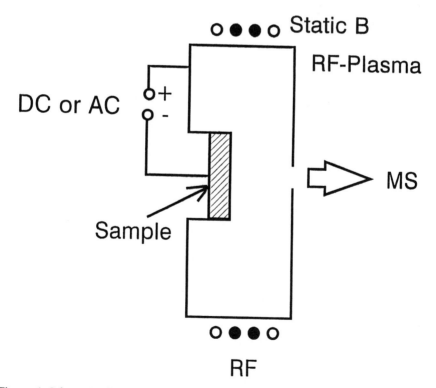

Figure 1 *Schematic diagram of an RF discharge used in sputtered neutrals mass spectrometry (SNMS)*

Principle. A schematic diagram of the commercially available version of an RF source used in the SNMS instrument is shown in Figure 1. An electron cyclotron resonance, in a static magnetic field generated by a set of Helmholtz coils, is used to produce an inductively coupled RF plasma under high vacuum conditions.[22] The main advantage of this arrangement consists in the fact that plasma generation is realized without any electrode, so that the physical properties of the sample, such as conductivity or dielectricity constant, will not influence the plasma, keeping the analytical procedure independent of the sample characteristics.

By connecting the sample to a high voltage, argon ions from the RF plasma are accelerated onto the sample surface, and hence the sample is atomized by sputtering. Depending on this voltage, the ion energy can be influenced over a wide range, and as a result the sputter rates can easily be varied. Only sputtered neutrals can overcome the potential barrier and diffuse into the plasma, where they are ionized by electron impact. By application of a very efficient ion energy filter, ions which are perhaps generated at the plasma potential can be effectively suppressed, so that ions representing the surface composition of the sample are dominant in the mass spectra.

Performance and Applications. A typical result obtained in in-depth analysis of surface layer systems by SNMS is shown in Figure 2. Analysis of a multilayer system made up from Cr and Ni layers, with steplike transitions, was used to

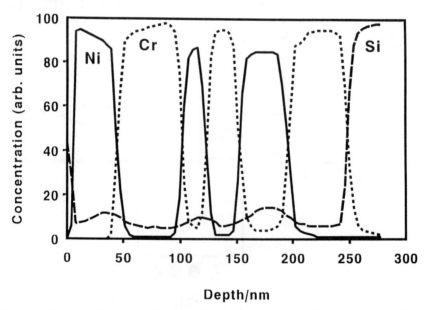

Figure 2 *Concentration versus depth profile of a multilayer sample measured with SNMS (successive layers of Ni 50 nm, Cr 50 nm, Ni 25 nm, Cr 25 nm, Ni 50 nm, and Cr 50 nm)*

demonstrate the depth resolution obtainable with SNMS. For this measurement, it should be realized that optimum depth resolution can only be achieved by keeping a critical material dependent voltage, as has been shown by Oechsner and Wucher.[23] The concentration scale was calculated from the intensities by use of relative sensitivity factors (RSF), which have been obtained by analysis of reference samples. The measured time scale was transformed to a depth scale using the concept of constant erosion rates. All single layers are well separated, and using the standard definition, a depth resolution[24] of about 9 nm can be calculated as the average from all interfaces. The lowest depth resolution ever achieved under optimum conditions was about 4 nm. It was shown, that this limit is caused physically by the sputter process itself.[25,26] Hence, the main application field of this ion source is the analysis of surface layers, on conducting or semi-conducting samples, with a thickness from the nm region to the μm region.

A selection of convincing applications of SNMS in in-depth and bulk analysis of conducting and semi-conducting samples is given in the references.[27-45] Additionally, the technique has been applied for analysis of coated metallic materials[46,47] and for analysis of organic layers like polymer interfaces[48] or organometallic microparticles for production of organic layers.[49] A round robin analysis of standard reference materials of steel and aluminium by SNMS users[50] showed that relative sensitivity factors (RSF) for the elements are constant for each instrument, but may differ significantly from one instrument to the other. The RSF values may be considered as comparative constants, and in this sense the technique can be considered to be free from matrix effects. Additionally, for most sample matrices, the RSF values are close to unity for a wide majority of elements, so that quantitative analysis without reference materials looks promising.

The versatility of SNMS as a tool for elemental analysis is demonstrated by further applications of which the analysis of particulate matter like pigments, outdoor aerosols, and welding fume should be mentioned here.[51] In recent work in our institute, SNMS was successfully applied for the in-depth analysis of powdered ceramic material[52] suspended on a metal foil, to study surface modification by different chemical treatments. This technique may be useful for evaluation of new developments in the production of ceramic materials.

Up to now the application of SNMS has been restricted to conducting and semi-conducting massive samples, because charging of non-conducting samples will build up retarding fields for ions at the sample surface and so terminate the sputter process. Nevertheless, analysis of glass samples was possible by applying a grid technique to overcome charging effects.[53] A more versatile tool to overcome charging effects is commercially available in the INA 3 instrument, simplifying the analysis of non-conducting samples considerably. With this device, the sample is connected to an RF generator. In the negative half cycles of the RF voltage, ions are bombarding the sample thus compensating the negative charge introduced by electron bombardment during the positive half cycles. Preliminary results demonstrated that this device may be really powerful, not only for analysis of non-conducting samples but for conducting samples too, because sputtering is more uniform.[54] As an alternative, the development of new RF sources for analysis of non-conductive materials by GDMS may also be promising.[55-57]

For completeness it should be mentioned here, that a review of SNMS for microanalysis is given elsewhere.[58]

Glow Discharge Sources

The capabilities of gas discharges as ion sources were described many years ago by Goldstein.[59] Gas discharges were used by the pioneers[60] of mass spectrometry (Aston,[61] Dempster,[62] and Thomson[63]). Nevertheless it took a very long time for a renaissance in the use of these sources, which began about 20 years ago, when considerable effort was devoted to the development of new techniques to overcome the limitations of SSMS.[64,65] Although Coburn was the first to revitalize GDMS, it was mainly the pioneering work of Harrison and co-workers, which demonstrated the analytical capabilities of glow discharges for mass spectrometry. GDMS has now reached a certain maturity for the direct analysis of conductive materials, as illustrated by the number of books and reviews on the subject.[66-80] A comprehensive review describing the role of the GD in AES and MS by Harrison[81] deserves a mention here. This high number of reviews does not mean that GDMS has reached stagnation. The exact opposite is true, because considerable research work is going on in different areas; especially atomization[82-85] and ionization[86-95] processes, which have been studied extensively. Although different commercial sources are available, optimization of the source design[96-101] is of special interest. Last, but not least, the problem of spectral interferences[91,102-107] and how to overcome them is important as never before for application of low resolution quadrupole based instruments.

Principle. The glow discharge is a self-sustaining, space-charge dominated discharge with high burning voltage between electrodes in a noble gas atmosphere of reduced pressure. In front of the cathode there is a strong voltage drop region, the so-called cathode fall, followed by the negative glow. The latter is a nearly field free region containing a quasineutral plasma, from which a bright light, the glow light, is emitted. In the cathode fall region, positive ions mainly of the discharge gas are accelerated towards the cathode (sample), where they impinge onto the surface resulting in atomization of the sample by sputtering. The sputtered atoms diffuse to the negative glow, where they are ionized in the plasma by electron impact and further ionization mechanisms. The negative glow produces a reservoir from which ions can be extracted through an ion exit aperture. Although a decoupling of the steps of atomization and ionization is realized in this source similar to SNMS, a certain influence of the sample matrix may be expected because the sample itself, different to SNMS, is an electrode of the discharge, so that heat conductivity, electron emission probability, resistance and other physical and chemical properties of the sample may significantly influence the discharge itself. The function of the sample as an electrode leads also to the requirement that the sample must have a certain conductivity.

In our laboratory developed GDMS system,[13] we made use of a modified Grimm type design for the source, as it is well known from emission spectroscopy. The sample is clamped against a cathode body and thus serves as the vacuum sealing part. An obstructed or hindered discharge is created by a special cylindrical anode tube, defining a surface area on the sample which is exposed to ion bombardment. The discharge is always operated in the anomalous mode. This arrangement is especially useful for bulk and surface analysis of flat samples.

Performance and Application. GDMS is an established technique for bulk analysis of conducting and semi-conducting solid samples with high detection power. Many applications have been reported in the literature, mainly applying the high resolution 'VG 9000' instrument for analysis of high purity metals,[108] refractory metals,[101,109-111] high purity Al,[112-115] semi-conducting materials in general,[16] and Si,[117,118] GaAs,[119] and Ga[120] in particular. Some special applications are the analysis of organic materials[121] and the analysis of starting materials for production of high temperature super conductors.[122] As already mentioned, MS has the advantage that isotopic information is included, which has been applied in an investigation of cold fusion experiments,[123] where isotope ratios of Pd have been measured with a precision better than 0.1%. As a medical application, the analysis of Pt drugs in urine has been reported.[124] Detection limits at low ng g^{-1} levels can easily be obtained for many elements. The ultimate performance has been obtained for determination of Th and U where detection limits down to below 0.01 ng g^{-1} have been realized in high purity Cu.[125] The determination of C, N, and O, which is of special interest in many cases, requires usually intensive cryo-cooling of the discharge cell. For analytical application of low resolution instruments, only a few results are available mainly concerning bulk analysis of conducting samples like steel,[126] refractory metals,[127] and Cu.[128] A problem of GDMS can be the very low sample consumption which gives rise to a high scatter of analytical results due to local inhomogeneities of the sample, which sometimes is mis-used for unqualified depreciation of the technique in general.[110]

For characterization of recent progress in GDMS, three different selected applications concerning surface analysis, the analysis of non-conductive samples, and of liquid residues will be discussed in more detail in order to demonstrate the versatility of GDMS, from which the expectation of an enlarging field of applications can be concluded.

Obviously there is an increasing interest for application of GDMS to surface and in-depth analysis of technical surface coatings, as it has also been the case with the Grimm type emission source in AES. Most technical layers have a thickness ranging from some hundred nm to several tens of μm. Traditional MS techniques for surface analysis like SIMS and SNMS require a rather long analysis time for such layers. Owing to considerably higher erosion rates of several μm min^{-1}, application of GDMS for surface analysis looks really

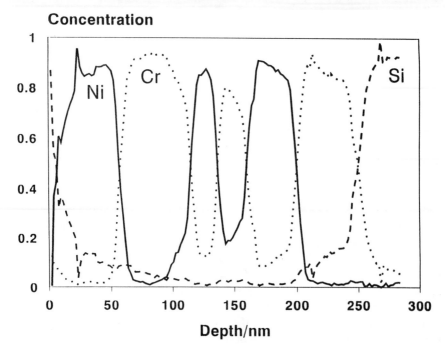

Figure 3 *Concentration versus depth profile of a multilayer sample measured with GDMS (successive layers of Ni 50 nm, Cr 50 nm, Ni 25 nm, Cr 25 nm, Ni 50 nm, and Cr 50 nm)*

promising, provided satisfactory depth resolution can be achieved. A necessary prerequisite for the latter is realization of plane crater profiles. In a model investigation of self-prepared multilayer systems,[129] we have shown that this requirement can be fulfilled if the operational conditions of the plasma are chosen adequately. The same Cr—Ni multilayer system, which was used for the SNMS analysis represented in Figure 2, was also analysed by GDMS, and Figure 3 shows the result obtained after conversion of the intensity *versus* time profile into the concentration *versus* depth profile, using a procedure, which is described in detail elsewhere.[129] The transformation of sputter time to depth has been done using a procedure developed by Oechsner *et al.* for SNMS.[23] All layers are well resolved, and the mean value for the depth resolution calculated to be about 12 nm, which is only a little bit worse than the SNMS result. With both techniques, it becomes obvious that the layers around 150 nm do not reach the full concentration, from which certain problems in the preparation of the layers must be concluded. In comparison to SNMS, a signal for Si appears from the first layer in GDMS. This must be attributed to some spectral interferences from residual gases, which obviously are suppressed in SNMS. Additionally, a small crater edge effect can be observed, demonstrated by an increasing Si signal in the last Cr layer. For analysis of thick technical layers, the depth resolution obtained by GDMS is satisfactory, so that with respect to the shorter analysis time, GDMS is very promising for this type of application.

Although the main application field of GDMS is the analysis of conducting and semi-conducting samples, non-conducting powder samples can also be analysed applying matrix modification by mixing and compacting with pure metal powders, as it is well known from AES.[130,131] The capabilities and limitations of this technique have been demonstrated by the example Al_2O_3.[132] The aspects of quantification due to the lack of standard reference materials in the case of ceramics have also been discussed, and a calibration was performed for some elements.

Another example of our work concerns the application of GDMS for micro analysis. For this purpose, a microvolume of the liquid to be analysed was pipetted onto the surface of a pure metal substrate. After cementation, the residue was exposed to GDMS analysis, and single ion monitoring profiles were registered.[133] The technique is well suited for analysis of Pt group elements in environmental liquid samples. Determination of these elements is a real challenge for modern analytical spectrochemistry, because Pt group elements are found in the exhaust gas from combustion engine catalysts.[134,135] Although they are released in very small amounts, a significant enrichment in the environment may occur over years. It was for this reason that performance of GDMS was investigated. Cautious drying of a solution containing Pt or Ir on a Cu substrate results in a cementation process covering the copper surface by a metallic layer of only monolayer thickness. For quantification, Au was used as an internal standard as can be seen in the example of the analytical signals obtained shown in Figure 4. The intensity *versus* time profiles of Pt and Au, with an absolute amount of 5 ng each, display nearly the same sputter dependence for both elements. The very high intensity of more than 10^6 cps demonstrates the extraordinary sensitivity of GDMS for microanalysis. Absolute detection limits in the low pg region could be realized easily for Pt group elements in aqueous solutions. For real life samples, matrix separation techniques like sorbent extraction will be necessary to remove the matrix and interfering reagents and to enrich the analyte, which will be necessary under certain conditions.[136] With a corresponding fully developed procedure, GDMS may become one of the most powerful techniques for microanalysis of liquid residues.

Inductively Coupled Plasma Sources

Inductively coupled plasmas as ion sources for mass spectrometry have a long history since they were already applied about 50 years ago.[137,138] In the early sixties, the ICP at atmospheric pressure was used for multi-element analysis[139] in AES. After the appearance of the first ICP-AES instrument on the market in 1974, it was about 10 years before the first commercial ICP-MS system became available, following the pioneering work of Gray and Houk in 1980.[140]

Principle. In the ICP source the plasma is generated by an oscillating high frequency electric field of an induction coil. Samples are introduced as an aerosol

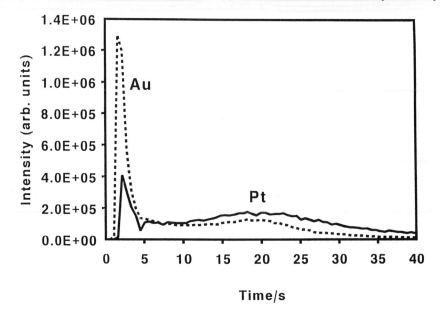

Figure 4 *Intensity versus time profile for a residue of a solution containing Pt and Au with a total amount of* 5 ng *each*

into the central channel of the argon plasma, with a high temperature of about 8000K, where fast evaporation, atomization, and ionization take place on the way through the plasma. The ions are extracted from the plasma at atmospheric pressure by means of a special interface, which is necessary to reduce the pressure by differential pumping to the working range of the mass analyser, which is nine orders of magnitude lower in comparison to the source. Among all the plasma sources used in combination with MS, this is the system that requires the highest quality vacuum technology.

Evaporation, atomization, and ionization are continuously taking place as the injected material passes through the plasma, so that there is not a clear decoupling between the steps. This is the reason why sample introduction procedures, as well as any techniques for conditioning the aerosol which finally is injected into the plasma, may have a considerable influence on the analytical performance and continue to be the subject of research. As a characterization of the state of the art in this field, some selected topics of recent progress concerning sample introduction techniques and aerosol conditioning will be discussed in the following sections. Additionally, the problem of spectral interferences will be discussed, with respect to the current trend towards application of high resolution in ICP-MS.

Performance and Application. As mentioned before ICP-MS is the most successful technique in elemental mass spectrometry up to now, as is demonstrated by the variety of books,[141] reviews,[142] and overviews of literature.[143] In particular, the excellent reviews of Lust,[144] McLaren,[145] and Hieftje et al.[146] should be mentioned, because they include hundreds of thematically ordered references. Some reviews deal with special applications, in particular for geochemistry.[147,148] A comparison with other atomic spectroscopy analytical techniques is given by Slavin[149] and with different instruments, by Denoyer.[150] An overview of fundamental research and applications can be found in the annual publication of *Atomic Spectrometry Update – Atomic Mass Spectrometry*,[151,152] which is a feature in the *Journal of Analytical Atomic Spectrometry*. The field displays a rapid increase in the number of publications, of which only part can be considered here.

Sample introduction systems. Many analytical applications in ICP-MS depend on the performance of a special sample introduction system, for which considerable progress has been made recently. One special aim of such work is the reduction of spectral and non-spectral interferences. For reduction of spectral interferences, the application of a desolvation system has proved useful.[153-155] Also application of electrothermal vaporization (ETV)[156-159] may reduce the appearance of molecules arising from solvent derived atoms. For reduction of non-spectral interferences, application of liquid chromatography for matrix separation can be useful to overcome the problem of sensitivity losses. In general, improvement of analytical figures of merit can be realized by application of new high performance sample introduction systems, which provide a higher nebulization efficiency and consequently improved detection limits.[160,161] Furthermore, the versatility of ICP-MS has been demonstrated by new application fields. A striking example is the direct analysis of solid samples without dissolution, by application of laser[162-165] or spark[166-168] ablation or by ETV.[169,170]

However, some severe disadvantages should not be concealed. The high cost of the equipment is a general disadvantage, as are the running costs and the expense of purchasing optional sample introduction systems for different applications. Furthermore, one should be aware that the main advantage of ICP-MS, the multi-element capability, may be reduced in significance if transient signals are generated, as is the case with laser ablation (LA) and with ETV, or when special element-sensitive procedures like gas and vapour generation[171-173] are applied. Last, but not least, it should never be forgotten that each special sample introduction system has been developed for a specific application, which may have the consequence that its performance is restricted significantly when used for different applications.

Some selected topics which are of greater interest in the context of our work will now be explained in more detail.

Aerosol conditioning and processing. Inspite of several new developments concerning sample introduction systems, pneumatic nebulizers have remained the workhorse in ICP-MS. Besides the nebulizer, a spray chamber is always part of the pneumatic nebulization unit. Its function is the removal of bigger droplets from the aerosol, which in a generalized sense, is a method for aerosol conditioning, as it changes significantly the droplet size distribution of the aerosol. This shows that aerosol conditioning of some sort is always employed in sample introduction to a plasma source, although one is usually not aware of it. The purpose of conditioning is mainly to influence a primary aerosol in such a way that introduction of the sample does not interfere with the plasma too much. In ICP-MS, analytical performance depends much more strongly, in comparison to AES, on the nature of the sample reaching the plasma. In most cases the sample is introduced to the plasma by droplets and sometimes by particles[174-176] or in special cases as a vapour, as mentioned before. Investigations of Olesik *et al.*[177] have shown, that the droplet diameter may significantly influence the plasma, so that as a consequence, the final analytical signal is changed by any changes in the processes of vaporization, atomization, and ionization in the plasma, because energy is consumed in these processes, and cooling of the plasma may be the result. Therefore, desolvation and cooling of the spray chamber are simple and effective approaches for influencing the droplet diameter and reducing energy losses due to a high water load.[178-181]

Desolvation is easily realized by a heated transfer line for evaporation of the droplets, followed by a cooling line where part of the evaporated solvent, mainly water, is condensed. It has been shown by Beauchemin *et al.*[182] that the heating part of this device may be sufficient to completely evaporate the aerosol droplets, giving an increase in sensitivity and a reduction of molecule formation, if only a high enough temperature is chosen. In a previous investigation,[183] we have shown that desolvation can significantly increase sensitivity and plasma stability. Likewise, removal of the solvent helps to reduce interferences significantly. On the other hand, a strong loading of the plasma with water is avoided which otherwise might lead to a strong intensity depression. It has been shown, in recent investigations,[184] that water addition to a dry aerosol can significantly reduce the sensitivity, and the same effect has also been observed for addition of hydrogen gas. From this point of view, dry aerosols are in general preferable in ICP-MS, and in the case of wet aerosols, efficient aerosol conditioning systems can be an effective means of improving the analytical figures of merit.

Electrothermal vaporization. As mentioned before, dry aerosols are preferable in ICP-MS, because the disturbance of the plasma is only low in this case. The highest efficiency in the conversion of an analytical sample to an aerosol can be obtained by application of ETV.[185-194] As a consequence, ultimate sensitivity and detection limits are achieved. Additionally, reduction of spectral interferences caused by solvents is possible due to removal of the solvent by evaporation at the drying step of the heating program. Furthermore, it is a method for analysis of small volumes of only a few µl, and hence it is useful for micro-analysis whenever the sample volume is limited, as for instance in medical or biological applications. As mentioned previously, a disadvantage of ETV is that transient signals must be used for the analysis, so that only a limited number of elements can be determined in a single run.

In our own work on the determination of Pt group elements with ETV, a tungsten oven[195,196] formed by a wire loop was used, as introduced for AAS by Berndt *et al.*[197] The main advantage of the tungsten wire loop unit is its very low expense of only some German Pfennigs for commercially available tungsten coils. In the case of the Pt group elements, the analysis is not limited by blank values. There are only interferences from WO^+ for some minor Pt isotopes. A power supply of only 150 W, which is modest in comparison to a demand of several kW for a graphite tube furnace, is sufficient to enable a full multi-step temperature program. Temperatures of up to 3000 °C can be reached in about 1 s, giving transient signals of about 1 s duration. Figure 5 shows an example of a calibration for Pt ranging from 0.1 ng to 10 µg. For this analysis, hydrogen has proved useful to increase the intensity by more than a factor of four. This was surprising, because it had been found before that hydrogen addition leads to an intensity loss for elements with ionization energies below 8 eV. From Figure 5, a linearity over more than five orders of magnitude can be achieved, and the detection limit was calculated to be in the low pg region. In comparison with microanalysis by GDMS, similar detection limits were obtained with both techniques, demonstrating the high efficiency of ETV sample introduction for ICP-MS.

Hydraulic High Pressure Nebulization (HHPN). Hydraulic High Pressure Nebulization (HHPN) has been introduced as a sample introduction technique for AAS and ICP-AES by Berndt.[198,199] Based on this work, we have adapted HHPN for sample introduction in ICP-MS.[200]

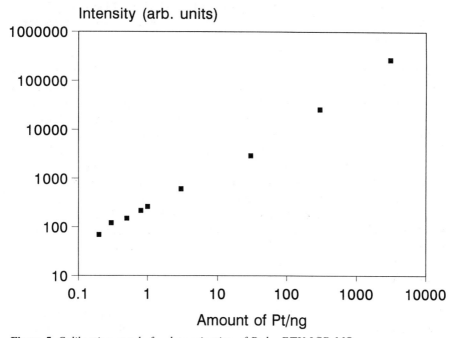

Figure 5 *Calibration graph for determination of Pt by ETV-ICP-MS*

In HHPN, a HPLC pump is used to force the analyte liquid through a nozzle with an exit diameter of about 10 μm, with a pressure of up to 400 bar. An impact bead is used to transform the size distribution of the resulting droplets so that finally a very dense aerosol, with droplet diameters below 4 μm, is generated. Samples are injected to a carrier flow by use of a sample loop with a volume between 10 μl and several ml. As with ultrasonic nebulization, a desolvation system for aerosol conditioning is indispensable for coupling to ICP-MS, because otherwise the high water load to the plasma can produce severe matrix effects or even instabilities in operation. In comparison to conventional pneumatic nebulization, HHPN increases the nebulization efficiency from 1–2% to about 30%. However, the gain in nebulization efficiency can only be fully exploited if a desolvation system is used to permit a corresponding gain in ion intensity. The mean value for the improvement in intensity was found as one order of magnitude for 23 elements investigated; only few elements, in particular B, show desolvation losses. Unfortunately, the gain in intensity could not always be converted to a corresponding improvement in detection limits, because for some elements, blank values were limited by contamination from reagents or the system itself.

Summarizing, the most obvious advantage of this method of sample introduction in ICP-MS is that it enables direct coupling of HPLC with a simultaneous increase in sensitivity. This is of considerable significance in the development of applications of PSMS in flow injection analysis (FIA)[201] matrix separation by liquid chromatography, and trace enrichment or chromatographic separation of interfering elements.[202] Furthermore, it may be considered for speciation analysis of environmental samples by use of LC, which is of increasing interest as can be seen from a recent review of this field, covering procedures for a variety of elements.[203]

The capabilities of ICP-MS with HHPN for speciation analysis of Cr, which had been investigated earlier for ICP-AES,[204] are shown in Figure 6. In this case, which is described in more detail elsewhere,[205] ion-pairing HPLC with a modified C_{18}-RP-column has been applied for on-line separation of the Cr(III) and the retained species Cr(VI) using tetrabutylammonium acetate (TBAA) as the complexing agent. Methanol–water was used as the eluent. Disturbances by strong carbon interferences on the main chromium isotope as well as the risk of clogging of the sampler or the skimmer orifice could be considerably reduced by oxygen addition to the aerosol gas and by application of a new desolvation system with two stage cooling.[206] Figure 6 shows chromatograms obtained in the analysis of a tap water, applying standard additions for calibration. From this figure, detection limits in the low ng ml^{-1} region can be estimated for real life samples. Besides Cr(III) and the hazardous species Cr(VI), a third species is present, which must be attributed to Cr, but remains unidentified so far.

High resolution. As discussed before aerosol conditioning or application of new high performance sample introduction systems are tools to overcome the problem of spectral interferences, but from a basis point of view, the application

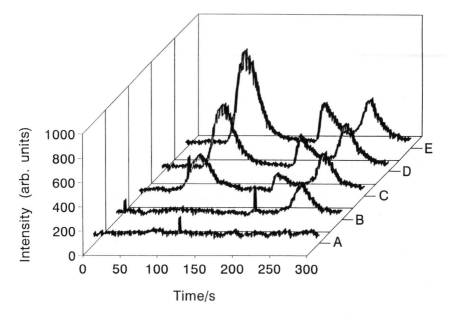

Figure 6 *Calibration for a tap water sample using standard additions of both Cr(III)[1] and Cr(VI)[2]: A: blank solution; B: sample; C: sample + 2.5 ng ml⁻¹ Cr(III) + 2.5 ng ml⁻¹ Cr(VI); D: sample + 5 ng ml⁻¹ Cr(III) + 5 ng ml⁻¹ Cr(VI); E: sample + 7.5 ng ml⁻¹ Cr(III) + 7.5 ng ml⁻¹ Cr(VI)*

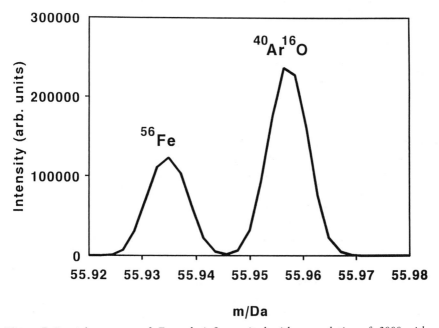

Figure 7 *Partial spectrum of Fe and ArO acquired with a resolution of 3000 with a laboratory prototype HR-ICP-MS instrument*

of higher mass resolution would be the best way to overcome problems arising from interferences.[207] The higher performance can of course only be achieved by higher expense, but it can be concluded from many discussions at conferences and also from the literature,[208] that the trend is going in this direction.

Recently, we have been working with a prototype of a new high resolution instrument from Finnigan MAT. A typical spectrum obtained in the high resolution mode of operation is shown in Figure 7, demonstrating that the well known interference of ArO^+ on the main isotope of Fe will no longer be a problem. It is likely that with this system, not only will most interferences be resolved, but also, the higher transmission achieved in comparison to a quadrupole instrument, will result in higher sensitivity. Additionally, the noise background is significantly reduced, and hence a significant improvement of the detection limits by more than one order of magnitude seems feasible.

5 Conclusion

Plasma source mass spectrometry displays rapid dynamic development in basic research, in instrumental performance, and in many applications for various analytical tasks. Automation of instruments, which is actually the main aim of different manufacturers, will improve the appeal of PSMS as a high performance tool for modern elemental analysis. With respect to the increasing demands in this field, no other technique can compete with PSMS in providing true multi-element analysis at extraordinary low detection limits. However, restrictions arise from stability problems and spectral and non-spectral interferences, and future work will mainly be devoted to solving these difficulties. Applications will continue to broaden in the bulk and surface analysis of conducting and non-conducting solids, especially high-tech ceramics,[209] and analysis of liquids of every kind arising from biology, medicine, geology, materials science, and last, but not least, environmental science.

Despite being enthusiastic about PSMS to a certain extent, the authors realize that other methods which promise high performance for analytical tasks should not be neglected. Each method has its limitations, and the increasing significance of analytical chemistry to the quality of life means that a selection of analytical tools will be required. It cannot be ignored that—perhaps rather soon—other methods will play a much more dominant role than today.[210] Total reflection X-ray fluorescence[211] is promising, as is laser spectroscopy[212] if technological barriers can be surmounted concerning the development of diode lasers which can be tuned over a large wavelength interval.

In its current state, PSMS is well established, but may nevertheless give a strong impetus for further work, because it continues to be a real challenge:

- for people in research to investigate new plasma sources for special applications with the aim to increase the versatility of PSMS;
- for manufacturers to design and develop new and cheap instruments with improved analytical performance and a higher degree of automation, including automatic optimization of operational conditions; and

- for all people applying analytical techniques, for whatever purpose, to make use of PSMS as a powerful tool with the aim of improving the quality of life.

References

1. G. Tölg and P. H. Garten, *Angew. Chem. Int. Ed. Engl.*, 1985, **24**, 485.
2. A. Montaser and D. W. Golightly, 'Inductively Coupled Plasmas in Analytical Atomic Spectrometry', VCH Publishers, New York, 1992.
3. K. Slickers, 'Die Automatische Atom-Emissions-Spektralanalyse', ed. K. A. Slickers, Gießen, Germany, 1992.
4. F. Adams, R. Gijbels, and R. van Grieken, 'Inorganic Mass Spectrometry', Wiley-Interscience, New York, Vol. 95, 1988.
5. F. Adams and A. Vertes, *Fresenius' Z. Anal. Chem.*, 1990, **33**, 638.
6. A. Benninghoven, F. G. Rüdenauer, and H. W. Werner, 'Secondary Ion Mass Spectrometry', *Ser. Chem. Anal.*, 86, John Wiley and Sons, Chichester, 1987.
7. B. Sundquist, A. Hedin, P. Hakansson, I. Kamensky, M. Salehpour, and G. Säwe, *Int. J. Mass Spectrom. Ion Processes*, 1985, **65**, 69.
8. G. Ramendik, J. Verlinden, and R. Gijbels, in 'Inorganic Mass Spectrometry', eds. F. Adams, R. Gijbels, and R. van Grieken, Wiley-Interscience, Vol. 95, 1988.
9. J. S. Becker and H.-J. Dietze, *Fresenius' Z. Anal. Chem.*, 1992, **344**, 69.
10. K. Ohls, *Fresenius' Z. Anal. Chem.*, 1987, **327**, 111.
11. B. W. Smith, N. Omenetto, and J. D. Winefordner, *Spectrochim. Acta, Part B*, 1984, **39**, 1389.
12. B. L. Bentz, C. G. Bruhn, and W. W. Harrison, *Int. J. Mass Spectrom. Ion Phys.*, 1978, **28**, 409.
13. N. Jakubowski, D. Stuewer, and G. Tölg, *Int. J. Mass Spectrom. Ion Processes*, 1986, **71**, 183.
14. A. Date and A. L. Gray, *Analyst (London)*, 1981, **106**, 1255.
15. D. A. Wilson, G. H. Vickers, and G. M. Hieftje, *Spectrochim. Acta, Part B*, 1987, **42**, 29.
16. R. S. Houk, Thesis, Iowa State University, Ames, USA, 1980.
17. N. Jakubowski, B. J. Raeymaekers, J. A. C. Broekaert, and D. Stuewer, *Spectrochim. Acta, Part B*, 1989, **44**, 219.
18. J. W. Coburn, *Rev. Sci. Instrum.*, 1970, **41**, 1219.
19. J. W. Coburn and E. Kay, *Appl. Phys. Lett.*, 1971, **18**, 435.
20. J. W. Coburn and E. Kay, *Appl. Phys. Lett.*, 1971, **19**, 350.
21. H. Oechsner and W. Gerhard, *Phys. Lett. A*, 1972, **40**, 211.
22. H. Oechsner, *Z. Physik*, 1970, **238**, 433.
23. A. Wucher and H. Oechsner, *Fresenius' Z. Anal. Chem.*, 1989, **333**, 470.
24. S. Hofmann, *Surf. Int. Anal.*, 1980, **2**, 148.
25. H. Oechsner, Festkörperprobleme XXIV, Vieweg Wiesbaden, 1984, p. 269.
26. S. Uhlemann, P. Weißbrodt, and D. Mademann, *Fresenius' J. Anal. Chem.*, 1993, **346**, 374.
27. H. Oechsner and W. Gerhard, *Phys. Lett. A*, 1972, **40**, 211.
28. H. Oechsner and W. Gerhard, *Surf. Sci.*, 1974, **44**, 480.
29. H. Oechsner, *Appl. Phys.*, 1975, **8**, 185.
30. H. Oechsner and E. Stumpe, *Appl. Phys.*, 1977, **14**, 43.
31. H. Oechsner and A. Wucher, *Appl. Surf. Sci.*, 1982, **10**, 342.
32. P. Beckmann, H. Oechsner, and H. Paulus, *Fresenius' Z. Anal. Chem.*, 1984, **319**, 851.

33. J. F. Geiger, M. Kopnarski, H. Oechsner, and H. Paulus, *Mikrochim. Acta*, 1987, **1**, 497.
34. H. Oechsner, *Vacuum*, 1987, **37**, 763.
35. H. Oechsner, *Scanning Microsc.*, 1988, **2**, 9.
36. H. Oechsner, *Thin Solid Films*, 1989, **175**, 119.
37. H. Oechsner, *Scanning Microsc.*, 1989, **3**, 411.
38. J. Sopka and H. Oechsner, *J. Non-Cryst. Solids*, 1989, **114**, 208.
39. J. Waldorf, H. Oechsner, H. J. Fuesser, and J. Mathuni, *Thin Solid Films*, 1989, **174**, 39.
40. A. Wucher, H. Oechsner, and F. Novak, *Thin Solid Films*, 1989, **174**, 133.
41. H. Oechsner, *Pure Appl. Chem.*, 1992, **64**, 615.
42. H. Oechsner, *Int. J. Mass Spectrom. Ion Processes*, 1990, **103**, 31.
43. J. Sopka and H. Oechsner, *J. Non-Cryst. Solids*, 1989, **114**, 208.
44. A. Wucher, H. Oechsner, and F. Novak, *Thin Solid Films*, 1989, **174**, 133.
45. H. Oechsner, *Scanning Microsc.*, 1989, **3**, 411.
46. K. H. Koch, D. Sommer, and D. Grunenberg, *Mikrochim. Acta*, 1990, **2**, 101.
47. K. H. Koch, D. Sommer, and D. Grunenberg, *Fresenius' Z. Anal. Chem.*, 1989, **333**, 447.
48. A. J. Gellmann, B. M. Naasz, R. G. Schmidt, M. K. Chaudhury, and T. M. Gentle, *J. Adhes. Sci. Technol.*, 1990, **4**, 597.
49. J. W. Bentz, H.-P. Ewinger, J. Goschnick, G. Kannen, and H. J. Ache, *Fresenius' Z. Anal. Chem.*, 1993, **346**, 123.
50. J. Bartella, R. Fuchs, J. Goschnick, and D. Grunenberg, *Fresenius' J. Anal. Chem.*, 1993, **346**, 131.
51. J. Goschnick, M. Lipp, J. Schuricht, A. Schweiker, and H. J. Ache, *J. Aerosol Sci.*, 1991, **22**, 835.
52. H. Jenett, S. Bredendiek-Kämper, and J. Sunderkötter, *Mikrochim. Acta*, 1993, **110**, 13.
53. J. Goschnick, D. Maas, J. Schuricht, and H. J. Ache, *Fresenius' J. Anal. Chem.*, 1993, **346**, 323.
54. D. Grunenberg, D. Sommer, and K. H. Koch, *Fresenius' J. Anal. Chem.*, 1993, **346**, 147.
55. D. C. Duckworth and R. K. Marcus, *Anal. Chem.*, 1989, **61**, 1879.
56. D. C. Duckworth and R. K. Marcus, *J. Anal. At. Spectrom.*, 1992, **7**, 711.
57. R. K. Marcus, *J. Anal. At. Spectrom.*, 1993, **8**, 935.
58. A. Wucher, *Fresenius' J. Anal. Chem.*, 1993, **346**, 3.
59. E. Goldstein, *Berl. Ber.*, 1886, **39**, 691.
60. K. T. Brainbridge and E. B. Jordon, *Phys. Rev.*, 1936, **50**, 282.
61. F. W. Aston, 'Mass Spectra and Isotopes', Longmans, New York, 1942.
62. A. J. Dempster, *Proc. Am. Phil. Soc.*, 1935, **75**, 755.
63. J. J. Thomson and G. P. Thomson, 'Conduction of Electricity through Gases', Vol. 1, Cambridge University Press, Cambridge, 1928.
64. W. W. Harrison and C. W. Magee, *Anal. Chem.*, 1974, **46**, 461.
65. B. N. Colby and C. A. Evans, *Anal. Chem.*, 1974, **46**, 1236.
66. W. W. Harrison, 'Inorganic Mass Spectrometry', eds. F. Adams, R. Gijbels, and R. van Grieken, John Wiley and Sons, New York, 1988, 85.
67. J. W. Coburn and W. W. Harrison, *Appl. Spectrosc. Rev.*, 1981, **17**, 95.
68. N. Jakubowski, D. Stuewer, and W. Vieth, *Fresenius' Z. Anal. Chem.*, 1988, **331**, 145.
69. D. Stuewer, in 'Application of Plasma Source Mass Spectrometry', eds. G. Holland and A. Eaton, Royal Society of Chemisty, Cambridge, 1990, 1.
70. W. W. Harrison, K. R. Hess, R. K. Marcus, and F. L. King, in 'Second. Ion Mass Spectrom., SIMS 5', Springer Ser. Chem. Phys., 1986, **44**, 75.
71. W. W. Harrison, K. R. Hess, R. K. Marcus, and F. L. King, *Anal. Chem.*, 1986, **58**, 341A.
72. W. W. Harrison and B. L. Bentz, *Prog. Anal. Spectrosc.*, 1988, **11**, 53.
73. R. J. Guidoboni and F. D. Leipziger, *J. Cryst. Growth*, 1988, **89**, 16.

74. W. W. Harrison, *J. Anal. At. Spectrom.*, 1988, **3**, 867.

75. W. W. Harrison, *Chem. Anal.*, 1988, **95**, 85.

76. F. L. King and W. W. Harrison, *Mass Spectrom. Rev.*, 1990, **9**, 285.

77. M. van Straaten and R. Gijbels, 'Application of Plasma Source Mass Spectrometry II', eds. G. Holland and A. Eaton, Special Publication No. 124, Royal Society of Chemistry, Cambridge, 1993, p. 130.

78. W. W. Harrison, *J. Anal. At. Spectrom.*, 1992, **7**, 75.

79. D. Stuewer, *Fresenius' J. Anal. Chem.*, 1990, **337**, 737.

80. D. Stuewer, in 'Analytiker Taschenbuch', eds. H. Günzler, R. Borsdorf, W. Fresenius, W. Huber, H. Kelker, I. Lüderwald, G. Tölg, and H. Wisser, Springer Verlag, Berlin, 1990, 127.

81. W. W. Harrison, C. M. Barshick, J. A. Klinger, P. H. Ratliff, and Y. Mei, *Anal. Chem.*, 1990, **62**, 943A.

82. M. van Straaten, Thesis, University of Antwerp, 1993.

83. M. van Straaten, A. Vertes, and R. Gijbels, *Spectrochim. Acta, Part B*, 1991, **46**, 283.

84. W. S. Taylor, J. G. Dulak, and S. N. Ketkar, *J. Am. Soc. Mass Spectrom.*, 1990, **1**, 448.

85. M. van Straaten, R. Gijbels, and A. Vertes, *Anal. Chem.*, 1992, **64**, 1855.

86. M. K. Levy, D. Serxner, A. D. Angstadt, R. L. Smith, and K. R. Hess, *Spectrochim. Acta, Part A*, 1991, **60**, 253.

87. C. M. Barshick and W. W. Harrison, *Mikrochim. Acta*, 1989, **3**, 169.

88. W. Vieth and J. Huneke, *Spectrochim. Acta, Part B*, 1990, **45**, 941.

89. W. Vieth and J. C. Huneke, *Spectrochim. Acta, Part B*, 1991, **46**, 137.

90. J. A. Klinger, C. M. Barshick, and W. W. Harrison, *Anal. Chem.*, 1991, **63**, 2571.

91. M. Saito, *Anal. Sci.*, 191, **7**, 541.

92. R. W. I. Smithwick, *J. Am. Soc. Mass Spectrom.*, 1992, **3**, 79.

93. R. L. Smith, D. Serxner, and K. R. Hess, *Anal. Chem.*, 1989, **61**, 1103.

94. K. R. Hess and W. W. Harrison, *Anal. Chem.*, 1988, **60**, 691.

95. D. Fang and R. K. Marcus, *Spectrochim. Acta, Part B*, 1990, **45**, 1053.

96. T. J. Loving and W. W. Harrison, *Anal. Chem.*, 1983, **55**, 1526.

97. G. Ronan, J. Clark, and N. Ketchell, *Mikrochim. Acta*, 1989, **3**, 231.

98. J. A. Klinger and W. W. Harrison, *Anal. Chem.*, 1991, **63**, 2982.

99. D. M. P. Milton, R. C. Hutton, and G. A. Ronan, *Fresenius' J. Anal. Chem.*, 1992, **343**, 773.

100. J. C. Woo, H. B. Lim, D. W. Moon, K. W. Lee, and H. J. Kim, *Anal. Sci. Technol.*, 1992, **5**, 169.

101. H. J. Kim, E. H. Piepmeier, G. L. Beck, G. G. Brumbaugh, and O. T. Farmer III, *Anal. Chem.*, 1990, **62**, 639.

102. J. A. Klinger, C. M. Barshick, and W. W. Harrison, *Anal. Chem.*, 1991, **63**, 2571.

103. J. A. Klinger, P. J. Savickas, and W. W. Harrison, *J. Am. Soc. Mass Spectrom.*, 1990, **1**, 138.

104. N. Jakubowski and D. Stuewer, *Fresenius' Z. Anal. Chem.*, 1989, **335**, 680.

105. J. A. Klinger, C. M. Barshick, and W. W. Harrison, *Anal. Chem.*, 1991, **63**, 2571.

106. F. L. King and W. W. Harrison, *Int. J. Spectrom. Ion Processes*, 1989, **89**, 171.

107. Y. Mei and W. W. Harrison, *Spectrochim. Acta, Part B*, 1991, **46**, 175.

108. R. Gijbels, *Talanta*, 1990, **37**, 363.

109. P. M. Charalambous, *Mikrochim. Acta*, 1987, **1**, 295.

110. P. Wilhartitz, H. M. Ortner, R. Krismer, and H. Krabichler, *Mikrochim. Acta*, 1990, **2**, 59.

111. D. Fang and P. Seegopaul, *J. Anal. At. Spectrom.*, 1992, **7**, 959.

112. G. Kudermann, *Fresenius' Z. Anal. Chem.*, 1988, **331**, 697.

113. L. F. Vassamillet, *J. Anal. At. Spectrom.*, 1989, **4**, 451.
114. N. E. Sanderson, P. Charalambous, D. J. Hall, and R. Brown, *J. Res. Nat. Bur. Stand.*, 1988, **93**, 426.
115. G. Kudermann, K. H. Blaufuß, C. Lührs, W. Vielhaber, and U. Collisi, *Fresenius' J. Anal. Chem.*, 1992, **343**, 734.
116. A. P. Mykytiuk, P. Semeniuk, and S. Berman, *Spectrochim. Acta Rev.*, 1990, **13**, 1.
117. D. M. P. Milton, J. Clark, D. Potter, and R. C. Hutton, *Anal. Sci.*, 1991, **7**, 1243.
118. D. M. P. Milton, R. C. Hutton, and G. A. Ronan, *Fresenius' J. Anal. Chem.*, 1992, **343**, 773.
119. R. J. Guidoboni and F. D. Leipziger, *J. Cryst. Growth*, 1988, **89**, 16.
120. W. Vieth and J. C. Huneke, *Anal. Chem.*, 1992, **64**, 2958.
121. R. Mason and D. Milton, *Int. J. Mass Spectrom. Ion Processes*, 1989, **91**, 209.
122. M. Tigwell, J. Clark, S. Shuttleworth, and M. Bottomley, *Mater. Chem. Phys.*, 1992, **31**, 23.
123. D. L. Donohue and M. Petek, *Anal. Chem.*, 1991, **63**, 740.
124. I. Evetts, D. Milton, and R. Mason, *Biol. Mass Spectrom.*, 1991, **20**, 153.
125. J. A. Dunlop, K. E. Ritala, J. R. Gibbard, R. Beauprie, B. Pouliquen, J. H. Reeves, J. C. Huneke, and W. A. Vieth, *J. Metals*, 1989, **41**, 18.
126. N. Jakubowski, D. Stuewer, and W. Vieth, *Anal. Chem.*, 1987, **59**, 1825.
127. N. Jakubowski, D. Stuewer, and W. Vieth, *Mikrochim. Acta (Wien)*, 1987, **1**, 302.
128. R. C. Hutton and A. Raith, *J. Anal. At. Spectrom.*, 1992, **7**, 623.
129. N. Jakubowski and D. Stuewer, *J. Anal. At. Spectrom.*, 1992, **7**, 951.
130. S. El Alfy, K. Laqua, and H. Massmann, *Fresenius' Z. Anal. Chem.*, 1973, **263**, 1.
131. H. Mai and H. Scholze, *Spectrochim. Acta, Part B*, 1986, **41**, 197.
132. J. C. Woo, N. Jakubowski, and D. Stuewer, *J. Anal. At. Spectrom.*, 1993, **8**, 881.
133. N. Jakubowski, D. Stuewer, and G. Tölg, *Spectrochim. Acta, Part B*, 1991, **46**, 155.
134. R. Eller, F. Alt, G. Tölg, and H. J. Tobschall, *Fresenius' Z. Anal. Chem.*, 1989, **334**, 723.
135. F. Alt, A. Bambauer, K. Hoppstock, B. Mergler, and G. Tölg, *Fresenius' J. Anal. Chem.*, 1993, **346**, 693.
136. M.-L. Lee, Thesis, University of Dortmund, 1993.
137. P. C. Thonemann, *Nature (London)*, 1946, **158**, 61.
138. H. Neuert, *Z. Naturforsch., Teil A*, 1949, **31**, 449.
139. S. Greenfield, I. L. Jones, and C. T. Berry, *Analyst (London)*, 1964, **89**, 713.
140. R. S. Houk, V. A. Fassel, G. D. Flesch, H. J. Svec, A. L. Gray, and C. E. Taylor, *Anal. Chem.*, 1980, **52**, 2283.
141. K. E. Jarvis, A. L. Gray, and R. S. Houk, 'Handbook of Inductively Coupled Plasma Mass Spectrometry', Blackie, London, 1992.
142. J. A. C. Broekaert, in 'Analytiker Tasschenbuch', eds. H. Günzler, R. Borsdorf, W. Fresenius, W. Huber, H. Kelker, I. Lüderwald, G. Tölg, and H. Wisser, Springer Verlag, Berlin, 1990, 127.
143. J. W. McLaren, *ICP Inf. Newsl.*, 1990, **16**, 371.
144. M. Lust, *At. Spectrosc.*, 1992, **13**, 29.
145. J. W. McLaren, *At. Spectrosc.*, 1992, **13**, 81.
146. G. M. Hieftje and L. A. Norman, *Int. J. Mass Spectrom. Ion Processes*, 1992, **118/119**, 519.
147. I. B. Brenner and H. E. Taylor, *Crit. Rev. Anal. Chem.*, 1992, **23**, 355.
148. A. R. Date, *Spectrochim. Acta Rev.*, 1991, **14**, 3.
149. W. Slavin, *Spectrosc. Int.*, 1992, **4** (1), 22.
150. E. R. Denoyer, *Atom. Spectrosc.*, 1991, **12**, 215.
151. J. R. Bacon, A. T. Ellis, A. W. McMahon, P. J. Potts, and J. G. Williams, *J. Anal.*

At. Spectrom., 1993, **8**, 261R.

152. D. W. Koppenaal, *Anal. Chem.*, 1992, **64**, 320R.
153. R. S. Houk, V. A. Fassel, and H. J. Svec, *Dyn. Mass Spectrom.*, 1981, **6**, 234.
154. J. W. Lam and J. W. McLaren, *J. Anal. At. Spectrom.*, 1990, **5**, 419.
155. R. Tsukahara and M. Kubota, *Spectrochim. Acta, Part B*, 1990, **45**, 581.
156. P. Hulmston and R. C. Hutton, *Spectroscopy (Eugene. Oreg.)*, 1991, **6** (1), 35.
157. J. M. Carey and J. A. Caruso, *Crit. Rev. Anal. Chem.*, 1992, **23**, 397.
158. D. C. Gregoire, M. Lamoureux, C. L. Chakrabarti, S. Al-Maawali, and J. P. Byrne, *J. Anal. At. Spectrom.*, 1992, **7**, 579.
159. R. D. Ediger and S. A. Beres, *Spectrochim. Acta, Part B*, 1992, **47**, 907.
160. A. Montaser, H. Tan, I. Ishii, S.-H. Nam, and M. Cai, *Anal. Chem.*, 1991, **63**, 2660.
161. D. R. Wiederin, F. G. Smith, and R. S. Houk, *Anal. Chem.*, 1991, **63**, 219.
162. E. R. Denoyer, K. J. Fredeen, and J. W. Hager, *Anal. Chem.*, 1991, **63**, 445A.
163. A. A. Heuzen, *Spectrochim. Acta, Part B*, 1991, **46**, 1803.
164. W. T. Perkins, R. Fuge, and J. G. Pearce, *J. Anal. At. Spectrom.*, 1991, **6**, 445.
165. J. Marshall, J. Franks, I. Abell, and C. Tye, *J. Anal. At. Spectrom.*, 1991, **6**, 145.
166. N. Jakubowski, I. Feldmann, B. Sack, and D. Stuewer, *J. Anal. At. Spectrom.*, 1992, **7**, 121.
167. S.-J. Jiang and R. S. Houk, *Spectrochim. Acta, Part B*, 1987, **42**, 93.
168. T. Hirata, T. Akagi, and A. Masuda, *Analyst (London)*, 1990, **115**, 1329.
169. N. Voellkopf, M. Paul, and E. R. Denoyer, *Fresenius' J. Anal. Chem.*, 1992, **342**, 917.
170. S. A. Darke, C. J. Pickford, and J. F. Tyson, *Anal. Proc. (London)*, 1989, **26**, 379.
171. D. S. Bushee, *Analyst (London)*, 1988, **113**, 1167.
172. U. Völlkopf, A. Günsel, and A. Janssen, *At. Spectrosc.*, 1990, **11**, 137.
173. D. T. Heitkemper and J. A. Caruso, *Appl. Spectrosc.*, 1990, **44**, 228.
174. J. G. Williams, A. L. Gray, P. Norman, and L. Ebdon, *J. Anal. At. Spectrom.*, 1987, **2**, 469.
175. L. Ebdon, M. E. Foulkes, H. G. M. Parry, and C. T. Tye, *J. Anal. At. Spectrom.*, 1988, **3**, 753.
176. J. R. Dean, L. Ebdon, and R. Massey, *J. Anal. At. Spectrom.*, 1987, **2**, 369.
177. J. W. Olesik and J. C. Fister III, *Spectrochim. Acta, Part B*, 1991, **46**, 851.
178. R. C. Hutton and A. Eaton, *J. Anal. At. Spectrom.*, 1987, **2**, 595.
179. G. Zhu and R. F. Browner, *J. Anal. At. Spectrom.*, 1988, **3**, 781.
180. J. W. Lam and J. W. McLaren, *J. Anal. At. Spectrom.*, 1990, **5**, 419.
181. R. Tsukahara and M. Kubota, *Spectrochim. Acta, Part B*, 1990, **45**, 581.
182. G. R. Peters and D. Beauchemin, *Spectrochim. Acta, Part B*, 1993, **48**, 1481.
183. N. Jakubowski, I. Feldmann, and D. Stuewer, *Spectrochim. Acta, Part B*, 1992, **47**, 107.
184. N. Jakubowski, I. Feldmann, and D. Stuewer, *J. Anal. At. Spectrom.*, 1993, **8**, 969.
185. W. L. Shen, J. A. Caruso, F. L. Fricke, and R. D. Satzger, *J. Anal. At. Spectrom.*, 1990, **5**, 451.
186. J. M. Carey, E. H. Evans, J. A. Caruso, and W. L. Shen, *Spectrochim. Acta, Part B*, 1991, **46**, 1711.
187. T. Etoh, M. Yamada, and M. Matsubara, *Anal. Sci.*, 1991, **7**, 1263.
188. N. Shibata, N. Fudagawa, and M. Kubota, *Anal. Chem.*, 1991, **63**, 636.
189. S. Al-Maawali and C. L. Chakrabarti, *Spectrochim. Acta, Part B*, 1992, **47**, 1123.
190. D. C. Gregoire, *Anal. Proc. (London)*, 1992, **29**, 276.
191. J. P. Byrne, C. L. Chakrabarti, D. C. Gregoire, M. Lamoureux, and T. Ly, *J. Anal. At. Spectrom.*, 1992, **7**, 371.
192. A. Ulrich, C. Huchulski, W. Dannecker, and U. Voellkopf, *Anal. Proc. (London)*, 1992, **29**, 282.

193. A. Ulrich, W. Dannecker, S. Meiners, and U. Voellkopf, *Anal. Proc. (London)*, 1992, **29**, 284.
194. J. P. Byrne, M. M. Lamoureux, C. L. Chakrabarti, T. Ly, and D. C. Gregoire, *J. Anal. At. Spectrom.*, 1993, **8**, 599.
195. N. Shibata, N. Fudagawa, and M. Kubota, *Anal. Chim. Acta*, 1992, **265**, 93.
196. R. Tsukahara and M. Kubota, *Spectrochim. Acta, Part B*, 1990, **45**, 779.
197. H. Berndt and G. Schaldach, in 'CAS-5, Proc. of Colloquium Atom. Spurenanalyse', ed. B. Welz, Bodenseewerk Perkin Elmer, Überlingen, 1989, p. 109.
198. H. Berndt, *Fresenius' Z. Anal. Chem.*, 1988, **331**, 321.
199. J. Posta and H. Berndt, *Spectrochim. Acta, Part B*, 1992, **47**, 993.
200. N. Jakubowski, I. Feldmann, D. Stuewer, and H. Berndt, *Spectrochim. Acta*, 1992, **47**, 119.
201. D. Beauchemin, *Analyst (London)*, 1993, **118**, 815.
202. D. Beauchemin, *Trends Anal. Chem.*, 1991, **10**, 71.
203. N. P. Vela, L. K. Olson, and J. A. Caruso, *Anal. Chem.*, 1993, **65**, 585A.
204. S.-K. Luo, H. Berndt, and J. Posta, *J. Anal. At. Spectrom*, to be published.
205. N. Jakubowski, B. Jepkens, D. Stuewer, and H. Berndt, *J. Anal. At. Spectrom.*, 1994, **9**, 193.
206. S. K. Luo and H. Berndt, *Spectrochim. Acta*, submitted for publication.
207. C.-K. Kim, R. Seki, S. Morita, S. Yamasaki, A. Tsumura, Y. Takaku, Y. Igarashi, and M. Yamamoto, *J. Anal. At. Spectrom.*, 1991, **4**, 205.
208. M. Sargent and K. Webb, *Spectroscopy Europe*, 1993, **5**, 21.
209. J. A. C. Broekaert, T. Graule, H. Jenett, G. Tölg, and P. Tschöpel, *Fresenius' Z. Anal. Chem.*, 1989, **332**, 825.
210. P. Boumans, *J. Anal. At. Spectrom.*, 1993, **8**, 767.
211. R. Klockenkämper, J. Knoth, A. Prange, and H. Schwenke, *Anal. Chem.*, 1992, **64**, 1115A.
212. H. Groll and K. Niemax, *Spectrochim. Acta, Part B*, 1993, **48**, 633.

Process Analysis Using Mass Spectrometry

J. H. Scrivens

ICI MATERIALS, PO BOX 90, WILTON RESEARCH CENTRE, WILTON, CLEVELAND TS90 8JE, UK

Summary

Among the criteria thought to be important for a technique to have widespread utility for process analysis are: ease of sampling; accuracy; reliability; straightforward maintenance; wide acceptability; simplicity; linearity of response; speed of analysis; versatility; and wide dynamic range. Mass spectrometry, in contrast, although widely used in laboratory work, is, when considered against these criteria: relatively expensive; needs secondary standards for mass and concentration calibration; does not have an established pedigree as a process analyser; is conceptually and instrumentally non-trivial; and requires gas samples at low pressure. The technique does however have a fast response, is capable of high precision measurements, and is versatile. These advantages are significant enough to warrant the serious consideration of mass spectrometry for process analysis. Significant developments have been made in the design and use of purpose built mass spectrometers in recent years and their contribution to the armoury of modern measurement systems is now established. The advantages and disadvantages of the technique will be discussed below and examples of current use given. The improvements that have led to the acceptance of the technique in process analysis are described and typical applications presented.

1 Introduction

Mass spectrometry was initially used for process analysis in the 1940s. By the 1950s the use was well documented.[1,2] The development of the gas chromatograph at this time[3] led to it replacing mass spectrometry for process analysis on the grounds of lower cost and greater reliability. The advent of stable, low cost, electronics, more reliable vacuum systems, and greater computing power has led to the gradual reappraisal of mass spectrometry for process analysis. The historical use of process mass spectrometry has been reviewed[4] covering the major application areas. The possible resurgence of the technique with purpose built equipment, giving much greater reliability and stability, has been discussed[5] and developments that improved the results that could be obtained outlined.[6] Mass spectrometry has now reached the stage where it is an established alternative for many forms of process analysis and indeed is the favoured technique for a number of applications.

2 Hardware

The basic mass spectrometry experiment consists of three stages. These are: ionization, separation, and detection. The sample is supplied to the instrument in the gaseous form at $< 200\,^{\circ}C$ and atmospheric pressure (or less). The pressure is then lowered to 1 Torr and then 10^{-6} Torr via a two stage pressure reduction system. The low pressure is required to avoid ion–molecule reactions occurring in the instrument. The two stage pressure reduction maintains the relative concentrations of species at the same level as in the process stream. Since many streams of interest do not fulfil these criteria a significant problem in obtaining a suitable sample can sometimes exist. Although mass spectrometry continues to enjoy a steady growth curve, with a wide diversity of sources, analysers, and detectors being developed and commercially available, process instruments, apart from a small number of special applications, are built around a very small number of options. These are an electron impact (EI) source, a single quadrupole or magnetic analyser, and either a Faraday cup or electron multiplier detector. The electron impact source makes use of an energetic beam of electrons (usually 70 eV) which impact on the sample of interest generating predominantly positively charged radical ions, which can further fragment to form product ions, which are often characteristic of the initial compounds. Although it is very difficult to predict from first principles what these fragments may be, libraries of spectra exist and little variation is seen from instrument to instrument when operated under the same conditions. Electron impact is a sensitive, although not an efficient, procedure which is relatively independent of experimental conditions and is extremely reproducible. After ionization and fragmentation in the source, the ions pass through an analyser which separates the ions according to their mass to charge ratio. For process mass spectrometry, two analysers have been largely favoured. These are the quadrupole and magnetic analysers.

The quadrupole makes use of a quadrant of four parallel hyperbolic or circular rods to which a varying radio frequency and d.c. voltage are applied. At a given field only ions of a certain mass undergo stable oscillation and pass through. By changing the d.c. component, ions of varying mass to charge ratio may be allowed to pass through the analyser to be detected. Quadrupole analysers are capable of high sensitivity and fast response and they are relatively inexpensive to manufacture. Other, less desirable, features include the relatively poor ability to resolve the peaks of very similar mass to charge ratios and the requirement for cleaning on a long-term basis.

When a charged particle is injected into a magnetic field, with velocity normal to the field, it obeys the equation:

$$m/z = B^2 R^2 / 2V$$

Where B is the magnetic flux density, R the radius of the magnetic field, V the accelerating voltage, and m and z are the mass and charge, respectively. In modern instruments, the masses to be studied are selected by varying the magnetic field although in earlier instruments voltage control or fixed collectors were often used. Magnetic analysers are less susceptible to contamination than quadrupoles, are more stable, and possess much greater powers of resolution. They are, however, more expensive and have a slower response time.

The detectors in common use are a Faraday cup, in which the ions strike a target and give up their charge, which is amplified, and an electron multiplier, in which the charge is increased via a series of dynodes and then amplified. The Faraday cup is linear and stable, but it is not very sensitive and has a slow response. The electron multiplier is much more sensitive and has a fast response. It is, however, much more expensive and is susceptible to ageing.

3 Applications

Mass spectrometry is used for process analysis in the nuclear, steel, coal, gas, glass, semi-conductor, and chemical industries with, in the latter case, ethylene oxide, vinyl chloride, and ammonia production among the processes monitored.[7-13] The technique has been used on-line with closed loop process control, but is more frequently used on-line as a measurement device to aid in process control decisions. Considerable use is also made of the technique for at-line or off-line measurements particularly in environmental applications. Detailed discussion of these important areas is outside the scope of this paper.

Table 1 shows a typical analysis stream measured by mass spectrometry, six components are measured in less than sixty seconds. The agreement with accepted concentrations is excellent. Mass spectrometry requires the provision of external standards to give accurate results and it is in the provision of these standards that most errors occur. Table 2 shows some long-term stability data on a similar process stream. These results show the excellent stability that can be obtained from modern instrumentation. No recalibration has been carried out between measurements. Ammonia conversion is an area where complete control of the process using mass spectrometric results is possible and has been achieved. This requires very fast and precise analysis. Table 3 shows some

Table 1 *Natural Gas Analysis by On-line Mass Spectrometry*

Component	Concentration/% (Actual)	Concentration/% (Mean)	Standard Deviation (Absolute %)
Methane	79.485	79.286	0.00709
Ethane	3	3.007	0.00575
Nitrogen	12.5	12.495	0.00614
CO_2	4.006	4.152	0.00645
Propane	0.75	0.768	0.00728
Butane	0.3	0.291	0.00194

Table 2 *Long-term Stability of Natural Gas Analysis by On-line Mass Spectrometry*

Component	Actual Vol.%	15 Days Vol.%	20 Days Vol.%	25 Days Vol.%	30 Days Vol.%
Helium	0.2	0.21	0.2	0.2	0.2
Nitrogen	41.57	41.57	41.53	41.54	41.57
CO_2	5.25	5.21	5.23	5.22	5.22
Methane	49.44	49.59	49.51	49.52	49.5
Ethane	1.78	1.77	1.77	1.77	1.76
Propane	1.02	1.02	1.01	1.01	1.01
n-Butane	0.44	0.42	0.42	0.42	0.42
i-Butane	0.31	0.32	0.33	0.32	0.32

Table 3 *Typical Results for On-line Analysis of Process Streams in an Ammonia Converter by Mass Spectrometry*

Component	Mean Vol.%	Standard Deviation Vol.%	Minimum Vol.%	Maximum Vol.%
Nitrogen	21.722	0.01908	21.645	21.791
Argon	3.838	0.00081	3.836	3.842
Hydrogen	66.778	0.01627	66.725	66.821
Methane	6.505	0.00406	6.491	6.528
Ammonia	1.157	0.00354	1.149	1.187
H/N Ratio	3.074	0.00329	3.064	3.082
$AR + CH_4$	10.343	0.0045	10.301	10.374

typical results from such a process. Four hundred results were obtained over a one hour period and the table gives the minimum and maximum results obtained. The combination of excellent precision with this speed of response gives results that could not be obtained with any other single technique.

The dynamic range is another feature of mass spectrometry which exhibits a significant advantage over other process analysis methods. Table 4 shows results of an ethylene oxide process stream analysed by mass spectrometry. Components from 77% to 1.2 p.p.m. are measured in the same stream at the same time with good precision. In order to justify the requirement for this precision we will consider the analysis in more detail.

The ethylene oxide process consists of two main reactions:

$$\text{Oxidation:} \quad 2C_2H_4 + O_2 = 2C_2H_4O$$

$$\text{Reduction:} \quad C_2H_4 + 3O_2 = 2CO_2 + 2H_2O$$

The important process measurements, in addition to the individual component concentrations, are the selectivity of the conversion and the oxygen and carbon

Table 4 *Typical Plant Data Obtained by On-line Mass Spectrometric Analysis of an Ethylene Oxide Process Stream*

Component	Mean (%)	Standard Deviation (%)	Minimum (%)	Maximum (%)
Nitrogen	77.113	0.01155	77.086	77.131
Oxygen	4.111	0.01336	4.095	4.1444
Ethylene Oxide	1.52	0.00904	1.499	1.532
CO_2	6.995	0.0109	6.972	7.011
Ethylene	10.169	0.01556	10.147	10.196
Ethane	0.076	0.00091	0.074	0.078
Argon	0.015	0.00051	0.014	0.016
CH_3CH_3Cl (p.p.m.)	1.288	0.02212	1.236	1.318
Selectivity	80.977	0.09891	80.824	81.182

balances. These are shown in Tables 5, 6, and 7. Six measurements of selectivity can be made. Square brackets indicate concentration, I and E refer to inlet and exit concentrations respectively. The initial requirements expected from the

Table 5 *Selectivity Estimates For Components in Ethylene Oxide Productions*

$$S(EO/O_2) = 6/[5 + (2\Delta O_2/\Delta EO)]$$
$$S(EO/CO_2) = \Delta EO/(\Delta EO + \Delta CO_2/2)$$
$$S(EO/C_2H_4) = \Delta EO/\Delta C_2H_4$$
$$S(O_2/CO_2) = (6\Delta CO_2 - 4\Delta O_2)/(5\Delta CO_2 - 4\Delta O_2)$$
$$S(O_2/C_2H_4 = (6 - 2\Delta O_2)/(5\Delta C_2H_4)$$
$$S(C_2H_4/CO_2) = (2\Delta C_2H_4 - \Delta CO_2)/(2\Delta C_2H_4)$$

Table 6 *Definition of Terms in Table 5*

$$S = \text{selectivity}$$
$$\Delta EO = p[EO]_E - [EO]_I$$
$$\Delta CO_2 = p[CO_2]_E - [CO_2]_I$$
$$\Delta C_2H_4 = [C_2H_4]_I - p[C_2H_4]_E$$
$$\Delta O_2 = [O_2]_I - p[O_2]_E$$

$p = $ volume reduction where

$$p = (100 + 0.5[EO]_I + [CO_2]_I)/(100 + 0.5[EO]_E + [CO_2]_E)$$

Table 7 *Oxygen and Carbon Balances in Ethylene Oxide Production*

$$OB = ((\Delta EO/2 + 3\Delta CO_2/2)/\Delta O_2) \times 100\%$$
$$CB = ((\Delta EO + \Delta CO_2/2)\Delta C_2H_4) \times 100\%$$

(OB = oxygen balance; CB = carbon balance. See Table 6 for other definitions).

analysis by the production personnel were that all measures of selectivity agree to within 0.5% and that the oxygen and carbon balances would be $100 \pm 1\%$. In order to evaluate whether or not this was experimentally possible the analysis was simulated using model data. The components assumed to be measured were nitrogen, argon, oxygen, methane, ethane, ethylene, carbon dioxide, and ethylene oxide. The mass spectral peaks recorded were m/z 14, 15, 16, 26, 27, 28, 29, 30, 32, 40, 42, 43, and 44.

Using matrix notation

$$I = AP$$

where i_m is the measured peak intensity at mass m, p_n the partial pressure of component n, and a_{mn} the peak intensity at mass m due to unit pressure of component n. (i, p, and a indicate matrix elements of I, P, and A.) The solution of this equation is:

$$P = A^{-1}I$$

where A^{-1} is the inverse matrix.

This is developed by calibration with pure components or mixtures of known composition. For the ethylene oxide case considered here $I = (13 \times 1)$, $P = (8 \times 1)$, and $A = (13 \times 8)$. An overdetermined set of equations results, which is easily solved to obtain the concentrations. An overdetermined set of data is used to improve the stability and the discrimination of the measurement. In order to probe the effect of experimental error on the data an error of 0.25% RSD was assumed in the measured peak intensities. This was propagated through the calculation assuming that this represented the mean of a Gaussian distribution. The solution to the overdetermined matrix solution was calculated to obtain the concentrations. The concentrations were then used to calculate the delta values and the selectivities and balances. The model result used in the calculations is shown in Table 8. The resulting selectivities and balances are shown in Table 9. As can be seen from the data it is not possible to meet the initial requirements of the production personnel. The data also shows that some measures of the selectivity of the reaction are more precise than others. This is obviously important to know when evaluating the performance of the catalyst used in the reaction. In order to compare this simulated data with actual production measurements three set of data were obtained, as shown in

Table 8 *Model Result Used in Calculations of Selectivity in Ethylene Oxide Production*

	Feed (%)	Product (%)	Delta (%)
Ethylene Oxide	0	2.3	2.3
O_2	8.12	5.33	2.79
C_2H_4	28.93	26.08	2.85
CO_2	5.47	6.56	1.09

Table 9 *Selectivity (S) and Balances (B) obtained From Model Results For Ethylene Oxide Production*

	Mean	Standard Deviation	% Relative Standard Deviation (% RSD)
$S(EO/O_2)$	80.84	0.86	1.06
$S(EO/CO_2)$	80.88	0.84	1.04
$S(EO/C_2H_4)$	80.79	3.48	4.31
$S(O_2/CO_2)$	80.86	1.77	2.19
$S(O_2/C_2H_4)$	80.85	1.18	1.46
$S(C_2H_4/CO_2)$	80.91	0.99	1.22
OB	99.88	3.18	3.18
CB	99.88	3.95	3.95

(Error of 0.25% RSD assumed on measured intensities; see Table 8 for model result data.)

Table 10 *Typical Ethylene Oxide Plant Analysis by On-line Mass Spectrometry*

	Set 1	Set 2	Set 3
$S(EO/O_2)$	79.69 (0.77)	79.89 (1.31)	79.86 (0.26)
$S(EO/CO_2)$	79.62 (0.21)	79.01 (1.87)	79.55 (0.74)
$S(C_2H_4/CO_2)$	80.24 (0.72)	79.45 (0.57)	79.45 (0.57)
$S(O_2/C_2H_4)$	80.92 (1.88)	79.73 (2.53)	79.73 (2.54)
$S(O_2/CO_2)$	79.50 (0.98)	79.01 (1.87)	79.01 (1.87)
$S(EO/C_2H_4)$	78.44 (3.12)	79.94 (5.86)	80.12 (5.26)
OB	100.26 (2.51)	101.15 (2.58)	100.67 (3.86)
CB	96.95 (2.86)	100.67 (5.73)	100.25 (7.04)

[Data is presented as the mean (standard deviation) for $n = 4$. S = Selectivity; B = Balance (see Table 9 for model results).]

Table 10, where each calculation is the sum of four measurements. The results are a good agreement with the simulated data (Table 9). These studies indicate that, when important process variables are derived via a series of calculations based on the individual concentrations of process component, the high precision afforded by the mass spectrometric measurements is necessary.

The improvement in precision of measurement obtained with modern mass spectrometric equipment derives from a number of factors. These include the following:

(i)   Source temperature control to $\pm 1\,^{\circ}C$. This gives much better stability of fragmentation.

(ii) Improved amplifier design giving improved linearity over wide ranges.

(iii) Improved sample inlet design reducing discrimination.

(iv) Increased speed of analysis giving the opportunity for averaging.

(v)   Development of purpose built process instruments to improve the reliability.

4 Future Developments

Mass spectrometry is now an accepted process analytical technique. The proven advantages of speed of analysis, good sensitivity, high precision, excellent dynamic range, and versatility outweigh, in many cases, the increased cost and complexity. Current remaining problem areas are in sampling, where improvements are needed in order to study liquid and solid samples, and in selectivity where tandem mass spectrometry methods and/or alternative techniques may offer advantages if they can be adapted for process use.

Acknowledgement

I should like to thank Fisons (VG Gas Analysis, Cheshire) for permission to use some of the process measurements presented in this paper.

References

1. C. F. Robinson, *Chem. Eng.*, 1951, **58**, 136.
2. B. W. Thomas, *Ind. Eng.*, 1953, **45**, 85A.
3. A. T. James and A. J. P. Martin, *Analyst (London)*, 1952, **77**, 915.
4. J. H. Scrivens, *Trans. Inst. Meas. Control*, 1985, **7**, 2.
5. J. H. Scrivens, *Vacuum*, 1982, **32**, 169.
6. J. H. Scrivens and J. C. Ramage, *Int. J. Mass Spectrom. Ion Processes*, 1984, **60**, 299.
7. P. Nicholas, *Met. Mater. (Inst. Met.)*, 1990, **6**, 647.
8. G. Dube, *Wiss. Z. Tech. Hochsch. Chem. 'Carl Schorlemmer' Leuna-Merseburg*, 1990, **32**, 30.
9. W. Grosse Bley, *Vak.-Tech.*, 1989, **38**, 9.
10. A. Choudhury, G. Brueckmann, H. Scholz, C. Smith, and B. Rogge, *Metall. Plant. Technol.*, 1988, **11**, 44.
11. 'Mass Spectrometry in Biotechnological Process Analysis and Control', (eds) E. Heinzle and M. Reuss, Plenum Press, New York, 1987.
12. V. H. Adams, *Am. Lab. (Fairfield Conn.)*, 1986, **18**, 72.
13. H. L. C. Meuzelaar, W. Windig, S. M. Huff, and J. M. Richards, *Anal. Chim. Acta*, 1986, **190**, 119.

Laser-induced Molecular Luminescence for Detection and Identification

C. Gooijer, R. J. van de Nesse, A. J. G. Mank, F. Ariese,
U. A. Th. Brinkman, and N. H. Velthorst

FREE UNIVERSITY, DEPARTMENT OF GENERAL AND ANALYTICAL
CHEMISTRY, DE BOELELAAN 1083, 1081 HV, AMSTERDAM, THE NETHERLANDS

Summary

To discuss the progress of laser-induced molecular luminescence in analytical science, it is appropriate to distinguish between the liquid and the solid state. Liquid state luminescence spectra are almost featureless; here lasers are involved because of the high irradiances they provide. Recent achievements concerning detection in liquid chromatography (LC) and capillary electrophoresis (CE) are discussed, i.e. laser excitation in the deep ultraviolet part of the electromagnetic spectrum, two-photon excitation utilizing pulsed lasers, and the potential of diode lasers in laser-induced fluorescence detection. For some classes of analytes in suitable solvents, highly resolved solid state spectra can be obtained under cryogenic conditions and molecular luminescence spectrometry can be utilized for identification purposes. Some examples are given, concerning applications of (laser-excited) Shpol'skii spectroscopy and fluoresence line narrowing spectroscopy.

1 Introduction

Molecular luminescence techniques, i.e. fluorescence, phosphorescence/lanthanide luminescence, and chemi/bioluminescence have received extensive attention in analytical chemistry.[1,2] The former two modes both require lamp (or laser) excitation; their main mutual difference is the radiative lifetime which is 10^{-7} to 10^{-9} s in fluorescence in contrast with 10^{-3} to 10 s in phosphorescence/lanthanide luminescence.[3,4] In chemi/bioluminescence instead of a light source a chemical reaction is responsible for the excitation process.[5]

Luminescence techniques are attractive because of their inherent sensitivity in comparison to absorption measurements. It is far more easy to observe a weak light in the dark emitted by the sample than to detect a small decrease in the intensity of an intense light. In principle, in photoluminescence, the sensitivity and the achievable detection limits can be further improved by utilizing lasers instead of conventional light sources for excitation, since lasers provide extremely high sample irradiances. The availability of lasers is even crucial for detection in micro-liquid chromatography (micro-LC) and capillary electrophoresis (CE) where extremely small sample volumes have to be analysed.[6] The emphasis of the present paper is on laser-induced fluorescence (LIF), although long-lived luminescence can also be successfully combined with (pulsed) laser excitation.

In comparison with absorption measurements, luminescence spectrometry is distinctly more selective, but nevertheless its identification power is limited. When considering fluorimetry, the absorption of ultraviolet or visible light causes few analytes (and possible interferents) to emit any fluorescence at all. In order to be detected, most species need to be chemically derivatized to fluorescent products. Unfortunately, the spectra recorded for such products in general do not reveal any characteristic information about the analyte.

For naturally fluorescent analytes, fluorimetry provides more selectivity than absorption measurements since two wavelengths can be selected independently (*i.e.* for excitation and emission) so that excitation and emission spectra can be recorded and techniques like synchronous scanning can be applied. Nevertheless the identification power is not very high since the spectra exhibit little structural detail. As a consequence, the other characteristic feature of laser radiation (in addition to its directionality), namely its high degree of monochromaticity, is hardly useful in fluorimetry. In view of the broad-banded spectra involved, it can be qualified as overkill.

However, under special conditions high-resolution spectra of naturally fluorescent analytes can be obtained, for instance in the cryogenic techniques Shpol'skii spectroscopy and fluorescence line narrowing spectroscopy.[6] In Shpol'skii spectroscopy the use of lasers has greatly improved the selectivity and sensitivity of the technique. In fluorescence line narrowing spectroscopy, sharp lines are obtained specifically as a result of laser excitation; here lamp excitation is not appropriate at all.

Without the intention of presenting an exhaustive treatment, in the following review some recent achievements of LIF detection in CE and LC are discussed first. Emphasis will be on: (a) excitation in the deep UV, for instance at 257 nm, which extends significantly the applicability of laser induced fluorescence to naturally-fluorescent analytes; (b) two-photon excitation utilizing pulsed lasers, which provides additional selectivity compared to normal excitation; and (c) the use of (cheap) diode lasers for LIF detection, which can only be applied in the red part of the electromagnetic spectrum. Concerning the latter point chemical derivatization of analytes is crucial and the role of chemistry will be illustrated. Recently, suitable near-infrared labels for thiol-groups have been developed.

In the second part of the paper, the analytical potential of high-resolution fluorescence spectrometry is illustrated. Extremely high selectivity is combined with high sensitivity, especially with laser-excited Shpol'skii spectroscopy, so that low-level identification of analytes in complex samples is possible without extensive sample treatment. This will be illustrated for the analysis of benzo(*a*)pyrene metabolites in fish bile, as part of a multidisciplinary study on the ecotoxicological effects of PAH contamination on the marine environment.

2 Laser-induced Fluorescence (LIF) Detection in Liquid Chromatography (LC) and Capillary Electrophoresis (CE)

Influence of Irradiance

Since in fluorescence the signal S is proportional to the sample irradiance I (in $W m^{-2}$) it is expected that the signal-to-noise (S/N) can be increased significantly by using laser-excitation instead of lamp-excitation. However, some points have to be noted.

First, in many practical examples of LIF detection, a significant background is encountered, caused by: (a) reflected and refracted laser radiation; (b) Rayleigh and (more importantly) Stokes–Raman scattering coming from the eluent; (c) last but not least, fluorescence background from various sources. S/N increases with I as long as the dominating source of noise in this background is shot noise: under this condition S increases in proportion to I and N in proportion to $I^{1/2}$. However, at high background levels, it is usually flicker noise, due to fluctuations in laser power, which dominates, so that N is proportional to I and further increase of irradiance is useless.

It should be realized that, in general, lasers are much more noisy than lamps.[7] By far, most applications of LIF detection in LC and CE have been realized with the He:Cd laser with a typical stability of 1–2%, and the argon-ion laser, with a typical stability of 0.5–1%. This explains why only for low background levels the shot-noise condition is fulfilled. Within this context recent instrumental improvements should be noted: stabilizing systems have become commercially available, providing a stability of 0.05% for the above lasers.[8] Furthermore the quality of diode lasers should be emphasized; their stability is better than 0.01%.

The second point to be noted is that the fluorescence signal S does not infinitely increase with I. At too high an irradiance, fluorescence saturation and/or photobleaching of the analyte plays a role. Mostly, the latter is important, depending on the analyte characteristics and the applied flow rate (*i.e.* the residence time in the laser-illuminated volume).[9] This is one of the reasons why in LIF detection for LC and CE, continuous wave (CW) lasers almost exclusively are used instead of pulsed lasers. The latter provide extremely high peak powers.

The consequence of the above discussion is that laser power plays a less important role in LIF detection than might be anticipated. In practice it is hardly advantageous to utilize powers higher than say 50 mW, provided that the line can be adequately focused.

Applicability

An obvious disadvantage of CW lasers in comparison to conventional light sources is that they only provide a small number of discrete lines. The He:Cd laser gives output at 325 and 442 nm; the output of the argon-ion laser depends on the laser type: while the air-cooled system only provides two visible lines at 488 nm and 514 nm, the water-cooled large frame type gives UV lines at 334, 355, and 364 nm, together with several visible lines ranging from 458 nm to 530 nm.[67]

As a consequence laser excitation is not always performed at the wavelength associated with the excitation maximum. More seriously, various naturally fluorescent analytes with the excitation maximum in the deep UV, readily measured by conventional fluorescence, cannot be excited and thus detected by LIF.

For this reason in LIF detection, chemical derivatization is even more important than in conventional fluorescence detection. Of course, existing procedures are followed as close as possible, although an additional requirement has to be met: the product should have an excitation maximum not too far from the available laser line. In the literature, a lot of effort has been devoted to this point. As an illustration, instead of using the well-known o-phthalaldehyde reaction for amines, in LIF the analogous naphthalene-2,3-dialdehyde reaction seems to me more appropriate, since the 442 nm He:Cd laser line is close to the excitation maximum of the reaction product.[10,11]

Detection Limits

It has been extensively shown in the literature that, for analytes labelled with a fluorescent tag, extremely low concentration limits of detection, CLODs, can be achieved (giving, in CE, extremely low mass limits of detection, MLODs, as injection volumes as small as 10 nl are common). Care should be taken here since different definitions are used in the literature. In some cases instead of CLOD, the minimal concentration at the detector cell, denoted as CDET, is reported; generally CDET is 5–10 times more favorable than CLOD owing to dilution of the injected plug during the elution process. Furthermore, instead of MLOD, which is the product of CLOD and the injection volume, sometimes the corresponding number of moles in the laser-irradiated part of the detector cell is calculated; this number is extremely low since it is obtained by multiplying CDET with a volume that can be as small as 10 pl.

Another point to emphasize here is that, frequently, the data reported apply to analytes labelled with a fluorescent tag. However in many examples the pre-column chemical derivatization reaction is performed at much higher analyte concentrations (for instance at the 10^{-6} M level).[12] In practice one has to deal with low analyte concentrations in complicated samples and derivatization reactions at analyte concentrations at the 10^{-10} to 10^{-9} M level usually give rise to many problems. In such cases it is the chemistry that limits the achievable CLODs.

Recent Achievements

In this section some examples of recent developments in LIF detection in separation techniques will be discussed without the intention of providing a rigorous review. Two examples concern naturally fluorescent analytes with excitation maximum in the deep UV. The third considers the use of visible diode lasers in LIF detection.

Deep UV Induced LIF. The possibility of excitation in the deep UV part of the electromagnetic spectrum would greatly enhance the applicability of LIF to naturally fluorescent analytes. Van de Nesse *et al.* have frequency-doubled the argon-ion laser 514 nm line, thus achieving a 257 nm line with a power of 5 mW.[13-15] With such a short-wavelength laser line a large number of analytes can be excited directly so that chemical derivatization can be avoided. A disadvantage of deep UV-induced LIF detection is the relatively high fluorescent background, caused by the fact that most fluorescent impurities are excited as well. One might expect also severe interferences from Stokes–Raman lines, as the intensity is proportional to λ^{-4}. However, the Raman lines are close to the 257 nm excitation line (see Figure 1), so they can readily be distinguished from fluorescence emission. Consequently, Raman scatter does not interfere with fluorescence detected at wavelengths greater than 295 nm.

Recently in our laboratory, a quantitative comparison was made between LIF and conventional fluorescence detection (in conventional-size LC) utilizing four UV-laser lines provided by the argon-ion laser, *i.e.* 257 nm, 293 nm (both obtained by frequency-doubling of visible laser lines), 334 nm, and 352 nm. The light intensity for the laser lines and a lamp inside the detector cell were determined by means of actinometry. At 257 nm, the intensity provided by the laser was 1.8 mW, about 100-fold higher than that provided by the pulsed Xe-lamp. Flicker noise coming from the laser was suppressed by simultaneous monitoring of the laser and fluorescence light, and ratioing the sample and reference signals. The effect of this procedure (applied to 293 nm induced LIF) is visualized in Figure 2. It was found that at 257 nm, the LIF detection limits are 10–20 times more favorable than in conventional fluorescence detection. However, the conventional mode has the advantage that the excitation and detection wavelengths can be optimized separately, so that the net advantage in detection limit is 5- to 10-fold.

It is known that LIF detection at 257 nm is also favorable for analytes that can be excited at wavelengths greater than 300 nm. Chromatographic detection limits for some PAHs obtained for LIF detection, are presented in Table 1. Comparing the data at 257 nm and 352 nm it should be realized that the power of the latter line is about 20 times higher. As expected LIF at 257 nm is more widely applicable than at 352 nm: at the lower wavelength pyrene and chrysene can be sensitively detected, while for the other compounds the CLODs are quite similar at both wavelengths. Three factors play a role here, which apparently largely compensate each other. For 352 nm excitation, the emission window has to be shifted to avoid detection of Raman scatter; this implies a partial loss

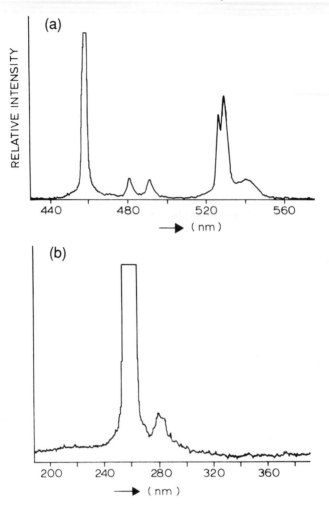

Figure 1 *A comparison between the Raman spectra of methanol/water (9:1, v/v) obtained at* 458 nm (a) *and* 257 nm (b) *laser excitation. From reference 13, with permission*

of fluorescence intensity. Secondly at 352 nm the molar absorptivities are lower. Thirdly, at 352 nm excitation, the laser intensity is higher and the fluorescence background about three times higher.

The results clearly underline the need to develop a continuous wave laser that is tunable in the deep UV. Tunability will not become possible in the near future. Nevertheless interesting developments in laser technology have been reported. Modifications of the plasma tube and mirrors have made 275.4 nm output feasible, while intracavity frequency doubling provides high power at 257 nm and 244 nm. Recently Lee and Yeung have shown the usefulness of 275.4 nm excitation for the LIF detection of underivatized proteins in CE.[16] This excitation wavelength is a much better match with the fluorescence–excitation

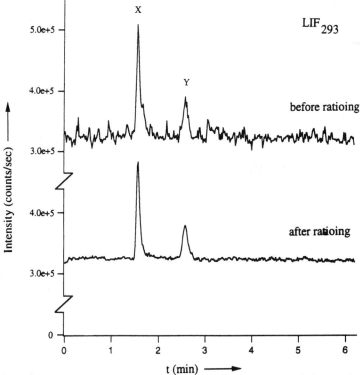

Figure 2 *Conventional-size LC chromatograms of two pharmaceutical compounds (denoted as X and Y) applying LIF-detection at 293 nm excitation (obtained by frequency-doubling of the 586 nm output of an argon-ion–dye laser combination) showing the effect of ratioing emission and excitation light. From reference 15, with permission*

Table 1 *Chromatographic concentration detection limits (10^{-10} M) from some model PAHs in a conventional size LC system, applying LIF-detection at four UV-excitation lines. The molar absorptivities (in $10^4 1 mol^{-1} cm^{-1}$) are given in parentheses; the excitation powers for 257 nm and 293 nm are determined within the detection cell (by actionometry); the other powers are measured with a power meter*

	257 nm[a] (1.8 mW)		293 nm[a] (1.5 mW)		334 nm[b] (14 mW)		352 nm[c] (37 mW)	
Anthracene	2.5	(1.1)	75	(0.0064)	3.0	(0.37)	3.0	(0.59)
Fluoranthene	30	(1.2)	250	(0.21)	10	(0.60)	9.0	(0.69)
Pyrene	10	(1.3)	40	(0.43)	1.5	(4.3)	260	(0.045)
Chrysene	2.5	(7.8)	20	(1.2)	220	(0.076)	900	(0.049)
Benzo[b]fluoranthene	5.5	(4.7)	12	(2.3)	4.5	(1.1)	4.0	(1.1)
Benzo[k]fluoranthene	2.0	(2.1)	1.0	(4.0)	1.5	(0.53)	3.0	(0.36)
Benzo[a]pyrene	2.5	(4.4)	2.0	(6.2)	4.5	(0.56)	3.0	(1.2)

[a] $\lambda_{em} = 400$ nm, maximum transmittance 54%, full width at half maximum (FWHM) 52 nm
[b] $\lambda_{em} = 410$ nm, maximum transmittance 36%, FWHM 29 nm
[c] $\lambda_{em} = 423$ nm, maximum transmittance 17%, FWHM 21 nm

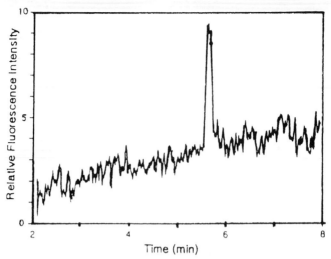

Figure 3 *Electropherogram of 5 × 10⁻¹⁰ M injected conalbumin applying LIF-detection at 275.4 nm excitation provided by a Spectra Physics Model 2045 argon-ion laser. At this wavelength, both the tryptophan and tyrosine residues are effectively excited; two (UG-1, Schlott) band pass filters were used to selectively pass the fluorescence. From reference 16, with permission*

maxima of most proteins (based on the fluorescence of tryptophan and tyrosine residues) than 257 nm. As illustrated by Figure 3, for 275.4 nm induced LIF a high sensitivity is reached. A crucial aspect of this study is that pre-column labelling is avoided. For samples of this type, such labelling gives rise to multiplet peaks in the electropherogram, because of differences in the extent of incorporation of the fluorescent tags in each protein molecule.

Two-photon Excitation. In two-photon excitation (TPE), visible laser light is used for excitation of UV-absorbing analytes, the fluorescence spectrum is on the short-wavelength side of the excitation and Stokes–Raman scattering does not play any role. In TPE, excitation of analytes is achieved by simultaneous absorption of two photons. A virtual electronic level is involved which has an energy lower than that of the first excited states. Since the lifetime of this virtual level is extremely short (about 10^{-15} s), the basic condition for TPE is that the two photons arrive at the same time. This explains why the excitation efficiency is much smaller than in conventional one-photon excitation (OPE). As shown in the literature, reasonable detection limits can be realized by utilizing pulsed lasers, despite the fact that TPE is extremely improbable.[17] This is due to the fact that the number of molecules excited per second in TPE is proportional to the laser power squared and pulsed lasers provide extremely high peak powers.

Pfeffer and Yeung reported a CLOD of 9×10^{-10} M for an oxadiazole (in a micro-LC system) utilizing a Cu vapour laser with 3 W average power, 20 kW peak power, and laser lines at 510 nm and 578 nm.[18]

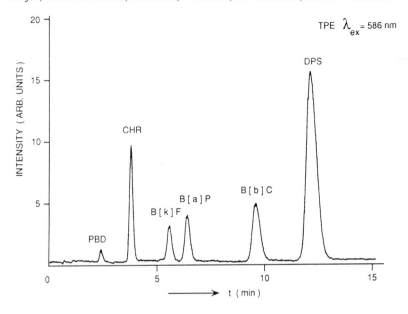

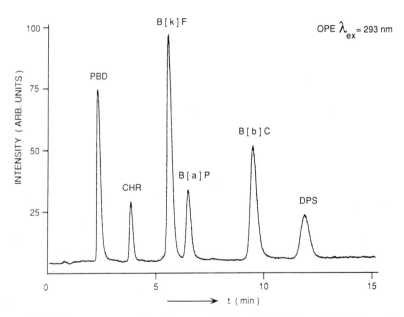

Figure 4 *LC chromatograms of a mixture of 2-phenyl-5-(4-biphenylyl)-1,3,4-oxadiazole (PBD), chrysene (CHR), benzo[k]fluoranthene (B[k]F), benzo[a]pyrene (B[a]P), benzo[b]chrysene (B[b]C), and 4,4'-diphenylstilbene (DPS), excited with 586 nm (TPE) and 293 nm (OPE). The fluorescence detection window was between 360 and 480 nm. The concentrations in the upper chromatogram (λ_{ex} = 586 nm) were 5×10^{-7} M for chrysene and 5×10^{-8} M for the other analytes; in the lower chromatogram (λ_{ex} = 293 nm) the concentrations for all analytes were 2.5-fold lower. From reference 19, with permission, copyright 1991 American Chemical Society*

In our laboratory we have used the same chromatography system to perform a direct comparison of TPE and one-photon-excitation (OPE) with an excimer-dye laser combination (without and with frequency doubling).[19] LC chromatograms for a number of model compounds excited at 586 nm and at 293 nm and utilizing a fluorescence window from 360 to 480 nm, are shown in Figure 4. The profiles of the chromatograms are clearly different. For example, whereas DPS gives the strongest peak in the upper chromatogram (excitation at 586 nm) its peak in the lower chromatogram (excitation at 293 nm) is relatively small; for PBD the reverse is true. Since the fluorescence-emission spectra associated with the two chromatograms are the same, the mutual difference is due to a difference in excitation. Obviously, the TPE efficiency at 586 nm cannot be directly related to the OPE efficiency at 293 nm. This indicates that there might be an interesting difference in selectivity for TPE and OPE.

Diode-laser Induced Fluorescence. As noted above, diode-lasers are extremely stable (0.01%) compared with He:Cd and argon-ion lasers. Furthermore they are cheap, have very long lifetimes and are extremely small. Recently, in addition to diode lasers emitting in the infrared (from 780 to 860 nm), visible diode lasers have become available, such as the In:Ga:Al:P laser providing 10 mW of 670 nm light.[20,21]

Obviously, long wavelength excitation has some inherent advantages in LIF detection: negligible fluorescence background and little photo decomposition. The Raman scatter intensity at long wavelengths is low but extended over a wide wavelength region.

However, there is a serious hindrance to the use of diode laser induced fluorescence detection (DIO-LIF). Only few analytes can both be excited at long wavelengths and produce naturally occurring fluoresence in the red or near infrared. Exceptions are phthalocyanines applied in photodynamic therapy. Therefore, chemical derivatization procedures have to be developed leading to covalently bound fluorophores detectable at 670 nm.

Such procedures have hardly been developed yet. Recently, Patonay and Antoine have considered the possible use of carbocyanines for this purpose, dyes that can be 'wavelength tuned' by modifying their structures.[22] In our laboratory we have developed a pre-column chemical derivatization of thiols, based on a sulfonated carbocyanine with one reactive iodoacetamido group[23]; attention was focused on determination of 2-mercaptobenzothiazole (MBT) (Figure 5). High fluorescence quantum yields in aqueous solutions were obtained enabling the application of reversed-phase LC. The detection limit of labelled MBT in the conventional size DIO-LIF-LC system was as low as 8×10^{-12} M. The detection limit of MBT was 1×10^{-9} M; lower concentrations are not labelled quantitatively.

A fascinating result was that in complex matrices such as spiked river water and urine samples, hardly any interference was obtained. This underlines the analytical potential of DIO-LIF and indicates the importance of developing chemical derivatization procedures for other functional groups.

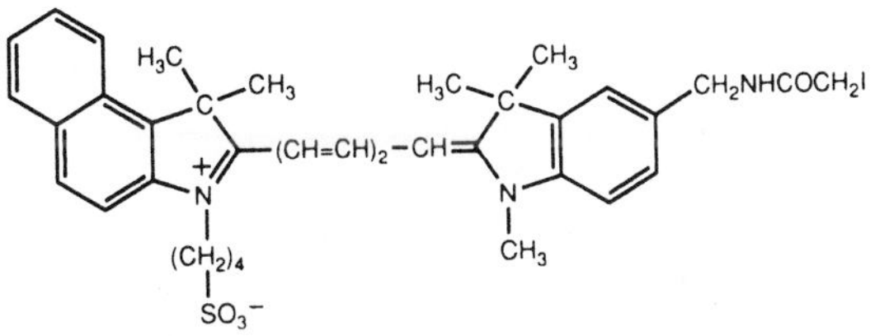

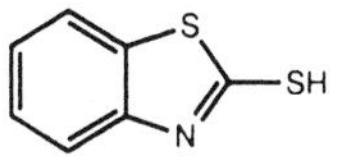

2-mercaptobenzothiazole (MBT)

Figure 5 *Structure of the sulfonated carbocyanine with an iodoacetoamido reactive group used for labelling of 2-mercaptobenzothiazole*[23]

3 High-resolution Fluorimetry

The Shpol'skii and Fluorescence Line Narrowing Effects

As noted above, the identification power of fluorimetry is strongly limited by the fact that under normal conditions in the fluorescence emission spectra, the vibrational structure is hardly resolved. The question should be raised why broad bands are observed in the spectra, whereas the well-known Jablonski energy diagram shows discrete energy levels and thus sharp vibronic transitions. The answer is that this diagram only applies for an isolated molecule. In the condensed liquid and solid phase, influences of the surrounding solvent cage on the molecule have to be accounted for. In particular, the energies of the electronic states (*e.g.* S_0, S_1, S_2, *etc.*) depend on the detailed cage structure whereas the vibrational levels are hardly influenced. It is of crucial importance that in the condensed phase the specific cage formation (and thus the S_0–S_1 energy distance) differs from molecule to molecule. Thus, for an ensemble of analyte molecules, a Gaussian distribution of narow lines will be observed resulting in featureless spectral bands, wherein the separate vibronic transitions are largely hidden, a phenomenon denoted as inhomogeneous broadening.[6]

Shpol'skii and fluorescence line narrowing (FLN) spectroscopy involve procedures which can reduce this broadening. In Shpol'skii spectroscopy an attempt is made to create equal analyte–solvent cage interactions for all the analyte molecules concerned; an example is given in Figure 6. It has been known since 1952, that this can be achieved for PAHs and PAH-derivatives in appropriate n-alkane solvents under cryogenic conditions. As the Shpol'skii effect is completely matrix-induced, conventional lamp excitation sources can be utilized to obtain high-resolution fluorescence spectra and lasers are not strictly needed. Nevertheless, as the long-wavelength part of the excitation

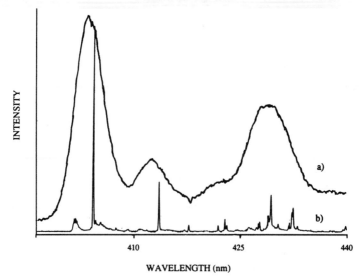

Figure 6 *Fluorescence emission spectra of benzo[k]fluoranthene illustrating the Shpol'skii effect. For both spectra the solvent is n-octane and xenon lamp excitation at 308 nm is performed; spectrum (a) was recorded at room temperature, for a concentration of 10^{-4} M; spectrum (b) was recorded at 26 K, for a concentration of 10^{-6} M. From reference 25, with permission*

spectrum also exhibits sharp lines, laser-excited Shpol'skii spectroscopy (LESS) provides much higher selectivity and sensitivity than conventional Shpol'skii spectroscopy.

Of course, in Shpol'skii spectroscopy there is little freedom in solvent choice. In FLN spectroscopy this freedom is much larger. In this technique the reduction of inhomogeneous broadening is laser-induced. It is based on the fact that out of the ensemble of analyte molecules only those molecules are excited that have a S_0–S_1 energy difference exactly equal to the photon energy at the laser wavelength applied. In an ideal situation, in the cryogenic solid state matrix, the excited population thus created remains intact and narrow fluorescence (vibronic) transitions are observed. This explains why FLN spectroscopy (though more widely applicable) is less sensitive than Shpol'skii spectroscopy: by far the largest fraction of analyte molecules is not excited by the laser line and thus does not show up in the spectrum.

Experimental Aspects

It is known from the literature, and frequently cited as an obvious disadvantage, that in Shpol'skii spectroscopy not only the choice of the solvent but also the preparation of the solid matrix (analyte concentration, rate of cooling, and freezing the sample) influence the spectral characteristics.[24] These experimental parameters determine the quality of the crystal lattice and possible segregation of sample constituents and thus the analyte–solvent cage interactions.

Recently, significant improvements have been realized by constructing the sample holders with materials guaranteeing high thermal conductivity and thus rapid sample solidification. Thus, for larger PAHs, in particular, completely reproducible spectra are obtained. Ariese *et al.* developed a sample holder in which four 10 µl samples can be handled simultaneously.[25] It is used in combination with a closed-cycle helium cryostat, an instrument that is easy to operate and does not consume helium.

In FLN and LESS, pulsed lasers such as the Nd-YAG and excimer lasers combined with dye lasers and frequency mixing and doubling techniques are very appropriate since they provide wavelength tunability. On the detection side, the introduction of multichannel instruments, based on intensified linear photodiode arrays (ILDA) or charge coupled devices (CCD) is a significant improvement. Compared with scanning instruments, the analysis time is greatly reduced. Also, each data point is equally affected by pulse-to-pulse fluctuations of the laser and by photochemical activity of the analyte.

Recent Achievements

LESS of PAH-metabolites. By far, most analytical applications of Shpol'skii spectroscopy are devoted to parent PAHs.[6,25,26] Recently, its applicability to PAH-metabolites in biosystems has been shown, a challenging analytical problem which is highly relevant because the analytical data can provide valuable information on the actual PAH uptake rate (biomonitoring).[26-28]

The metabolite, 3-hydroxy-benzo[*a*]pyrene (3OH-B*a*P), was determined in fish bile (of the flatfish species flounder) by LESS. Of course the problem is that hydroxy-PAHs are less compatible with n-alkane matrices because of their increased polarity compared with parent PAHs. As a result 3OH-B*a*P could only be directly identified for fish injected with B*a*P, a rather unrealistic situation. The problem could be solved by invoking a (rapid, practical, and straightforward) methylation reaction, adopted from Weeks *et al.*[29] In contrast to the OH-B*a*Ps, the corresponding CH_3O-B*a*Ps provide good Shpol'skii spectra with an increase in fluorescence intensity of about 20-fold due to improved host–guest compatibility; an example is given in Figure 7. The detection limit of 3-methoxy-B*a*P obtained by lamp excitation is 5×10^{-10} M; in LESS performed at 418.36 nm, it is as low as 5×10^{-12} M. Considering the sample treatment steps involved and applying laser excitation, the detection limit of 3OH-B*a*P in the original bile sample was 2×10^{-10} M or 0.05 ng ml^{-1}, low enough to detect bile in fish living in relatively clean areas (see Figure 7).

FLN Spectroscopy of Analytes on TLC-plates. As indicated above, in FLN-spectroscopy the solvent choice is less stringent than in Shpol'skii spectroscopy.[6] An interesting recent finding is that FLN is even applicable to analytes on various types of thin layer chromatographic (TLC) plates.[30] In other words it can be applied after performing TLC separation. Even more wide perspectives can be opened, since liquid chromatography and TLC can

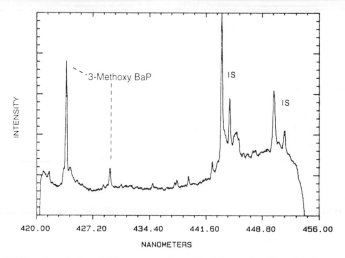

Figure 7 *LESS of methylated bile extract from flatfish species flounder living in a clean area. Selective laser excitation is performed at 418.36 nm; the presence of 3-methoxybenzo[a]pyrene is obvious. The peaks indicated as IS are from the internal standard perylene d_{12}. From reference 27, with permission*

be coupled under certain conditions (current research is underway).[31,32] The effluent from the LC-column is deposited on a TLC plate moving with a constant speed so that an 'immobilized' chromatogram is obtained. Efficient eluent elimination and conservation of chromatographic integrity during the deposition process is of utmost importance here.

References

1. 'Luminescence Techniques in Chemical and Biomedical Analysis', eds. W. R. G. Bayens, D. de Keukelaire, and K. Korkides, Marcel Dekker, New York, 1991.
2. 'Molecular Luminescence Spectroscopy, Part 3', ed. S. G. Schulman, Wiley, New York, 1993.
3. C. Gooijer, R. A. Baumann, R. W. Frei, and N. H. Velthorst, *Prog. Anal. Spectrosc.*, 1987, **10**, 573.
4. C. Gooijer, M. Schreurs, and N. H. Velthorst, in 'HPLC Detection—Newer Methods', ed. G. Patonay, VCH Publishers, New York, 1992, Chapter 2, p. 27.
5. C. Gooijer, N. H. Velthorst, and R. W. Frei, in 'Selective sample handling and detection in HPLC, Part B', eds. K. Zech and R. W. Frei, Elsevier, Amsterdam, 1989, Chapter 6, p. 216.
6. J. W. Hofstraat, C. Gooijer, and N. H. Velthorst in 'Molecular Luminescence Spectroscopy, Part 3', ed. S. G. Schulman, Wiley, New York, 1993, Chapter 9, p. 323.
7. For details on commercial systems see *Laser Focus World, Buyers Guide*, 1991, **26**.
8. X. Xi and E. S. Yeung, *Anal. Chem.*, 1990, **62**, 1580.
9. R. A. Matthias, K. Peck, and L. Stryer, *Anal. Chem.*, 1990, **62**, 1786.
10. M. C. Roach and M. D. Harmony, *Anal. Chem.*, 1987, **59**, 411.
11. S. C. Beale, Y. Z. Hisch, D. Wiesler, and M. Novotny, *J. Chromatogr*, 1990, **499**, 579.

12. S. Wu and N. J. Dovichi, *J. Chromatogr*, 1989, **480**, 141.
13. R. J. van de Nesse, G. Ph. Hoornweg, C. Gooijer, U. A. Th. Brinkman, and N. H. Velthorst, *Anal. Chim. Acta*, 1989, **227**, 173.
14. R. J. van de Nesse, G. Ph. Hoornweg, C. Gooijer, U. A. Th. Brinkman, N. H. Velthorst, and S. J. van der Bent, *Anal. Lett.*, 1990, **23**, 1235.
15. R. J. van de Nesse, G. Ph. Hoornweg, C. Gooijer, U. A. Th. Brinkman, N. H. Velthorst, and B. Law, *Anal. Chim. Acta*, 1993, **281**, 373.
16. T. T. Lee and E. S. Yeung, *J. Chromatogr.*, 1992, **595**, 319.
17. M. J. Wirth and H. O. Fatumnbi, *Anal. Chem.*, 1990, **62**, 973.
18. W. D. Pfeffer and E. S. Yeung, *Anal. Chem.*, 1986, **58**, 2103.
19. R. J. van de Nesse, A. J. G. Mank, G. Ph. Hoornweg, C. Gooijer, U. A. Th. Brinkman, and N. H. Velthorst, *Anal. Chem.*, 1991, **63**, 2685.
20. T. Imasaka, A. Tsukamoto, and N. Ishibashi, *Anal. Chem.*, 1989, **61**, 2685.
21. A. J. G. Mank, H. Lingeman, and C. Gooijer, *Trends Anal. Chem.*, 1992, **11**, 210.
22. G. Patonay and M. D. Antoine, *Anal. Chem.*, 1991, **63**, 321A.
23. A. J. G. Mank, E. J. Molenaar, H. Lingeman, C. Gooijer, U. A. Th. Brinkman, and N. H. Velthorst, *Anal. Chem.*, 1993, **65**, 2197.
24. J. W. Hofstraat, I. L. Freriks, M. E. J. de Vreeze, C. Gooijer, and N. H. Velthorst, *J. Phys. Chem.*, 1989, **93**, 184.
25. F. Ariese, 'Shpol'skii Spectroscopy and Synchronous Fluorescence Spectroscopy', Ph. D. thesis, Free University, Amsterdam, 1993.
26. F. Ariese, C. Gooijer, N. H. Velthorst, and J. W. Hofstraat, *Anal. Chim. Acta*, 1990, **232**, 245.
27. F. Ariese, S. J. Kok, M. Verkaik, G. Ph. Hoornweg, C. Gooijer, N. H. Velthorst, and J. W. Hofstraat, *Anal. Chem.*, 1993, **65**, 1100.
28. F. Ariese, S. J. Kok, C. Gooijer, N. H. Velthorst, and J. W. Hofstraat, in 'PAH: synthesis, properties, analysis, occurrence and biological effects', eds. P. Garrigues and M. Lamotte, Gordon and Breach, London, 1993, pp. 761–768.
29. S. J. Weeks, S. M. Gilles, R. L. M. Dobson, S. Senne, and A. P. D'Silva, *Anal. Chem.*, 1990, **62**, 1472.
30. J. W. Hofstraat, M. Engelsma, R. J. van de Nesse, C. Gooijer, U. A. Th. Brinkman, and N. H. Velthorst, *Anal. Chim. Acta*, 1987, **193**, 193.
31. R. J. van de Nesse, G. J. M. Hoogland, J. J. M. de Moel, C. Gooijer, U. A. Th. Brinkman, and N. H. Velthorst, *J. Chromatogr.*, 1991, **552**, 613.
32. R. J. van de Nesse, I. H. Vinkenburg, R. H. J. Jonker, C. Gooijer, U. A. Th. Brinkman, and N. H. Velthorst, 'FLN spectroscopy as off-line identification method for narrow-bore column liquid chromatography', *Appl. Sci*, in press.

Imaging Analytical Chemistry: Trends in Imaging and Image Processing for Micro, Nano, and Surface Analysis

M. Grasserbauer, G. Friedbacher, H. Hutter, and M. Leisch[1]

INSTITUT FÜR ANALYTISCHE CHEMIE, TECHNISCHE UNIVERSITÄT WIEN, GETREIDEMARKT 9, A-1060 WIEN, AUSTRIA
[1]INSTITUT FÜR FESTKÖRPERPHYSIK, TECHNISCHE UNIVERSITÄT GRAZ, PETERSGASSE 16, A-8010 GRAZ, AUSTRIA

Summary

The rapid progress in high technology constantly poses new challenges for analytical chemistry and prompts the development of new techniques and procedures. Imaging Analytical Chemistry is evolving as a new field. This review summarizes methodological developments and applications in three frontier areas: three-dimensional stereometric analysis with SIMS; three-dimensional nanoanalysis with field ion mass spectrometry; and atomic resolution analysis with STM and AFM.

1 Introduction—Imaging in Science and Technology

Imaging and image processing are widely spread in modern science and technology. Satellite imaging has opened up new dimensions for local, regional, and global observations creating a new discipline, namely geoinformatics. Imaging in medicine has revolutionized diagnosis, particularly through more recent developments like NMR-tomography. In material science new imaging techniques are arising, like X-ray tomography based on the use of synchrotron radiation leading to three dimensional structural representations of material objects and their inner structure.

In analytical investigations of materials, imaging techniques have for a long time played a major role if one considers the widespread use of light microscopy, and scanning or transmission electron microscopy. In surface analysis, imaging is a common feature now for AES, XPS, and SIMS. However in the past few years new qualities in analytical imaging have been added:[1]

- the third dimension—leading to three-dimensional analysis;
- implementation of elaborate mathematical techniques for processing images leading to substantial improvements of the analytical figures of merit and to an increase in the useful analytical information that can be extracted from the data; and
- access to the atomic dimension in surface imaging.

'Imaging Analytical Chemistry', which designates the correlation of spatial and chemical information, is important because the information obtained is relevant for the material properties. It is rapidly developing due to progress in computer science and the influx from other more advanced areas like satellite cartography. Imaging Analytical Chemistry is a subdiscipline of Analytical Chemistry because it has its own 'philosophy' and specific combination of knowledge and technology combining chemistry, physics, mathematics, informatics, and computer science in general.

2 Recent Developments

This paper will review current significant trends in Imaging Analytical Chemistry and will demonstrate the new aspects mentioned earlier based on work performed in the authors' laboratories.

Three-dimensional Stereometric Analysis with SIMS

Many material properties are derived from the three-dimensional distribution of their components. Often trace elements influence these properties strongly, as in semiconductors, high performance metals, or composites. To characterize these materials fully, it is necessary to obtain information not only on the chemical identity of these trace elements and their components (phases), but also on stereometric features such as size, distance, and homogeneity in three-dimensional space. SIMS is ideally suited for this task since the laterally resolved secondary ion signals can be measured during sputter removal of the material, thus yielding signals with *n* chemical (number of masses or elements measured) and three spatial dimensions. Surface and subsurface structures up to a depth of 10 (or more) µm can be investigated.

Secondary ion *microscopy* offers a spatial multiplex advantage, since all the picture elements (PIXELS) of a surface are obtained simultaneously allowing short recording times for one element (typically 1 s) and consequently the sputtering of a surface layer of some 10 µm in 1 h. Representative three dimensional distributions of components of a material are therefore obtained. The lateral resolution of such a system is about 1 µm, the depth resolution approximately 5 nm.

We have developed a system, which is based on the recording of ion micrographs obtained on a double channel plate secondary ion detector (offering single ion registration) with a CCD camera.[2,3] The sequentially obtained ion micrographs are fed into an ITI 151 image processor. There the collection of the images, their mathematical treatment, and visualization of the distribution of the components is performed in conjunction with a 486-type PC operating at 66 MHz. Typically 100 MByte of data are obtained in one image depth profile. This is about 1000 times more than in a conventional spatially non-resolved depth profile, thus substantially more useful analytical information is generated.

System Description. The special features of the system are:

- 16-bit dynamic range through storage of the highest 8-bit intensity and marking of position;
- linear range of CP 10^6 through automatic gain adjustment;
- optimized hardware and all new peripherals, *e.g.* 20-bit magnetic field control, new counting system, control unit for sample position, control unit for HV of CP, bargraph;
- new interface; and
- scanning generator with 100 nm accuracy in beam positioning for nanoanalysis with a fine focus ion source preserving the high transmission of the mass spectrometer.

The major image processing operations implemented include:

- colour encoding;
- background subtraction and linearization of the channel plate response;
- smoothing through moving average;
- contour enhancement through Laplace filtering;
- reconstruction of images to obtain 1D profiles (line scans, depth profiles), 2D sections, and 3D cylinders or cubes;
- statistical processing of images; and
- image correlation.

Analytical Information. A large amount of analytical information can be gained, particularly:

- qualitative elemental (isotopic) identification;
- semiquantitative identification of phases with relative sensitivity factors (RSFs);
- spatial correlation of elements (isotopes);
- number of phases within analytical volume;
- individual and mean distances between precipitates;
- size of phases (mainly through depth profiles); and
- total volume fraction of phases through total ion intensities and RSFs.

Application for Materials Research. 3D-analysis has proved to be particularly useful in the study of the influence of trace elements on material properties.[4–10] This aspect is demonstrated here by two examples:

(i) Characterization of nucleation centers for CVD diamond deposition.[8,9]

The influence of substrate surface preparation on diamond nucleation is a major research topic when diamond films are produced by chemical vapor deposition (CVD). The chemical nature of the substrate, the polished material, its grain size, and the resulting surface roughness all influence diamond nucleation and film growth. In this context the analytical task is to characterize potential polishing residues on the surface of a substrate and their reaction with the substrate at elevated temperatures (*ca.* 900 °C) occurring during diamond deposition. Of special interest as polishing material is diamond itself, which could form nucleation sites for CVD-diamond.

3D-SIMS has been successfully applied to prove that residues of diamond are present at the surface of a substrate after polishing (Figure 1). Their area density (typically 20 to 30 mm^{-2}) can be determined as well as their size from the retrospective depth profiles extracted from the 3D information (typically 0.1–0.7 µm). Due to their low number and small size a characterization of these residues was not possible by other techniques. Only 3D-SIMS provided the necessary detection power and spatial resolution to detect and characterize the diamond residues by number and size.

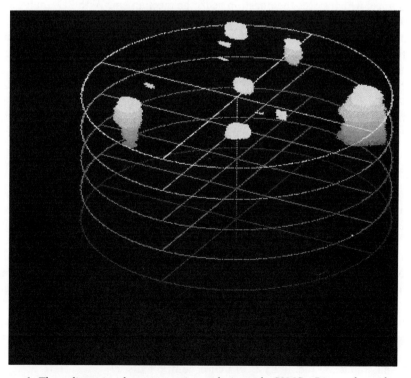

Figure 1 *Three-dimensional stereometric analysis with SIMS. Diamond residues on molybdenum after polishing with 1 µm powder. Images of atomic carbon signal*

The investigation of the reaction of diamond with the substrate during heating is a challenging task, because carbon in two different chemical forms, namely diamond and carbide have to be distinguished. This was achieved through an elaborate measurement and evaluation procedure, which is based on the measurement of cluster ions (C_6 for diamond and MoC for the molybdenum carbide), on the evaluation of their local intensities taking into account the size of the particles detected and a spatial correlation of these ion micrographs shown in Figure 2a and 2b.

Figure 2a shows the SIMS micrographs of a molybdenum surface after polishing with 1 μm diamond powder and heat treatment (1000 °C for 3 h: (a) C_6-cluster ions from diamond residues, (b) MoC-cluster ions and contours of the locations of the C_6-clusters in Figure 2b. These results show, that on molybdenum substrates diamond residues were partially dissolved during the heat treatment, but that a fraction of the diamond residue was still present after heat treatment and can provide nucleation sites for diamond deposition.

(ii) Characterization of the influence of trace elements on the corrosion properties of materials.

High purity chromium is a corrosion resistant material used in the chemical industry. Impurities in the μg g^{-1}-range drastically reduce the corrosion resistance, which poses the question as to the mechanism. With 3D-analysis the distribution of significant trace elements like Na, K, Mg, and O can be characterized. Figure 3 shows a comparison of the distribution of sodium in a material of reduced corrosion resistance (a) and in one of high quality (b). In the material of poor quality this element, whose bulk concentration has been determined to be less than 5 μg g^{-1}, is enriched at the grain boundary creating a local electrochemical potential and thus enabling the corrosive attack. In the high quality material an enrichment at a grain boundary cannot be observed. Sodium in this case forms local nanoenrichments distributed statistically across the bulk, thus causing no harm.

Image Combination and Correlation. A major effort in the further development of 3D-analysis is directed towards the combination of images obtained with different methods and consequently different information content to overcome the inherent limitations of the individual technique. A combination of high synergistic power is SIMS plus Electron Probe Micro Analysis (EPMA), since EPMA is the most important method for quantitative microanalysis and high resolution microstructural imaging (BSE-, SE-micrographs) of solid specimens, but suffers from a rather poor detection power (typically 0.01–0.1% within the analysed volume). SIMS, on the other hand, has an extremely large detection power (ng g^{-1}–μg g^{-1}), but suffers from a lower spatial resolution in images and from problems in quantification for heterogeneous materials and is therefore not well suited for the analysis of major components.

The combination of microstructural and elemental images obtained with the two techniques on one specimen must be based on a spatial fitting procedure

(a)

(b)

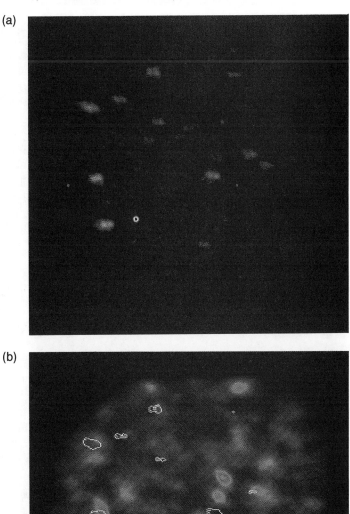

Figure 2 *SIMS micrographs of a molybdenum surface after polishing with* 1 µm *diamond powder for the specific detection of diamond residues and carbide formation;* (a) C_6-*cluster ions from diamond residues,* (b) *MoC-cluster ions indicating the formation of carbide phases from diamond particles under deposition conditions of CVD-diamond coatings (Steiner et al.[9])*

(a)

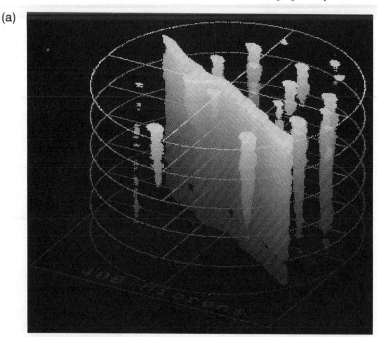

(b)

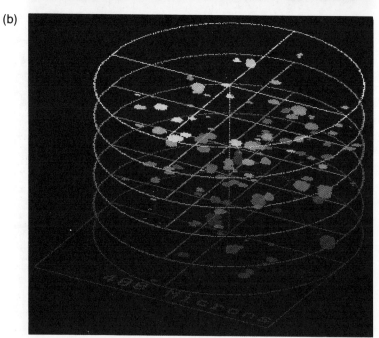

Figure 3 *Three-dimensional stereometric analysis. Distribution of sodium (bulk concentration less than $5\,\mu g\,g^{-1}$) in high purity chromium obtained with ion microscopy in combination with image processing. (a) Above: material with reduced corrosion resistance; (b) Below: high quality material. The total sputtered depth is $7\,\mu m$ (Wilhartitz et al.[6])*

in order to match the positions analysed on the same specimen, but sequentially and with different orientation and position of the actually analysed area. Characteristic features of the sample seen with both techniques are selected to fit the images. Due to the fact that all images are inherently distorted, non-linear algorithms based on weighted least-square-fitting are applied in a two dimensional mode.[10,11] The distribution of chromium in a technical solder obtained with EPMA and with SIMS is shown in Figure 4. The micrograph obtained with EPMA can be quantified with high accuracy, while the SIMS images allow study of the distribution of light elements, trace constituents, or nanophases. This latter task can be achieved either through measurements performed with high resolution scanning instruments using liquid metal ion sources (resolution limit *ca.* 100 nm) or through spatial correlation and classification of low resolution (*ca.* 1 µm) secondary ion micrographs obtained in the ion microscope mode.

Image Classification for Phase Identification. Phase identification is based on the measurement of atomic and cluster ions which under certain conditions represent the original nearest-neighbour relationship of the elements in the material. By correlation of the ion images of the different elements through image classification techniques based on the use of scatter diagrams and Principal Component Analysis, the spatial correlation of elements can be detected and determined in an objective unbiased manner.[10,12]

The first step is the calculation of a scatter diagram showing the positions of the measured secondary ions in the two images compared in the *x*- and *y*-axes with a colour encoded intensity scale in the *z*-axis. If the two images are identical the scatter diagram would yield a 45° straight line. If the images are only partially correlated, they appear in other aeas of the scatter diagram, *e.g.* the boron phase lies in the field with a high intensity for boron secondary ions and a low or negligible intensity for the other component. The secondary ion distributions corresponding to a certain chemical phase can be displayed automatically after selection of an area in the histogram. Different areas correspond to different phases, which then can be represented in the classified image in different colours. The detection and two-dimensional representation of chromium borides, other boron containing phases, and pure chromium phases in the solder discussed in the previous paragraph are shown in Figure 5.

Principal component analysis is a transformation of these scatter diagrams with the goal to determine the number of images that are linearly independent. Chemical features evolve in a more pronounced manner than in the simple image classification described above. By combination of *n* images (representing different elements) in *n*-dimensional space, all possible correlations are established and the existing phases automatically classified. New phases not suspected before can now be detected. The example in Figure 6 is based on PCA of 12 images of different atomic or cluster ions taken from the solder yielding 12 score images, which are ordered by decreasing grey level variance corresponding to the decreasing information content. Figure 6 shows the first three principal components employing a minimum distance algorithm and the spatial phase

SIMS Cr   EPMA Cr

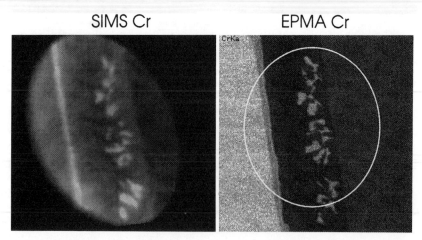

Figure 4 *Spacial correlation of images obtained with different techniques: mathematically fitted micrographs of the chromium distribution obtained with EPMA and SIMS in a soldering layer system consisting of chromium/solder (Ni-Fe-Cr-Se-B)/steel (Nikolov[11])*

distribution obtained from these three components. The major chemical phases are identified and their distribution is shown: in this case the chromium and chromium boride phase, then a nickel boride and iron boride phase (which are the 'other boron containing phases' of Figure 5), and nickel silicide phase.

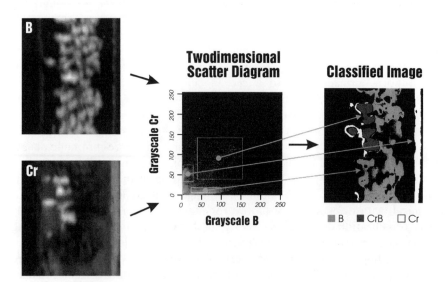

Figure 5 *Chemical correlation of secondary ion micrographs by means of image classification: ion micrographs of Cr and B, the scatter diagram indicating the correlation of these elements, and the spatial distribution of the chemical entities (phases) found (Latkoczy[12])*

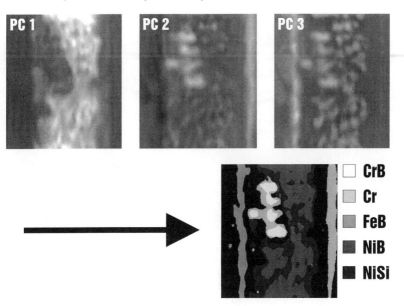

Figure 6 *Chemical correlation of secondary ion micrographs by means of Principal Component Analysis (PCA); top: the first 3 principal components, bottom: spatial distribution of the phases found (Latkoczy[12])*

Maximum Entropy Methods for Image Processing. In any analytical imaging technique an inherent permanent loss of information occurs due to the limited lateral resolution of the method. For SIMS the lateral resolution is either determined by the beam diameter (for scanning probes) or by the stigmatic aberrations of the lenses of the ion microscope. The fundamental question arises whether it is possible to recover part of the information loss through processing of the images.

A promising method for this purpose is the deconvolution of micrographs applying the Maximum Entropy Method (MEM) developed by Gull and Skilling,[13,14] which is used to reconstruct incomplete and noisy data in radio astronomy and technological imaging. The MEM occupies a special position among signal processing techniques, because it provides the statistically most probable solution for all kinds of data, which are positive and additive. It can in principle be applied to spectra, depth profiles, and images.

The principle of MEM is the following: an assumed 'true' distribution of an element is used as a hypothesis. It is convoluted with a response function of the instrument (causing the loss of lateral resolution) for maximum correspondence (ideally identity) with the measured ion micrograph. If the differences between the assumed and measured micrograph are smaller than the standard deviation of the PIXEL intensities the hypothetical distribution has the highest probability of trueness. This hypothesis fulfils the principle of maximum entropy because it contains the lowest number of possible structures (minimal information), which are necessary to explain the experimental data set.

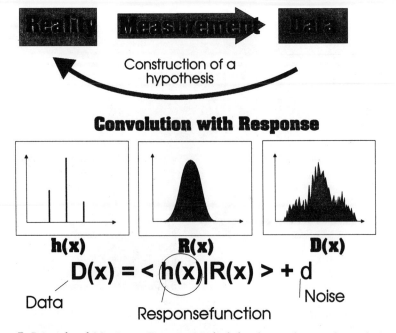

Convolution with Response

$$D(x) = < h(x)|R(x) > + d$$

Data   Responsefunction   Noise

Figure 7 *Principle of Maximum Entropy Method for deconvolution of signals (Ritter, Hutter, and Grasserbauer[15])*

The principle can best be demonstrated with the example of the deconvolution of mass spectra (Figure 7). The line shape in a SIMS spectrum is a convolution of the theoretical line shape and the instrumental response. So the measured signal $D(x)$ consists of an original undisturbed signal ('hypothesis' $h(x)$), which is falsified by an instrumental function (response $r(x)$). In addition, overlapping peaks exhibit statistical fluctuations of intensity ('noise'). In the case of registration of low secondary ion intensities due to a low concentration of the element of interest, the signal is characterized by Poisson statistics of small integer values.

MEM enables the inference of a physically 'real' hypothesis back from measured data. This results in practice in a spectral deconvolution and a mathematical increase of mass resolution. For example Figure 8 shows the secondary ion mass spectrum of a phosphorus doped silicon sample.[15] At mass 31 the spectrum measured with a mass resolution of 800 is broad and shows virtually no structure. We do know however that the spectrum actually consists of two overlapping peaks (^{31}P at $m = 30.97376$ and ^{30}SiH at $m = 30.98158$).

The measured spectrum can be deconvoluted with an appropriate response function obtained from measurement of a mass peak, which is not disturbed by an interfering species (in our case ^{28}Si) yielding the original undisturbed spectra ('hypothesis'). The interfering peaks are well separated and can be evaluated in a quantitative manner. For the problem of analysis of phosphorus in silicon MEM allows to increase the mass resolution of SIMS by a factor of 20. This means that low resolution spectra or profiles can be measured and

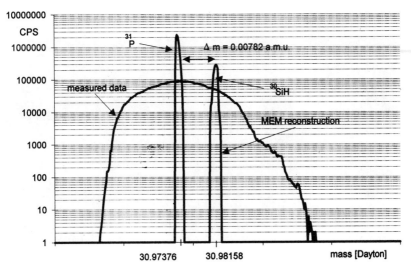

Figure 8 *Maximum Entropy Method for the deconvolution of mass spectra. Measured low resolution (M ΔM = 800) mass spectrum of a phosphorus doped silicon sample at mass 31 and calculated high resolution (M ΔM = 10 000) spectrum exhibiting a clear separation of the interfering ions ^{31}P and ^{30}SiH (Ritter, Hutter, and Grasserbauer[15])*

then deconvoluted. This procedure has the advantage that the high transmission of the mass spectrometer at a low mass resolution (typically 800) is conserved and the intensity loss for measurement at the necessary resolution of 4000 to separate the interfering peaks, which is typically a factor of 50, is not encountered. Consequently, lower detection limits can be obtained.

For deconvolution of images the response function of the instrument is the Gaussian profile of the beam or of the distortion of a PIXEL due to the stimatic imaging aberrations. The main advantage of MEM for image processing is the potential to deconvolute noisy data. Since the signal intensities in the individual PIXELs are low (less than 256 counts, often only a few counts), MEM offers probably the best potential for this purpose. An increase of the lateral resolution of ion microscopic images by at least a factor of three is obtained as it is shown in Figure 9. The main disadvantage of MEM is the rather time consuming mathematical processing, which makes the use of a workstation adviseable. The basic potential for enhancement of the spatial resolution can be used advantageously for the study of phenomena occurring in small (sub-micrometer) dimensions, *e.g.* segregation of trace elements to grain boundaries in polycrystalline materials.

Limitations of SIMS. Although SIMS has evolved as one of the most powerful techniques for micro and surface analysis during the last decade it exhibits severe limitations with respect to many analytical problems in material science. The major limitations are:

SIMS Micrographs of Cr

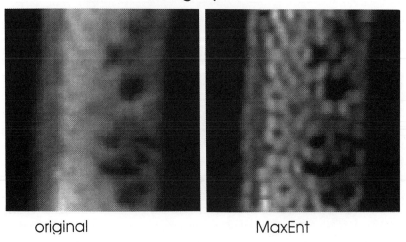

original   MaxEnt

Figure 9 *Enhancement of spatial resolution of SIMS micrographs with Maximum Entropy Method. Left: measured image (chromium distribution in solder), right: deconvoluted image exhibiting a substantially improved lateral resolution (Ritter, Hutter, and Grasserbauer[15])*

- spatial resolution: *ca.* 10 nm in depth, lateral 1 μm for microscopy, and *ca.* 100 nm for scanning SIMS using liquid metal ion guns;
- disturbance of atom positions by the primary ion impact (sputtering artefacts); and
- difficult quantitation.

These limitations can largely be overcome by development of a technique, which uses a gentle process for the generation of analytical ions and which exhibits maximal lateral resolution. This technique is based on the combination of field desorption of species from a sample as ions, their positional imaging using the principle of field ion microscopy, and their mass spectrometric identification and quantitative registration. It is called 'Atom Probe' or more appropriately 'Field Ion Mass Spectrometry'.

Three-dimensional Nanoanalysis with Field Ion Mass Spectrometry

The field ion mass spectrometer (atom probe) is a combination of the field ion microscope and a time-of-flight mass spectrometer with a single ion detection sensitivity. In the last few years, a new class of atom probe instruments has been developed which combines this single atom sensitivity mass spectrometry with position sensing. The combination of mass and position information allows the original chemical variations present in a material to be reconstructed in three dimensions on a nearly atomic scale. The first instrument of this type was the position sensitive atom probe (POSAP) developed at Oxford.[16] Other instruments of this kind are currently in use like the optical atom probe (OAP) by M. Miller[17] or the tomographic atom probe developed by Blavette *et al.*[18] Following the concept of the optical atom probe, a version of a three dimensional atom probe has been developed by M. Leisch.[19]

A schematic diagram of our three-dimensional optical atom probe is shown in Figure 10. The sample, shaped to a very fine needle, is mounted on a finger, cooled by liquid nitrogen. Structural features on atomic scale can be observed on the channel-plate screen (field ion image), when the surface is imaged with field ionized noble gas ions (*e.g.* He or Ne). Analysis and chemical imaging is performed by controlled field desorption of individual surface atoms by means of a nanosecond high voltage pulse (without imaging gas, in UHV). The mass to charge ratio is calculated from the time-of-flight from the tip to the channel-plate and allows chemical indentification of the desorbed species. The time measurement is performed by an eight channel time-to-digital converter. The short light pulse, generated when an ion strikes the detector, is picked up by the CD video camera and defines the x and y coordinates. In the present configuration (flight path 210 mm) usually an area of about 10 to 20 nm in diameter (depending on tip curvature) is located within the acceptance angle of the instrument.

Since the ionization efficiency for all species on kink sites on the surface is 100% (no matrix effects) quantitation is obtained simply by counting the ions detected. To get an error free result it is necessary that not more than one ion strikes the detector for each desorption pulse. By removal of atomic layer by atomic layer the material can be analysed in depth. The sample is kept at low temperature, surface diffusion is minimized, and no disturbance of the atoms occurs when surface species are field desorbed, contrary to sputtering as in SIMS. From the data sets: mass to charge ratio, x, y position, and n atomic layer a 3D reconstruction of the material analysed can be obtained on nearly atomic level.

To illustrate new possibilities for application of this sensitive 3D analytical technique an analysis on nanoprecipitates in a high speed steel sample (bulk composition (in wt.%): 1.3% C, 3.8% Cr, 4.3% Mo, 5.2% W, and 0.8% N) is presented.

The 3D reconstruction, depicted in Figure 11a represents a volume analysed of 12 nm diameter and 10 nm in depth. For clarity only certain added elements and their carbides and nitrides are represented in the CD plot. The existence of several small precipitates can be observed. The precipitates predominantly contain chromium, vanadium, and their nitrides and carbides (grey). Additionally molybdenum (dark grey) is found embedded or in close proximity to these clusters. Tungsten atoms (black) are more randomly distributed. This particular sample had been heat treated in nitrogen, atomic nitrogen found is presented in light grey.

At this point it should be mentioned that surface atoms are field desorbed mainly as doubly or triply charged ions. For a clear separation of different isotopes high mass resolution is needed. The mass resolution of this wide angle instrument is limited by the energy spread of HV-pulse desorbed ions (about $100 \, M \, \Delta M^{-1}$). Vanadium (51 amu) lies within the isotopes of Cr (50–54 amu).

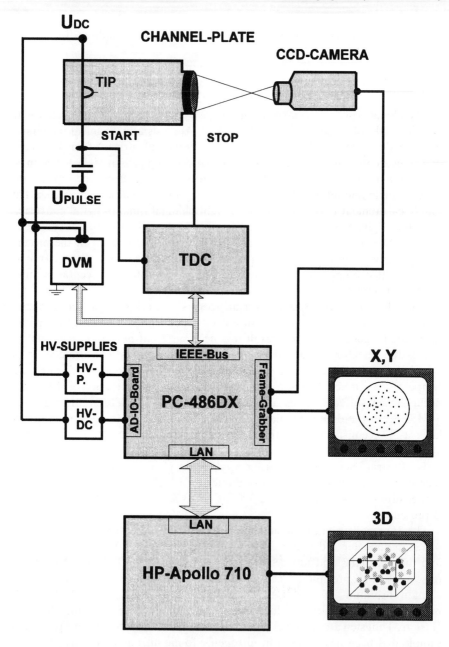

Figure 10 *Schematic diagram of the 3D optical atom probe. Area of analysis typically 12 to 15 nm in diameter. Time of flight measurement by a 8 channel time to digital converter (mass of ion). CCD-camera for position sensing. PC controls the instrument. 3D reconstruction is performed in the graphics-workstation (HP) connected (Leisch[19])*

(a)

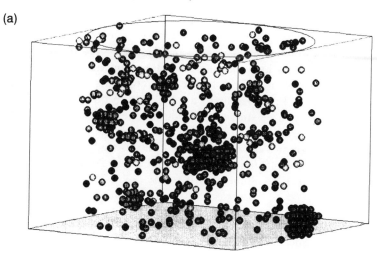

(b)

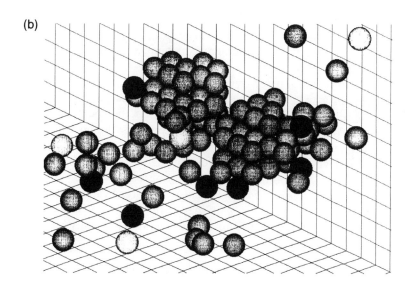

Figure 11 (a) *Three-dimensional reconstruction of a small volume of a high speed steel (diameter of analysed cylinder = 12 nm, 10 nm in depth) containing several small precipitates with near spherical shape. Cr, V, Cr—X, V—X (X = C,N) are represented as grey balls, molybdenum in dark grey, tungsten in black, and atomic nitrogen in light grey. Other elements added as well as Fe are not represented for clarity (Leisch[19]); (b) Three-dimensional reconstruction of a precipitate in a high speed steel sample with high magnification (plotted in a simple cubic lattice), represents the ultimate frontier in 3D-nanoanalysis. Again Cr, V, Cr—X, V—X (X = C,N) are represented as grey balls, molybdenum in dark grey, tungsten in black, and atomic nitrogen in light grey. Other elements added as well as Fe are not represented for clarity*

For a better separation of these elements located in the precipitates, complementary random depth profiling analyses in an energy compensated instrument, as carried out by Fischmeister, Karagöz, and Andren,[20] are required.

Figure 11b shows a high magnification 3D reconstruction of an individual Cr—X, V—X (X = N and C) precipitate that represents the ultimate frontier in 3D nanoanalysis and physical limits attainable with this method. Clusters and small precipitates can be studied on an atomic level, analysed quantitatively, and important microstructural parameters like average size and mean distance between the precipitates can be determined. In this study, about 60 precipitates with an average size of 1 nm in diameter and about 20 precipitates with 2 nm in size were found in a volume of $10\,000\,nm^3$.

Limitations of 3D-FIMS. As one of the most powerful techniques for 3D analysis in a nanometer scale it exhibits limitations in respect of many analytical problems in material science. The major limitations are:

- sample preparation: preparation of specimen tips with radii below 100 nm from certain materials is extremely difficult or time consuming (ion milling);
- electrical conductivity: high voltage pulse desorption works only on electrically conductive materials, semiconductors are accessible only with laser pulse desorption;
- operating conditions: analyses have to be carried out in ultra high vacuum, problems with materials with high vapour pressure;
- mass resolution: mass resolution limited by the energy spread of high voltage pulse desorbed ions, no energy compensation in a wide angle type instrument;
- analysed volume: only small volumes can be analysed into depth (tip radius becomes too large after desorption of 300 to 500 atomic layers), many samples necessary to get statistically significant results; and
- measuring time: restriction of desorption rate (maximum 1 ion per desorption pulse) leads to long measuring times (typically 6 h for a volume of $1000\,nm^3$).

Affected by the low mass resolution of the imaging instrument the quantitative separation of elements with an overlap in the isotope distribution represents the major limitation in practical use. But this can be overcome by complementary random depth profiling analysis in an energy compensated atom probe.

In-situ Atomic Resolution Analysis of Surfaces (Chemical Nanoscopy)

Atomic resolution imaging of surfaces in real space has been made possible by the invention of the scanning tunnelling microscope by Binnig and Rohrer.[21] STM has become a major technique in many physics laboratories and has produced amazing and astonishing results, *e.g.* the observation of nucleation of metal films on a semiconductor substrate, a process of crucial importance in the production of microelectronic devices. Now about 1000 scientific publications per year document the large potential of the STM for science and technology.

For the chemist the second invention by Binnig, the development of Atomic Force Microscopy (AFM), made together with Gerber and Quate at Stanford University[22] might be even more important. AFM measures the attractive and repulsive forces between a fine tip and the surface atoms of a material. AFM images—in many cases with atomic resolution—have been obtained for a variety of metals, semiconductors, and insulators including, *e.g.* mica, graphite, multi-quantum-well structures, halides, and carbonates.[23-30]

Operation Principle of AFM. In the AFM a sharp tip mounted on a soft lever is scanned across the sample surface by means of piezoelectric translators, while the tip is in contact with the surface. The force acting on the tip changes according to the sample topography resulting in a varying deflection of the lever, which in commercial instruments is measured by means of laser beam deflection off a micro fabricated cantilever and subsequent detection with a double segment photo diode. Today cantilevers with extremely low force constants of less than 0.1 N m^{-1} and resonance frequencies of more than 100 kHz can be produced, which allow imaging with forces in the nN and sub-nN range. These forces are about 10 000 to 100 000 times lower than the force of gravity introduced by a fly (1 mg) sitting on a surface.

Figure 12 shows a SEM micrograph of a silicon nitride cantilever with the integrated pyramidal tip (base: $4 \mu\text{m} \times 4 \mu\text{m}$, height $4 \mu\text{m}$), which is commonly used with commercial instruments. The nominal radius of curvature is 20 to 50 nm. In the ideal case the tip would be terminated by a single atom.

Analytical Information. The basic information provided by the AFM is that of surface topography. The following properties, make the AFM a useful tool for the analytical chemist:

- imaging can be performed from the 100 µm range down to the atomic scale;
- the images contain direct depth information;
- *in-situ* measurements under liquids allow investigation of, *e.g.* dissolution, crystallization, electrochemical processes, and other surface reactions; and
- both conductors and insulators can be studied without tedious sample preparation (no coating is necessary).

Examples. Figure 13 shows a SEM and an AFM image of an IR-transparent fibre used for biosensors. While the surface appears smooth in the SEM image, AFM reveals a surface topography with a root mean square (rms)-roughness of 3 nm.

An example for the technical significance of surface topography is, *e.g.* electroluminescence displays (ELSs). In these devices light emission strongly depends on surface roughness and grain structure. Figure 14 demonstrates the influence of the deposition temperature on the roughness of a SrS–EL film

Figure 12 *Atomic Force Microscopy: SEM micrograph of an AFM silicon nitride tip. The pyramidal tip (base: 4 µm × 4 µm, height: 4 µm) is integrated in the cantilever. The nominal radius of curvature of the apex is 20 to 50 nm. In the ideal case the tip is terminated by a single atom*

(thickness: 500 nm) prepared by means of atomic layer epitaxy (ALE) from organic Sr-compounds and H_2S. Deposition at 300 °C leads to a layer with a rms-roughness of 15 nm and with a grain size in the range of 50 to 100 nm, while a temperature of 400 °C leads to a coarser structure with a rms-roughness of 33 nm and a grain size of 150 to 200 nm. This kind of information is useful for the optimization of the deposition process.

Figure 15 shows an example for the *in*-situ imaging capabilities under liquids. Here a sensitive hydrated glass sample could be imaged readily under paraffin after polishing. The products of first corrosion reactions are visible especially along the polishing riffles. The analytical potential of imaging sensitive surfaces under protective liquids are currently being studied for a number of other systems. Another advantage of AFM experiments performed under liquid is the reduction of the force acting on the sample surface by a factor of approximately 100 (from *ca.* 100 nN to 1 nN in most cases). This is especially important for applications on soft systems like biomaterials and other systems that can be destroyed by high forces (*e.g.* molecular resolution imaging of domains in Langmuir–Blodgett films).[31]

AFM is not only *the* topographical probe for non-conducting specimens, but offers another fascinating potential: due to the fact that no electrical currents

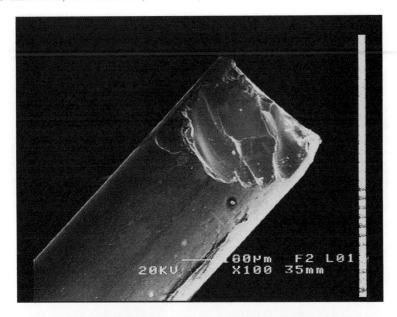

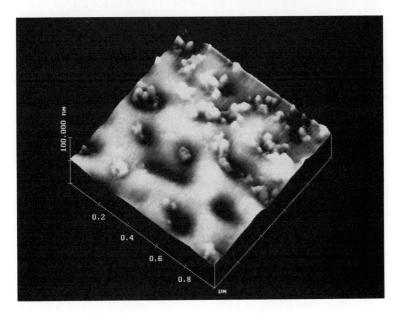

Figure 13 *High resolution topographical imaging of surfaces with AFM. Micrograph of the surface of an IR-transparent fibre showing a roughness of* 3 nm *(bottom) in comparison to a SEM image of the same fibre (top). AFM image size:* 1 μm × 1 μm, *depth scale:* 100 nm *from black to white*

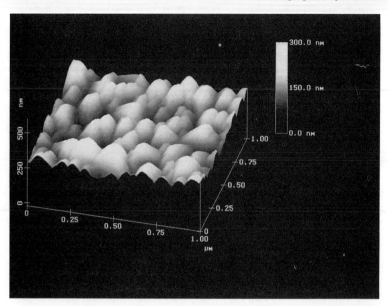

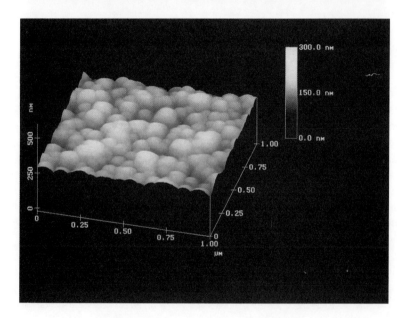

Figure 14 *Surface roughness and grain structure of SrS−EL-layers produced with atomic layer epitaxy (film thickness = 500 nm). Top: layer deposition at a substrate temperature of 300 °C; rms-roughness = 15 nm, grain size = 50 to 100 nm. Bottom: A deposition temperature of 400 °C leads to a coarser structure of the EL-film; rms-roughness = 33 nm, grain size = 150 to 200 nm. Image size: 1 μm × 1 μm, depth scale: 300 nm from black to white*

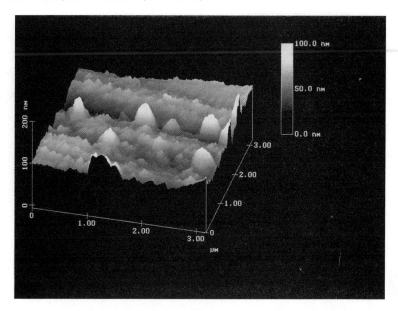

Figure 15 *AFM micrograph of a hydrated glass surface imaged under paraffin. Corrosion products mainly along the polishing riffles can be observed. Image size: 3.3 μm × 3.3 μm, depth scale: 100 nm from black to white*

are needed the *in-situ* observation of surface reactions is facilitated and often possible with atomic resolution. This permits the study of corrosion reactions at grain boundaries, precipitation from solution, as well as chemical surface processes *in-situ* and in real time. AFM will provide a new important tool for the analytical chemist and give access to the new field of chemical nanoscopy.

With AFM the removal of single atomic layers can be observed and described quantitatively.[26,27] In principle it is possible to study such processes with atomic resolution as shown in Figure 16, but the removal of individual atoms and atomic layers is often too fast in relation to the speed of image aquisition. Methods for artifically slowing down surface reactions are presently being developed. At this point it should also be emphasized, that these results on alkaline halide samples show the potential of high resolution analysis of insulators, which allowed determination of atomic spacings with an accuracy of better than 10% (0.56 ± 0.05 nm for the Cl—Cl spacing in NaCl, 0.66 ± 0.06 nm for the Br—Br spacing in KBr).

In addition to the topographical imaging capabilities AFM also bears the potential of obtaining chemical information about surfaces. One option is the force *versus* distance curves, that can be recorded with an AFM.[32] These curves describe the force interaction between the AFM tip and the sample surface, and allow the determination of the 'pull-off-force' acting when the tip is removed from the sample surface. Figure 17 shows a comparison of two force curves, one recorded on a silicon sample covered with its native oxide layer, the other

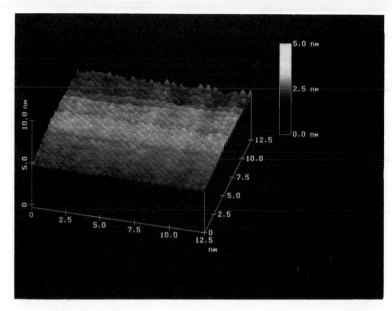

Figure 16 *In-situ atomic resolution analysis of surfaces (chemical nanoscopy). In-situ observation of the reaction of the surface of NaCl with absolute ethanol. Lateral atomic resolution could be achieved with the AFM under these conditions. The line corresponds to atomic steps, which move across the surface with a velocity of approximately 100 nm s^{-1} (Prohaska et al.[27]). Image size: 12.5 nm × 12.5 nm, depth scale: 5 nm from black to white*

one etched with HF (hydrogen terminated). Although no change in topography could be observed (see also Figure 17), for the oxidized surface a pull-off-force of 1.9 nN and for the etched surface a pull-off-force of <0.6 nN has been determined, which shows the influence of the chemical composition of the sample surface on the force–distance curve. Further thorough study in this field will be necessary but these preliminary results look very promising.

3 Conclusion

Imaging Analytical Chemistry is developing at a high pace. For imaging and image processing in surface analysis the demands arising particularly from High Technology are steadily rising, prompting analytical innovations. The new techniques that have been developed often reach out to the very limits of physics, *e.g.* when individual atoms are observed, when single ions are detected, or monolayers on the surface of materials are selectively analysed. Surface analysis is moving more and more into the chemistry laboratory as the study and design of surface chemical processes increases and surface chemical properties of materials become even more important.

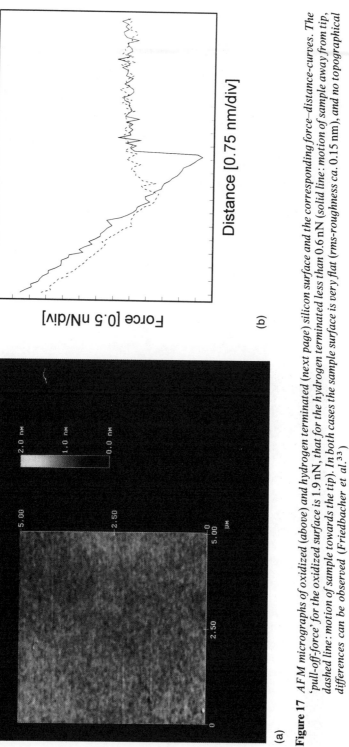

Figure 17 *AFM micrographs of oxidized (above) and hydrogen terminated (next page) silicon surface and the corresponding force–distance-curves. The 'pull-off-force' for the oxidized surface is 1.9 nN, that for the hydrogen terminated less than 0.6 nN (solid line: motion of sample away from tip, dashed line: motion of sample towards the tip). In both cases the sample surface is very flat (rms-roughness ca. 0.15 nm), and no topographical differences can be observed (Friedbacher et al.[33])*

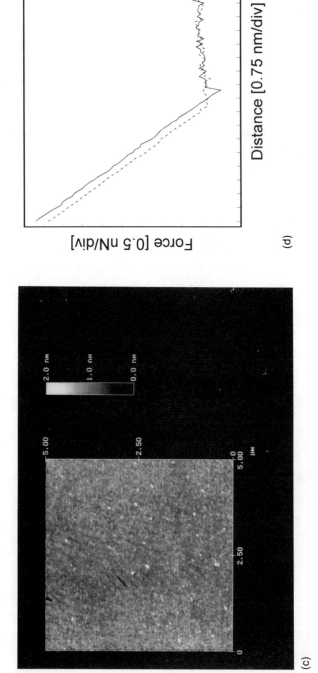

Figure 17 *Continued*

Acknowledgements

Support for the research activities on which this paper is based by the following institutions is gratefully acknowledged: Austrian Scientific Research Council (projects 7495, 9065, S5902, S6205, and S6206), Austrian Industrial Research Council (projects 2/275 and 2/293), Jubilee Funds of the Austrian National Bank and the City of Vienna, Metallwerk Plansee, Reutte, and Wacker Chemitronic, Burghausen. The authors thank Kurt Piplets for his valuable contributions to the development of SIMS-hardware and measurements.

References

1. M. Grasserbauer, G. Friedbacher, H. Hutter, and G Stingeder, *Fresenius' J. Anal. Chem.*, 1993, **346**, 594.
2. H. Hutter and M. Grasserbauer, *Mikrochim. Acta*, 1992, **107**, 137.
3. H. Hutter and M. Grasserbauer, 'Development and Application of a New Imaging System for SIMS', SIMS VIII-Proceedings of the Amsterdam Conference, John Wiley, Chichester, 1992, p. 533.
4. S. Gara, H. Hutter, G. Stingeder, C. Tian, H. Führer, and M. Grasserbauer, *Mikrochim. Acta*, 1992, **107**, 149.
5. M. Grasserbauer, *Pure Appl. Chem.*, 1992, **64**, 485.
6. P. Wilhartitz, G. Leichtfried, H. P. Martinz, H. Hutter, A. Virag, and M. Grasserbauer, 'Application of 3-D SIMS for the Development of Refractory Metal Products', Proceedings of the 2nd European Conference on Advanced Materials and Processes, Cambridge, 1991, eds. T. W. Clyne and P. J. Withers, The Institute of Materials, London, 1992, Vol. 3, 323.
7. P. Wilhartitz, R. Krismer, E. Garber, H. Hutter, M. Grasserbauer, H. M. Ortner, W. F. Müller, and W. Wegscheider, *Int. J. Refract. Hard Mater.*, 1992, **11**, 235.
8. R. Steiner, G. Stingeder, M. Grasserbauer, R. Haubner, and B. Lux, *Diamond Diamond Rel. Mater.*, 1993, **2**, 958.
9. R. Steiner, G. Stingeder, H. Hutter, M. Grasserbauer, R. Haubner, and B. Lux, *Fresenius' J. Anal. Chem.*, in press.
10. H. Hutter and M. Grasserbauer, *Chemometrics for Surface Analysis*, in press.
11. St. Nikolov, Diploma Thesis, Sofia University and Technical University Vienna, 1992.
12. Chr. Latkoczy, Diploma Thesis, Technical University Vienna, 1994.
13. S. F. Gull and G. J. Daniell, *Nature (London)*, 1978, **272**, 686.
14. S. F. Gull and J. Skilling, *IEEE Proc.*, 1980, **131F**, 646.
15. M. Ritter, H. Hutter, and M. Grasserbauer, *Fresenius' J. Anal. Chem.*, 1994, **349**, 186.
16. A Cerezo, T. J. Godfrey, and G. D. W. Smith, *Rev. Sci. Instrum.*, 1988, **59**, 862.
17. M. K. Miller, *Surface Sci.*, 1991, **246**, 428.
18. D. Blavette, A. Bostel, J. M. Sarrau, B. Deconihout, and A. Menand, *Nature (London)*, 1993, **363**, 432.
19. M. Leisch, *Fresenius' J. Anal. Chem.*, 1994, **349**, 102.
20. H. F. Fischmeister, S. Karagöz, and H. O. Andren, *Acta Metall.*, 1988, **36**, 817.
21. G. Binnig, H. Rohrer, Ch. Gerber, and E. Weibel, *Phys. Rev. Lett.*, 1982, **49**, 57.
22. G. Binnig, C. F. Quate, and Ch. Gerber, *Phys. Rev. Lett.*, 1986, **56**, 930.
23. D. Rugar and P. Hansma, *Phys. Today*, 1990, **43**, 23.
24. S. N. Magonov, *Appl. Spectrosc. Rev.*, 1993, **28**, 1.
25. J. Frommer, *Angew. Chem.*, 1992, **104**, 1325.

26. T. Prohaska, Diploma Thesis, Vienna University of Technology, 1992.
27. T. Prohaska, G. Friedbacher, and M. Grasserbauer, *Fresenius' J. Anal. Chem.*, 1994, **349**, 190.
28. G. Friedbacher, D. Schwarzbach, P. K. Hansma, H. Nickel, M. Grasserbauer, and G. Stingeder, *Fresenius' J. Anal. Chem.*, 1993, **345**, 615.
29. G. Friedbacher, P. K. Hansma, D. Schwarzbach, M. Grasserbauer, and H. Nickel, *Anal. Chem.*, 1992, **64**, 1760.
30. G. Friedbacher, P. K. Hansma, E. Ramli, and G. D. Stucky, *Science*, 1991, **253**, 1261.
31. L. M. Eng, H. Fuchs, S. Buchholz, and J. P. Rabe, *Ultramicroscopy*, 1992, **42–44**, 1059.
32. A. L. Weisenhorn, P. Maivald, H.-J. Butt, and P. K. Hansma, *Phys. Rev. B.*, 1992, **45**, 11226.
33. G. Friedbacher *et al.*, *Appl. Surf. Sci.*, submitted.

Neural Networks and Fuzzy Logic for Analytical Spectroscopy

M. Otto

INSTITUTE FOR ANALYTICAL CHEMISTRY, TU BERGAKADEMIE FREIBERG,
LEIPZIGER STRASSE 29, D-09599 FREIBERG, GERMANY

Summary

Applications of neural networks are reviewed with respect to parameter estimation, classification, and clustering in analytical spectroscopy. Examples are given for multicomponent analysis in atomic and molecular spectroscopy, for identifications of spectra as well as for classification and clustering of infra-red or mass spectra. The combination of neural and fuzzy approaches is demonstrated with the Fuzzy Associative Memory for generating rules in infra-red spectra interpretation systems.

The new branch in computer science called **soft computing** is being exploited more and more for data evaluation in Analytical Spectroscopy. Soft computing includes methods, such as, neural networks, genetic algorithms, probabilistic reasoning, and chaos or fuzzy logic. The large amount of data recorded in a spectrum favours knowledge processing methods based on numerical approaches rather than on symbolic procedures. Therefore, statistical multivariate methods are increasingly supplemented by methods based on neural networks and on fuzzy logic.

This paper will mainly concentrate on neural network applications in analytical spectroscopy as well as neuro-fuzzy methods where combinations of fuzzy logic and neural nets have been explored. The principal usage of fuzzy methods in spectroscopy have been described elsewhere.[1]

Neural networks are now studied in most of the chemometrically oriented laboratories. Very often, however, the results are not compared to the performance of existing statistical methods so that the results are judged over-optimistically. In addition one sometimes finds unfair comparisons between neural and statistical methods where, *e.g.* in NIR-spectroscopy, scaling or transformations of input spectra are not allowed as a preliminary to statistical data analysis.

Compared to statistical analysis much improvement has to be made in neural network data evaluation with respect to interpretation aids. The presently dominating 'black box' usage of neural networks will not convince the experienced spectroscopist who applies highly sophisticated tools for judging the quality of calibration or spectra classification.

In principle, neural networks can be used for the same objectives as statistical methods, *i.e.* for parameter estimation, pattern recognition, and cluster formation. In spectroscopy this translates into solving problems in multicomponent analysis or modelling of spectrum–structure relationships, in spectra identification as well as in classification and interpretation of spectra.

The present review will discuss problems for the above objectives. A major aim of the paper is to give hints for extracting more detailed knowledge out of the neural network applications. Finally the link between neural networks and fuzzy methods will illustrate the emerging field of neuro-fuzzy approaches in analytical spectroscopy.

1 Introduction

Basic introductions to neural networks in analytical chemistry are available[2–4] hence it is only necessary to introduce the terminology for discussing the different network approaches in analytical spectroscopy.

At a minimum, a neural net consists of an input layer for presenting the spectrum and a second layer that is used as a competitive layer in case of unsupervised learning or it serves as an output layer for supervised training of categories of spectra or of chemical components. For more complicated mapping problems one or several hidden layers with different numbers of neurons are added (Figure 1). Each neuron receives input from one or several neurons, aggregates this input usually as linearly weighted sum, transforms it into the output signal, and processes the output to one or several neurons (Figure 2).

Mathematically these processes can be summarized as follows:

neuron aggregation

$$\text{Net}_j = \sum_{i=1}^{n} x_i w_{ij}$$

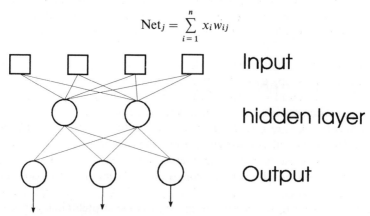

Figure 1 *Neural Network architecture exemplified with the multi-layer perception*

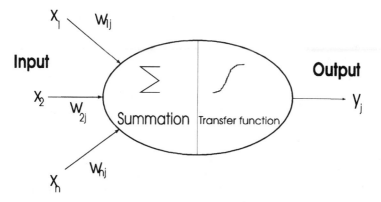

Figure 2 *Model of an artifical neuron representing the input x, the weights w, and the output y*

and signal transformation and output

$$y_j = f(\text{Net}_j)$$

where Net is net input signal; x_i is input i; w_{ij} is synaptic weight between input i and output j, and y_j is output j.

Learning laws are based on either associative learning, such as the Hebbian correlation learning, or on competitive learning schemes which will be demonstrated later, in connection with spectra clustering.

The most frequently applied network models,[5] *i.e.* the Hopfield and Hamming net, the Bidirectional Associative Memory (BAM), the Backpropagation and Counterpropagation network, the Kohonen network, and more recently Grossberg's binary version of Adaptive Resonance Theory networks (ART1) have been explored also in analytical spectroscopy.

The following spectral methods have been studied in connection with neural networks: optical atomic emission spectrometry;[6,7] UV-molecular spectroscopy;[8-10] NIR-,[11-15] IR-,[16-23] ^{1}H-NMR-,[24] and ^{13}C-NMR-[25,26] spectroscopy; mass spectrometry;[27,28] X-ray fluorescence spectroscopy;[29] and molecular 2D-fluorescence spectroscopy.[30] The main objectives of these studies will be detailed in the following sections.

2 Multicomponent Analysis and Modelling of Spectrum–Structure Relationships

Multicomponent analysis is based on calibrating spectra for component concentrations by means of linear algebra tools, such as Multiple Linear Regression (MLR), Principal Component Regression (PCR), Partial Least Squares Analysis (PLS), or the method of Alternating Conditional Expectations (ACE). The estimated calibration parameters correspond to the trained weights in the neural network approach.

As long as the relationship between original or transformed signals and concentrations can be described by a linear model the network approach does not give any significant advantage over the existing statistical methods. Linear models are those for all signal–concentration-dependences that can be modelled by means of linear algebra, *i.e.* curved dependences can also be treated if the input signals and/or output concentrations are properly transformed preliminary to the network training. In addition, by using multiwavelength spectroscopy non-linear signal–concentration-dependences observed in plots of signals recorded at one wavelength can often be described by a linear model if additional wavelengths are chosen (Figure 3). Therefore, to find intrinsic non-linear signal–concentration-dependences in analytical spectroscopy is more difficult than is stated.

The first example studied in our laboratory was concerned with the calibration using the overlapping lines of arsenic and cadmium in ICP-Atomic Emission Spectroscopy.[6,7] It was shown that with a minimum network consisting of the input and output layer connected by one layer of synaptic weights, by using the backpropagation algorithm, and by applying linear signal transfer functions the weights were (within computational noise) identical to the statistically estimated regression parameters, *e.g.* by using MLR based on singular value decomposition.[7]

If the component concentrations are assigned to the output neurons, y, and the calibration spectra to the input neurons, x, the weights correspond to regression parameters as estimated in inverse calibration models:

$$y = Wx$$

If it is required to relate these weights to the original spectra it can be shown that they are identical to the pseudo-inverse of the signals of the individual components, k_{ij}, *i.e.*[7]

$$y = K^+ x$$

In principal, the linear model approach is also applicable for other spectroscopic calibrations as, for example, for NIR-spectroscopy. If the spectroscopic data are properly pretreated by the most suitable transformation and scaling procedures significant improvements will not be found with neural network

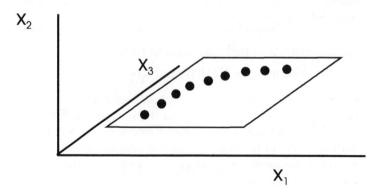

Figure 3 *Modelling of a dependence that is curved with respect to variables x_1 and x_2 but linear in three-dimensional space*

calculations. If it is difficult, too laborious, or because of lack of qualification of the operator impossible to find the best data preprocessing procedure, neural networks can be used to calibrate and predict concentrations in a comparable precise and robust manner as is feasible with statistical methods and as has been demonstrated in the NIR-range, *e.g.* for analysing fat in meat.[12] The special architecture of the networks used is based on direct connections between input and output neurons in order to estimate the linear part of the data. As transfer function a linear or linear threshold function is applied. In one hidden layer a minimum of neurons, often only a single neuron, is arranged with a sigmoid transfer function for modelling non-linear parts of the data. Apart from presentation of the original spectral data to the net orthogonalized data obtained from principal component analysis can be advantageously input to the network.

Pronounced curvature in signal–concentration-dependence has been found in remission spectroscopy of powders in the UV-range[10] to establish a quality control procedure for direct analysis of headache tablets. The UV-spectra recorded on the solid tablet were used for multicomponent analysis of the pharmaceutical substances and also for the tablet matrix. The UV-spectra of caffeine in mainly cellulose as a support material are given in Figure 4 and used for calibration and prediction of the two components. As shown by the data

Absorbance (apparent)

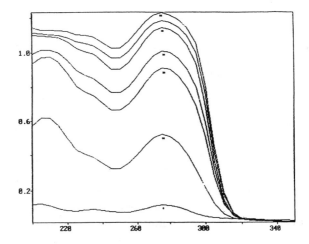

Wavelength, nm

Figure 4 *Remission spectra of caffeine tablets recorded with a M500 Zeiss two-channel UV/VIS-spectrometer*

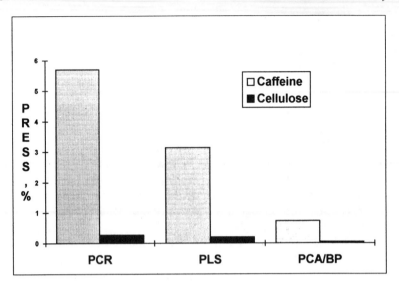

PRESS (Predictive Residual Error Sum of Squares):

$$\% \, PRESS \; = \; \sqrt{\frac{\sum\limits_{i=1}^{m} (c_i - c_{actual})^2}{\sum\limits_{i=1}^{m} c_{actual}^2}} * 100$$

Figure 5 *Predictions for 50.0 mg caffeine tablets based on the calibration methods PCR, PLS, and a neural network with PCA scores as input and backpropagation as learning method (PCA/BP)*

in Figure 5 the classical PCR or PLS analysis with the untransformed data gave obviously worse prediction results than those from the neural network based on PCA scores input data and the backpropagation learning algorithm (PCA/BP).

If calibration can be based on a variety of calibration standards a different network architecture, *e.g.* the counterpropagation network introduced by Hecht–Nielsen[31] may be useful. This network consists of the input layer for the spectra, the competitive Kohonen-layer, and the output component concentration layer designed as Grossberg outstar neurons (Figure 6).

In the Kohonen-layer the input calibration spectra are clustered by a competitive learning algorithm as known from the Kohonen network. The process is to find the winning neuron j by comparing the input vector at time t, $\mathbf{x}(t)$, with the corresponding weight vector $\mathbf{w}_j(t)$ by using a distance measure

A. Calibration standards

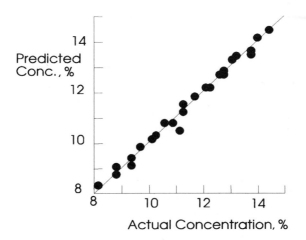

B. Counterpropagation Network

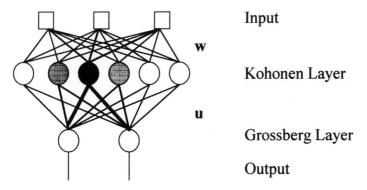

Input

W

Kohonen Layer

u

Grossberg Layer

Output

Figure 6 *Spectroscopic multicomponent analysis based on many calibration standards (A) using the feedforward only counterpropagation network (B)*

such as the euclidian vector norm:

$$\|\mathbf{w}_j(t) - \mathbf{x}(t)\| = \min_i \|\mathbf{w}_i(t) - \mathbf{x}(t)\|$$

Update the winning neuron $\mathbf{w}_j(t)$;

$$\mathbf{w}_j(t + 1) = \mathbf{w}_j(t) + \eta_t[\mathbf{x}(t) - \mathbf{w}_j(t)]$$
$$\mathbf{w}_i(t + 1) = \mathbf{w}_i(t) \quad \text{for } i \neq j$$

where η_t is the learning coefficient.

After training of the Kohonen weights, the weights are fixed and the weights in the Grossberg layer are now adjusted by means of an associative learning law:

$$u_{ij}(t + 1) = u_{ij}(t) + \eta_t(-u_{ij} + \text{Out}_j)z_i$$

where j is the considered neuron in Grossberg layer; z_i is the activation of Kohonen-neurons; and Out_j is the actual Output in the Grossberg Layer.

If there is only one winning neuron in the Kohonen-layer the value for z is equal to 1 for the winning neuron and 0 for all the remaining neurons. This, however, would mean that in future prediction applications only the calibration samples could be precisely measured. In order to interpolate between concentrations of calibration standards more than one neuron has to be activated in the Kohonen-layer. This activation is usually derived from the reciprocal distance between the weight and input vector of the considered neuron. As a result, the network will be able to recognize exactly all identical or almost identical spectra as used during the calibration process. The performance for estimating intermediate concentrations is comparable to the backpropagation approach.

Other applications of neural networks as parameter estimators are known, e.g. for modelling chemical shifts in ^{13}C-NMR spectroscopy.[26] A fully-connected three-layer network was trained to predict chemical shifts for keto-steroid compounds using 13 descriptors as input information to the network. Because of non-linear data relationships, the network performed approximately twice as well as linear regression analysis.

3 Identification and Classification of Spectra

Neural networks can be used for classification if the output layer represents categories rather than continuous values as in the examples for multicomponent analysis. If each presented spectrum is related to one output category the network can be used for spectra identification. If a spectrum is related to several output categories the network approach serves the purpose of relating structural parts or fragments to the input spectrum as applied in IR- or mass spectroscopy.

The application of neural networks for **identification of spectra** has been described for UV-spectra based on the adaptive BAM[8] as well as for UV-[9] or ^{1}H-NMR spectra[24] by means of the backpropagation multilayer perceptron. In the NMR-example, unfortunately no comparison was made with existing methods of library search. In an exhaustive study for identifying UV-spectra[8] we compared the performance of the correlation coefficient with a multilayer perceptron trained by the backpropagation algorithm. As can be seen from the data given in Figure 7 the advantage with the neural network approach is a higher tolerance to noisy spectra, i.e. if spectra with 1% relative noise have to be identified the network reveals correct classification by 100% compared to the usually used correlation coefficient with 93.3% correctly classified spectra. This improvement is also observed for spectra of still higher noise characteristics (cf. Figure 7).

Contrary to the correlation coefficient a problem with the neural net approach is the identification of spectra not used in the training process. Usually, the

Correlation Coefficient

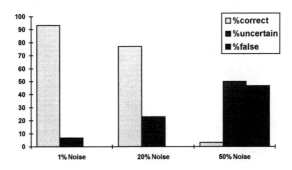

Backpropagation Net

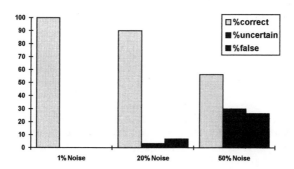

Figure 7 *Comparison of identification of 200 UV-spectra by means of the correlation coefficient and a backpropagation multilayer perceptron*

network classifies them as false positives. Therefore, identification by using neural networks has to be improved in the future by incorporating a novelty detector into the network as known from the ART-design.[5] The novelty detector should detect untrained spectra and open a new category to adapt the novel information to the network weights.

With the backpropagation-network only noise in the intensity axis of the spectrum can be taken into account. An additional problem in UV-spectroscopy, however, is if the spectrum shift depends on the pH-value or the actual solvent or solvent mixture composition. To tolerate such systematic changes the BAM-network was applied.[8] This network can be run for spectra identification as an auto-associative network where the input spectrum is simultaneously presented at the output (Figure 8).

The simple BAM enables binary input–output-patterns to be recognized. In order to account for systematic spectral deviations a BAM with continuous data input is used in its adaptive form—the Adaptive Bidirectional Associative

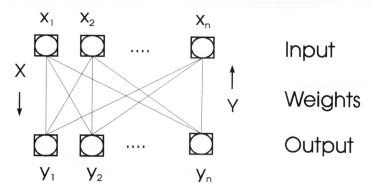

Figure 8 *Bidirectional Associative Memory for Encoding UV-Spectra in an auto-associative network*

Memory (ABAM). Apart from Hebbian learning the weights, w, as well as the activations in the input x, and output, y, are changed dynamically according to the following equations:

Short Term Memory (STM)

$$\dot{x}_i = -x_i + \Sigma_j S(y_j)w_{ij} + I_i$$

$$\dot{y}_j = -y_j + \Sigma_i S(x_i)w_{ij} + J_j$$

where x_i, y_j are passive decay terms, I_i, J_j represent constant external inputs and $S(x_i)$ and $S(y_j)$ stand for signal functions of x and y, respectively.

Long Term Memory (LTM)

$$\dot{w}_{ij} = -w_{ij} + S(x_i)S(y_j)$$

Instead of encoding the intensities wavelength by wavelength the two-dimensional intensity–wavelength plot is digitized (Figure 9A). If the adaptive version of the BAM is used the systematically shifted spectra are trained as fuzzy functions (Figure 9B). This improves recall significantly compared to the above described spectra interpretation approach.

For identification of an unknown spectrum the library spectra are compared with the final output of the network via intersection of the crisply encoded library spectrum with the fuzzy output of the ABAM.[8]

Applications of neural networks for **classification problems** have been intensively studied for the interpretation of IR- and mass-spectra. The first example for IR-spectra interpretation was by Munk *et al.*[16,17] who studied the backpropagation network for classification of 36 functional groups. As input the spectra digitized in discrete intervals are presented and at the output the presence or absence of a functional group is labelled by 1's or 0's as target output. Apart from this pioneering work the results for correct classifications are not good enough to be implemented in interpretation systems for routine use. As long as the network

A.

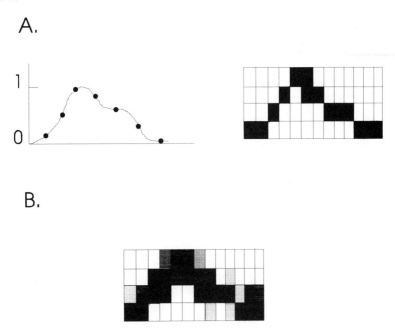

B.

Figure 9 (*A*) *Input to the adaptive BAM wavelength by wavelength* (*left*) *and discretized as spectrum* (*right*). (*B*) *Training of similar spectra as a fuzzy function*

is used for intepretation of unknown spectra that contain similar sets of fragments the classification runs smoothly. However, if an unknown spectrum of a multi-functional compound is presented to the network that could not be trained with examples of such combinations of functional groups classification usually fails. To overcome these problems, in later developments, the training set was expanded with simulated spectra or the spectrum structure relationship is not trained once but by decomposing the problem into specific sub-problems with small, dedicated neural networks.[19] In case of small systems for specific problems some improvement has been shown. Further research will show whether this approach can be extrapolated to more general systems.

The idea to substructure the interpretation network follows from Curry and Rumelhart's MSnet.[27] This network classifies molecular fragments based on mass-spectra. A backpropagation network was designed consisting of 493 input spectral features that were related to 36 top-level classes of structural fragments. The top-level classes are then subdivided into subclasses, *e.g.* for the carboxyl group 22 subclasses are formed. The performance of the network approach was in general superior to the existing expert system STIRS.[27]

Another application of the classification type is that of recognizing spectroscopic peaks in IR-spectroscopy.[21] Here part of the spectrum is presented to the network input and related to one output neuron that activates if eventually a peak has been detected.

4 Formation of Spectral Clusters

Grouping of spectra by neural networks may be used to find relationships between spectra and chemical structures or fragments of chemical structures. Examples have been developed for grouping of mass spectra[28] as well as for IR-spectra clustering.[18,20] The approach is based on unsupervised learning by means of a competitive network. The spectra are presented as inputs to the network and trained *versus* a competitive layer that acts like a vector quantizer.

One example of a competitive layer has been given earlier with the Counterpropagation network (*cf.* Figure 6B). To represent the clusters visually the competitive layer is usually arranged in two-dimensions as introduced in the Kohonen-network.[34] The special feature of the Kohonen-net is its ability to preserve topological neighbourhood in the patterns presented. If a geometrical shape, *e.g.* a cross, is trained this shape can be found on the output Kohonen layer again (Figure 10). To realize this feature the competitive neurons fire in dependence on a neighbourhood function, such as the mexican hat function.

In connection with grouping IR-spectra the neighbourhood feature might have the advantage to depict chaining of subsets of structural fragments.[20] A problem with mapping IR-spectra onto a Kohonen-layer arises from the high number of input neurons. Typically an IR-spectrum is digitized by 1400 points. If a library of about 3000 spectra is to be trained on a relatively fast working station for a 20×20 Kohonen network the estimated training time is approximately 450 days.[20] Therefore, training of the Kohonen-network should be either based on a parallel computing architecture[20] or by input of the Fourier or Hadamard transform coefficients to the network.[18]

Several possibilities have been examined with the unsupervised Kohonen-network spectra clustering method.

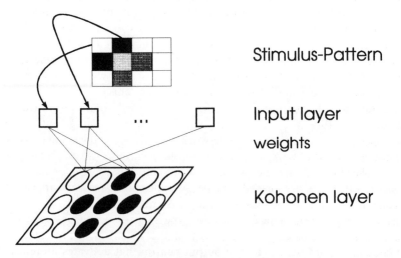

Figure 10 *Kohonen network for mapping a cross as stimulus pattern onto the Kohonen layer*

- Spectra clustering can be found for spectra that contain a certain fragment. Often the fragment clusters fall apart into several clusters.
- A distinction between 'strong' and 'weak' fragments can be made in the sense that narrow and more scattered clusters are found, respectively.
- Interpretation of the weights for a *specific output neuron* reveals typical spectra the network has derived from the training patterns. These typical spectra represent both well known spectral characteristics and additional bands that are important for fragment differentiation.
- The weights related to *one input neuron* represent information about the importance of different spectral regions. If contour maps of those weights are overlayed by the winning neurons in the output layer representing a certain functional group the importance of different spectral regions can be evaluated.[18]

Present developments have culminated in the building of modular neural networks in order to substructure the molecular fragments in a decision-tree hierarchy.[20]

5 Neuro-fuzzy Methods

In connection with fuzzy set theory neural networks are nowadays applied to generate fuzzy rules, to fine-tune membership functions or they can be designed for use in approximated reasoning systems.[1]

Generation of fuzzy rules is feasible by means of Kosko's Fuzzy Associative Memories (FAM).[32] Given the variable sets X and Y. Then, if $A, A_1, A_2, \ldots, A_n$ and $B, B_1, B_2, \ldots, B_n$ are subsets of X and Y, respectively, the general m rules can be derived:

$$\text{IF } X = A_1 \quad \text{THEN } Y = B_1$$
$$\text{IF } X = A_2 \quad \text{THEN } Y = B_2$$

$$\text{IF } X = A_m \quad \text{THEN } Y = B_m$$

Given: $X = A$ reasoning about Y is possible by Zadeh's compositional rule of inference[35] in the usual manner. Representation of the situation by using Kosko's FAM is depicted in Figure 11. The input A passes the different rules and the resulting B's are aggregated and often additionally weighted to form the output B of the FAM.

As an example consider the spectral range between 1750 and 1490 cm^{-1} for interpretation of amine vibrations. The position and intensity of the different bands can be described by fuzzy sets 'low', 'medium', 'high', *etc.* as given in Figure 12. If the substitution type of the amine, *i.e.* a primary, secondary, or tertiary amine, is to be inferred from the spectral bands in the considered region a set of rules of the following form has to be formulated:

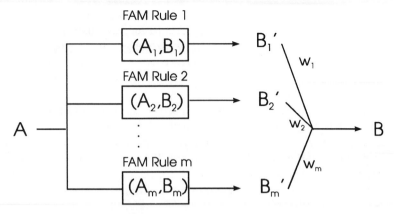

Figure 11 *FAM–Fuzzy Associative Memory. Reasoning about B is performed by input of the fuzzy set A to all rules and aggregating the result by a suitable operator, e.g. the maximum operator*

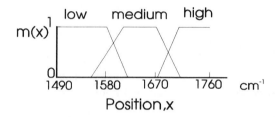

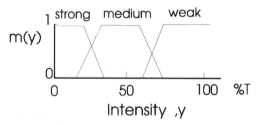

Figure 12 *Representation of the variables position and intensity in IR-spectroscopy by fuzzy sets*

IF Position = *LOW*   AND Intensity = *WEAK* THEN Substituent = *prim. Amin*
IF Position = *LOW*   AND Intensity = *WEAK* THEN Substituent = *sec. Amin*
IF Position = *LOW*   AND Intensity = *WEAK* THEN Substituent = *tert. Amin*
IF Position = *MEDIUM* AND Intensity = *WEAK* THEN Substituent = *prim. Amin*
IF Position = *MEDIUM* AND Intensity = *WEAK* THEN Substituent = *sec. Amin*
⋮

In principle, for the three levels of each position, intensity and substitution type a rule base of 27 individuals is obtained. These rules are interpreted as

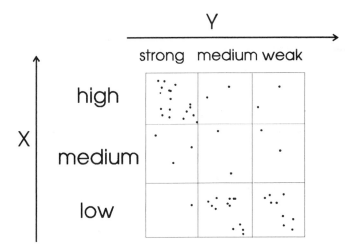

Figure 13 *FAM cells representing the antecedents position, X, and intensity, Y, and depicting trained sample data*

FAM cells as demonstrated in Figure 13. If training data are presented to the FAM the data will be distributed among the different FAM cells. Rules are then derived from data clusters. A frequency distribution approach would be based on representing the sample data in a FAM cell in a histogram and weighting the importance of different FAM cells by the number of sampled data. In order to smooth the data and to handle also dynamic systems adaptive quantization of the training data is performed on the basis of a competitive neural network.[32] The properly digitized input and output data are presented to the network as individual vectors for every sample spectrum with one competitive layer as given in Figure 14. The number of neurons in the competitive layer corresponds here to the number of FAM cells. Clustering of the synaptic

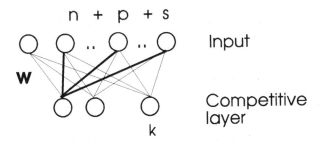

Figure 14 *Training of n position, p intensity, and s substituent values in a competitive network consisting of k synaptic vectors*

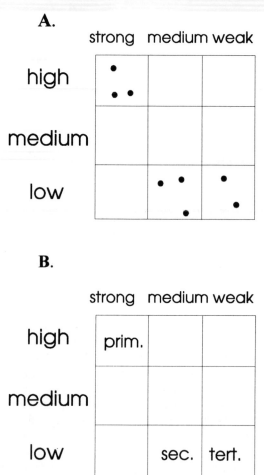

Figure 15 *Clustering of the synaptic weights by a competitive network (A). Derived rules for the FAM cells (B)*

weights reveals the results as given in Figure 15A. Finally the following rules could be derived from the neural network training of the FAM (*cf.* Figure 15B):

IF Position = *HIGH* AND Intensity = *STRONG* THEN Substituent = *primary amine*
IF Position = *LOW* AND Intensity = *MEDIUM* THEN Substituent = *secondary amine*
IF Position = *LOW* AND Intensity = *WEAK*   THEN Substituent = *tertiary amine*

The described principle can be extended to train additional antecedents, such as the band width, and consequences if conclusions are to be drawn with respect to several substituents.

Tuning of membership functions—the characteristic functions of fuzzy sets—by a neural network is possible if the transfer function is interpreted as membership function:

$$m(x) = f(\text{Net}) = f(wx + \theta)$$

where $m(x)$ is the membership function over the universe, X; Net is the Net signal of neuron; $f(\ )$ is a function, *e.g.* the sigmoid function; and w, x, and θ are weight, input, and threshold of the neural network.

Thus, the linguistic terms 'low' or 'high' can be represented by a single neuron. For description of the level 'medium' at least two neurons are necessary to model the increasing and decreasing branch of the membership function (*cf.* Figure 12).

In the area of analytical spectroscopy fine-tuning of membership functions has been demonstrated in an expert system for interpretation of X-ray fluorescence spectra.[33]

6 Pros and Cons

Applications of neural networks and fuzzy logic in analytical spectroscopy represent an exciting field of current chemometric research. The main advantages of neural networks over existing chemometric data evaluation techniques comprise their high level of **fault tolerence** as well as the feasibility to work **without explicitly described models**.

It should be recognized, however, that present state of neural network research is rather in its infancy. Compared to biologic neurons there are several drawbacks to be overcome within the next years. As a matter of fact ascribing to neural networks the ability to 'learn from examples' is not true yet. If one considers the learning process of, *e.g.* a backpropagation network then the conclusion of Bezdeck should be quoted that 'feedforward/backpropagation is a low-level computational model and nothing more (elegant)'.[36] Also the initial idea of introducing neural networks systems for parallel processing of information is seldom exploited because actual networks are simulated as sequentially working software algorithms. Also the choice of learning parameters may become a laborious step in developing real world applications in spectroscopy, so that methods for pre-estimations of the network parameters are highly desired. Most of the applications up to now are based on local rather than global solutions. For example, the IR-spectra interpretation studies concentrate on illustrative cases of individual models so that there is a long way to go to arrive at global interpretation systems.

It is considered that the adaptation of neural networks to new environments, *e.g.* adaptation of calibration models to different spectrometers or of spectrum interpretation systems to novel libraries, will be an important research topic in the coming years.

Acknowledgement

The author thanks the German Bundesministerium für Forschung und Technologie for supporting this work.

References

1. M. Otto, *Anal. Chim. Acta*, 1993, **283**, 500.
2. B. J. Wythoff, *Chemometrics Intell. Lab. Sys.*, 1992, **18**, 115.
3. J. Zupan and J. Gasteiger, *Anal. Chim. Acta*, 1991, **248**, 1.
4. L. M. C. Buydens and P. J. Schoenmakers, 'Intelligent Software for Chemical Analysis', Elsevier, Amsterdam, 1993.
5. J. A. Freeman and D. M. Skapura, 'Neural Networks—Algorithms, Applications, and Programing Techniques', Addision-Wesley, Reading, 1991.
6. C. Schierle, M. Otto, and W. Wegscheider, *Fresenius' J. Anal. Chem.*, **343**, 1992, 561.
7. C. Schierle and M. Otto, *Fresenius' J. Anal. Chem.*, 1992, **344**, 190.
8. U. Hörchner and M. Otto, 'Application of fuzzy neural networks to spectrum identification', in 'Software Development in Chemistry 4', ed. J. Gasteiger, Springer, Berlin, 1990, 377–384.
9. C. R. Mittermayr, A. C. J. H. Drouen, M. Otto, and M. Grasserbauer, *Fresenius' J. Anal. Chem.*, 1993, submitted for publication.
10. R. Janka and M. Otto, 'Quantitative Analysis of Powders in the UV-VIS Range', PittCon 93, Atlanta, Abstract No. 260P.
11. J. R. Long, V. G. Gregoriou, and P. J. Gemperline, *Anal. Chem.*, 1992, **62**, 1791.
12. C. Borggaard and H. H. Thodberg, *Anal. Chem.*, 1992, **64**, 545.
13. A. Bos, M. Bos, and W. E. van der Linden, *Anal. Chim. Acta*, 1992, **256**, 133.
14. L. McDermott, 'Applications of neural networks for near infrared calibration development', Pittsburgh Conference, New Orleans, 9–12 March 1992, Abstract #176.
15. T. Naes, K. Kvaal, T. Isaksson, and C. Miller, *J. Near Infrared Spectroscopy*, 1993, **1**, 1.
16. E. W. Robb and M. E. Munk, *Mikrochim. Acta*, 1990, **I**, 131.
17. M. E. Munk, M. S. Madison, and E. W. Robb, *Mikrochim. Acta*, 1991, **II**, 505.
18. M. Novic and J. Zupan, *Vestn. Slov. Kem. Drus.*, 1992, **39**, 195.
19. J. R. M. Smiths, P. Schoenmakers, A. Stehmans, F. Sijstermans, and G. Kateman, *Chemometrics Intell. Lab. Syst.*, 1992, **18**, 27.
20. W. J. Melssen, J. R. M. Smits, G. H. Rolf, and G. Kateman, *Chemometrics Intell. Lab. Syst.*, 1993, **18**, 195.
21. B. J. Wythoff, S. P. Levine, and S. A. Tomellini, *Anal. Chem.*, 1990, **62**, 2702.
22. U.-M. Weigel and R. Herges, *J. Chem. Inf. Comput. Sci.*, 1992, **32**, 723.
23. A. Bruchmann, H.-J. Götze, and P. Zinn, *Chemometrics Intell. Lab. Sys.*, 1993, **18**, 59.
24. J. U. Thomsen and B. Meyer, *J. Magn. Reson.*, 1989, **84**, 212.
25. V. Kvasnicka, *J. Math. Chem.*, 1991, **6**, 63.
26. L. S. Anker and P. C. Jurs, *Anal. Chem.*, 1992, **64**, 1157.
27. B. Curry and D. E. Rumelhart, *Tetrahedron Comput. Method.*, 1990, **3**, 213.
28. H. Lohninger, 'Classification of mass spectral data using neural networks', in 'Software Development in Chemistry 5', ed. J. Gmehling, Springer, Berlin, 1991.
29. M. Bos and H. T. Weber, *Anal. Chim. Acta*, 1991, **247**, 97.
30. A. L. Allanic, J. Y. Jézéquel, and J. C. André, *Anal. Chem.*, 1992, **64**, 2618.
31. R. Hecht-Nielsen, *Appl. Optics*, 1987, **26**, 4979.
32. B. Kosko, 'Neural Networks and Fuzzy Systems', Prentice-Hall, London, 1992.

33. B. Walczak, E. Bauer-Wolf, and W. Wegscheider, *Mikrochim. Acta* [Wien], 1994, **113**, 137.
34. T. Kohonen, 'Self-organization and Associative Memory', Springer, New York, 1988.
35. L. A. Zadeh, *Inf. Control*, 1965, **8**, 338.
36. J. C. Bezdeck, *Int. J. Approx. Reasoning*, 1992, **6**, 85.

The Role of Analytical Methods in the Scientific Examination of Paintings at The Getty Conservation Institute

D. C. Stulik

THE GETTY CONSERVATION INSTITUTE, 4503 GLENCOE AVENUE, MARINA
DEL RAY, CA 90292, USA

Summary

Analysis of paintings brings important and needed information to art historians
and art restorers who study, research, and work on paintings. Pigment
identification and paint layer stratigraphy analysis are important for understanding
artists' techniques. Polarized light microscopy, X-ray diffraction analysis, and
electron microprobe analysis are important analytical techniques used by
conservation scientists for pigment identification. The latest advances in
analytical instrumentation and methodology also makes it possible to focus on
the identification of binding media and the organic portion of the paint layer.
New developments in absolute dating techniques make it possible to determine
the age of paint layers using radiocarbon dating. Analytical results achievable
using advanced analytical methodology are discussed and related to questions
frequently asked by painting restorers and art historians when restoring or
researching paintings.

1 Introduction

The analysis of paintings is a challenging application of modern analytical
chemistry to the study of objects which are usually treated only by art restorers
and researched by art historians.

In centuries prior to the Industrial Revolution and the development of
manufactured paints, a painter was not only an artist, but also a colourist. By
mixing pigments with selected binding media (the material which holds the
pigment together and bonds the paint to a support), the painter personally, or
with the help of his assistants, made his own paints. This personal experience
with paint preparation gave an artist a great understanding of artist's materials
and their properties. Artists often followed a variety of traditional paint recipes.
However, differences in paint formulations proliferated as painters experimented

with a multiplicity of binding media, seeking that special combination which would give their paints the desired optical and handling properties.

After the introduction of collapsible paint tubes in 1841 and the development of the paint industry in the 18th and 19th centuries, artists became detached from the process of paint manufacturing and most of them lost the motivation to learn the skills of paintmaking. There was, however, a major advantage in this technical development, artists became more creative. The availability of collapsible paint tubes allowed them to leave their studios and develop new styles of painting. Without these advances in artist's materials, there would not have been movements such as Impressionism. The diminishing knowledge of artist materials by painters had, on the other hand, some serious negative effects. Using incorrect materials, working with poorly tested paints, and experimenting with paint formulations without an intimate knowledge of the possible consequences sometimes had disastrous effects on the longevity of the paintings. One notable painter infamous for his disastrous experimentation was Sir Joshua Reynolds (1723–1792), who was the first President of the Royal Academy. He is known for using unstable pigments which quickly faded, painting with bitumen which did not dry, and for using binding media which were responsible for cracking the paint layer. Even during Sir Reynold's lifetime, Sir Horace Walpole made jokes about Reynold's technique saying that: 'Reynolds should be paid in annuities only for so long as his pictures last'.[1] Many great paintings created during the last two hundred years are a challenge for museum curators and keep an armada of paint restorers working overtime.

There are two important reasons for analysing paintings. A painting restorer, before embarking on the cleaning or restoration of a painting, needs to know the materials used in the painting in order to design a working strategy which will not damage the work of art. The knowledge of pigments and binding media used in the creative process can also assist an art historian in his or her provenance and authentication studies and in detailed studies of painting techniques.

2 Sampling

The analysis of a painting usually starts with proper sampling. Based on questions asked by restorers or art historians, the conservator or conservation scientist removes a sample of picture varnish, paint layer, ground, or a sample for a paint layer cross-section. Taking a sample for analysis is a critical point in which the interest of art historians and conservation scientists differ. The art historian would like to have all questions concerning the materials used in the painting answered without touching the painting. On the other side, the analytical chemist knows that the quality of his answers depends, to a great extent, on the quality of the samples available for analysis and on the number of samples allowable to offset sampling errors due to inherent heterogeneity of the paint layer. The actual sampling strategy usually represents a workable compromise between these two fundamental approaches. Samples are removed using a sampling needle or scalpel and, whenever possible, the samples are taken from areas at the edge of the painting or areas where sampling does not interrupt the integrity of the painting. If samples for a paint cross-section have to be taken, the conservator or conservation scientist looks for an already damaged area or for existing cracks where the samples can be taken without causing further extensive damage and in such a way that the untrained eye would not notice any difference.

Samples, including cross-sections, are almost invisible to the naked eye and the amount of sample available for complete chemical analysis is almost always well below 1 mg. Based on this amount of sample, the analytical chemist has to answer several crucial questions concerning the identification of pigments and binding media, and paint layer stratigraphy.

3 Identification of Pigments

The analysis of samples from paintings usually starts with the identification of pigments. Pigment chronology, such as that shown in Figure 1, shows when different pigments were used and, specifically, when new pigments were introduced to the artist's palette. As can be seen in Figure 1, some pigments have been used from Paleolithic cave paintings to recent times. Some others were introduced after being discovered or developed by alchemists and chemists. The authentication of paintings is sometimes based on the identification of pigments. For example, if titanium white (TiO_2), which was introduced as an artist pigment after 1920, is found in a supposedly mediaeval painting, it might indicate that the painting is modern in origin (a painting made in the mediaeval

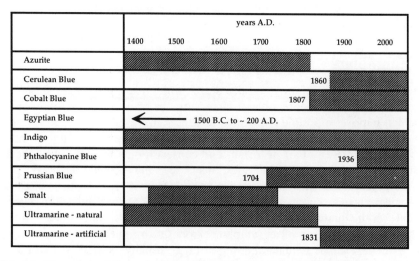

PIGMENT CHRONOLOGY - BLUE PIGMENTS

	years A.D.						
	1400	1500	1600	1700	1800	1900	2000
Azurite							
Cerulean Blue					1860		
Cobalt Blue					1807		
Egyptian Blue	← 1500 B.C. to ~ 200 A.D.						
Indigo							
Phthalocyanine Blue						1936	
Prussian Blue				1704			
Smalt							
Ultramarine - natural							
Ultramarine - artificial					1831		

Figure 1 *Pigment chronology of selected blue pigments. Dates of new pigment introduction are shown and time periods of major use of the pigment are shaded*

style, a 20th century copy of a mediaeval painting, or a forgery). However, to condemn a painting only because titanium white was found in one sample would be wrong. The presence of the titanium white may be due to 20th century restoration or retouching. In such a case, more samples of white pigment would be needed from different locations in the painting to find out if some other white pigments were also used and where in the 3D structure of the painting they are present. The need for resampling usually finds the full support of art historians, for whom the genuine nature of the paintings is a primary concern.

Information about pigment particle size and particle morphology is also important. To distinguish a natural from an artificial ultramarine is a relatively easy task despite the fact that both have the same chemical composition. The natural ultramarine is a pigment prepared from the semiprecious stone, *Lapis lazuri*, and it appears as large blue crystals under the microscope. In contrast, the particles of artificial synthetic ultramarine are very small, usually smaller than 1 μm.[2]

The preparation of some pigments has also undergone changes throughout the ages. For example, a beautiful red pigment, vermilion (HgS), can be prepared from the mineral cinnabar or it can be synthesized chemically using the dry method (heating mercury and sulfur together), or by using a wet chemistry procedure (solution precipitation). All three methods were used in different times and different geographical areas. The morphology of vermilion particles prepared using different methods can be differentiated with a scanning electron microscope.

Today a whole battery of instrumental methods exists which can be used to identify inorganic pigments in a paint sample. *X*-Ray fluorescence (XRF) is a powerful analytical method for art research because, with suitable instrumentation, the analysis can be conducted non-destructively, without touching the painting, *a fact that pleases every museum curator* (Figure 2). Many pigment-related questions can be answered using XRF, but the method has some serious limitations which preclude relying only on these types of experiments. Very often, pigments in the paint layer are mixed to achieve the desired paint tint and hue and only some pigments are used in their pure form. Additionally, many pigments used in artist paints are adulterated with fillers and extenders which are usually various inert inorganic materials. Moreover, the paint layer of some mediaeval paintings is composed of many individual layers which can have quite different pigment composition such as ground layer, underdrawing, preparatory layer, series of paint layers, and multiple glazes. Primary *X*-rays, from an *X*-ray tube, penetrate the whole paint layer and *X*-ray fluorescence radiation generated within the irradiated spot (about 1 mm in diameter) carries information about all elements present within the irradiated area of the paint layer, regardless of its stratigraphy. In such a case, XRF results have to be regarded as preliminary results which need confirmation from other methods, which usually require actual removal of samples.

When samples are taken from a painting for identification of pigments, a standard procedure, used by analytical chemists of The Getty Conservation Institute (GCI), starts with examination by polarized light microscopy (PLM).

Figure 2 *X-Ray fluorescence spectrometry using 1 mm in diameter X-ray beam and energy dispersive X-ray spectrometer is an ideal non-destructive tool for pigment identification*

In a series of simple tests, pigments can be identified based on their colour, particle size, particle morphology, birefringence, pleochroism, and anomalous polarization colours.[3]

A well-trained microscopist who has a good collection of historical and modern standard pigment samples is able to identify most pigments used in paintings and answer more than 80% of pigment related questions. For example, PLM was used to identify wheat starch as a component of red glazes over gilding on baroque polychromed structures from Brazil. The starch grains showed a typical black cross pattern when studied using PLM (Figure 3).

In the case of non-conclusive PLM results, electron microprobe analysis (EMPA) and X-ray diffraction (XRD) can be used to support and confirm the PLM results. The qualitative EMPA analysis can be used to identify all metal and most non-metallic elements present within the sample. This helps to resolve identification problems concerning mixed pigments, fillers, and extenders and to provide important input data for subsequent XRD analysis. The quantitative EMPA analysis helps to differentiate pigments which have the same elemental composition, but different concentrations of each element. For example, realgar (As_2S_2) and orpiment (As_2S_3) pigments both contain arsenic and sulfur, but in different proportions. The high lateral resolution of EMPA analysis is advantageous in studying the pigment stratigraphy of paint cross-sections.

XRD analysis provides crystallographic data to supplement quantitative data on the elemental composition of pigment particles obtained from EMPA measurements. Detailed study of already mentioned titanium white pigment is one of these cases for which the need for XRD analysis can be very well

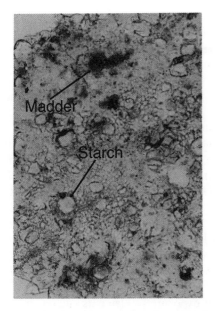 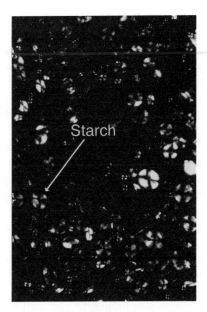

Figure 3 *Identification of a wheat starch as a component of red glazes over gilding on baroque polychromed sculptures from Brazil using polarized light microscopy (PLM). The starch grains show a typical black cross pattern*

documented. Titanium white is known to exist in nature as the minerals brookite, anatase, and rutile. Artificially prepared titanium dioxide, introduced after 1920, was available only in the anatase form. A change in titanium dioxide production technology immediately before 1939 resulted in the availability of the pigment in the rutile form.[4] Therefore, the identification of the crystalline structure of titanium dioxide in a paint sample using XRD analysis can provide important information for painting authentication.

4 Binding Media Analysis

Identification of paint media and varnishes is the next step in painting analysis. Although it is usually a much more difficult task than identifying pigments, identifying binding media is just as important, if not more so.

Artists' techniques used through the centuries (if we do not count revolutionary changes in the artist's palette in the 19th and 20th centuries) differed with respect to binding media more than with respect to pigments. It is the paint medium that determines the technique of painting. Binding media are not as diverse as pigments, but they are much more complex compounds.[5] They are all organic substances and most of them are inherently complex natural products (*e.g.* egg used as a binder in tempera paintings contains proteins, fats, sugars, vitamins, sterols, dyes, water, inorganic salts, *etc.*) and many have been treated to separate impurities (animal glue) or to improve their properties (prepolymerization of raw linseed oil to produce a more viscous stand oil).

The chemical make-up of binding media can also be a function of its primary source (different concentrations of various saturated and unsaturated fatty acids in linseed, poppy seed, and walnut oils). In some cases, even factors such as geographical location, weather variation, and seasonal changes can be responsible for changes in the chemical composition of natural products used as binding media (*e.g.* the chemical composition of plum-tree gum collected during the spring in California is slightly different than the chemical composition of plum-tree gum collected during autumn in Sweden).

A number of artists used mixed binding media techniques, such as oil–Copal resin–Venice turpentine painting medium. Different parts of a painting might also be painted using different binding media. Some painters, working with linseed oil-based paints, used poppy seed oil to prepare white and blue paints because it yellowed less. Other works of art may be built up using multiple layers of different binding media: animal glue ground, oil emulsion paint layer, oil glazes, and natural resin varnish. Works of art may also have been treated many times by other artists, restorers, and conservators throughout their history. Many of these treatments may have involved additional materials or chemicals which could alter the original composition of the paint layer. Organic materials are also well known to age with time and exposure to elements such as light, oxygen, and pollutants. All these factors make analysis of binding media extremely challenging.

Similar to pigment analysis, there are several individual analytical methods or their combinations, which can be used to identify binding media in paint samples.[6,7] The process of binding media identification starts with orientation tests which should provide information on the class of binding media used in the painting. This might seem to be an easy task and some analytical chemists who visit museums might ask if it is necessary to do it at all when all paintings in museums and galleries have the painting medium clearly identified (oil, tempera, *etc.*) along with the title and artist's name. It is interesting to note that the identification label and reality are sometimes quite different and binding media analysis might bring surprises even to well-trained conservators and curators (such as the reclassification of many 'oil' paintings).

Three analytical methods are very useful when identifying the class of binding medium: Organic Elemental Analysis (OEA), Fluorescence Microscopy (FM), and Fourier Transform Infrared (FTIR) Spectrometry.

Organic elemental analysis (OEA) provides quantitative information on the amounts of carbon, hydrogen, nitrogen, sulfur, and oxygen in organic material. The OEA is not a new method, but only recent advances in instrument design and a drastic reduction of sample size required for analysis has allowed its application in art research.[8] Modern elemental analysers are capable of providing quality analytical data from samples containing as little as 100 µg of organic matter. Detection of nitrogen in a paint sample indicates the presence of proteinaceous binding media. Measuring C/N concentration ratios can even allow preliminary sorting of proteinaceous binding media to specific subclasses (animal glue, egg, casein) as indicated in Figure 4.

The advantage of using fluorescence microscopy for binding media classification is that it works with paint cross-sections which are also used for pigment

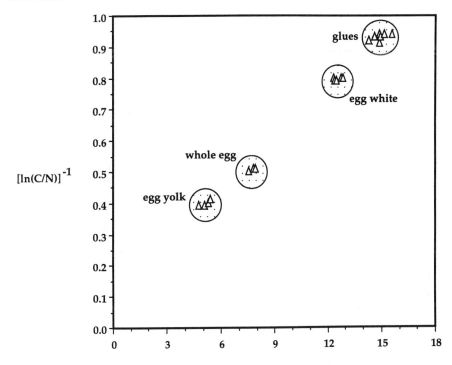

Figure 4 *Organic Elemental Analysis can be used for identification of the binding media class. The % N versus* $[\ln(C/N)]^{-1}$ *plot helps distinguish different proteinaceous binding media*

stratigraphy studies and EMPA analysis. Using the same cross-section for several analytical tests minimizes the time of the whole analytical procedure. Fluorescence microscopy applied to art research evolved from medical research and staining techniques used in clinical pathology. The use of reactive molecular probes[9] allows the attachment of 'fluorescent tags' to functional groups which are specific to given binding media. Other methods use selective adsorption or staining using fluorescing molecules. A fluorescent microscope equipped with a broad range UV source in combination with properly selected excitation and emission filters allows the analyst to determine if a given binding medium is present and which layers of the paint structure contain the same binding medium. Fluorescence microscopy can even deal with complex mixed binding media. Under favourable circumstances, as low as 0.1 wt% of a specific binding medium can be detected in mixed binding media paint samples. The analytical chemist has to be cautious when interpreting fluorescence microscopy images, because the presence of some pigments interferes with the analysis and, under unfavourable conditions, binding medium cannot be positively detected at concentrations as high as 20 wt%.

More sophisticated analysis of binding media in paint cross-sections can be achieved using FTIR microscopy. An infrared spectrum is a characteristic pattern of absorption bands which are related to various functional groups in the analysed organic material.[10] Modern FTIR microscopes provide the ability to analyse selected sample areas as small as $10 \times 10\,\mu m$. In combination with a computer controlled X–Y stage, the FTIR microscope can be used to map selected areas of the sample and provide information about the concentration of different functional groups across the sample. Such maps make it easier to interpret FTIR analyses of cross-sections, especially when mixed binding media have been used. An optical photomicrograph and infrared concentration maps of a paint cross-section from *Adoration of the Magi* by Mantegna are shown in Figure 5.

It would be ideal to analyse binding media in a paint cross-section using the FTIR reflection method which does not need any special sample preparation. However, because reflected radiation contains both specular and diffuse radiation and cross-sections samples are highly heterogeneous and contain various amount of particulate inorganic matter, spectra from FTIR reflection measurements are very difficult to interpret. The more feasible approach is to use a transmission method of analysis, but in such cases, a thin-section sample ($2–10\,\mu m$ thick) has to be prepared from a cross-section using a precise microtome.[11] Another analytical option is to use the newly developed attenuated total reflection (ATR) lens which solves the problem of reflection measurement and allows analysis of sample areas as small as $5 \times 5\,\mu m$. A major reason why FTIR is not able to serve as an ultimate tool for binding media analysis is its relative low sensitivity. The Getty Conservation Institute studies focusing on quantitative FTIR analysis of binding media mixtures have shown that the generally reported detection limit for FTIR (2–5%) does not apply when dealing with binding media mixtures in the presence of pigment. More realistic detection limits are greater than 10 wt% for most binding media. This is not good enough for analysis of major and minor binding media components in complex multicomponent samples. Also, the differentiation between binding media within one class needs more involved analytical procedures than FTIR.

Once major components of the complex binding medium have been identified, the next task is to identify all minor components and exact identification of the binding medium within its class. Usually, more than one analytical method has to be used to answer related questions and reconfirm analytical findings.

For example, when drying oil has been used as a painting medium, it might be important to determine the plant source of the oil. This question relates closely to artist's technique; the working style and techniques of most artists developed with time and experience. Some artists experimented heavily, always looking for new recipes and methods to achieve desired optical and structural effects in their work. Others mastered their medium and did not want to risk the consequences of further experimentation. Drying oil used for artist paints can be pressed from linseeds, poppy seeds, or black walnuts. Linseed oil was the most common, but we also know that, for example, Claude Monet used poppy seed oil in many of his paintings. The use of different types of oil could also be a function of local availability. Walnut oil was more popular in Flanders and Germany than in Italy.

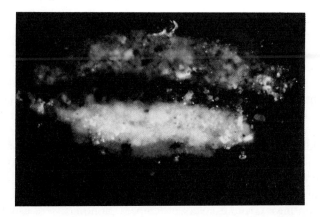

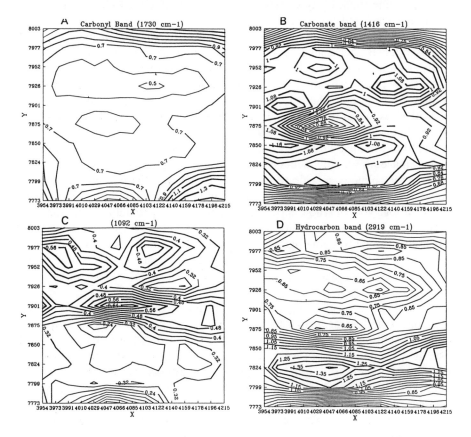

Figure 5 *Optical photomicrograph and infrared concentration contour maps for a painting cross-section from* 'Adoration of the Magi' *by Andrea Mantegna. Functional groups mapped are*: *A–carbonyl band* (1730 cm^{-1}); *B–carbonate band* (1416 cm^{-1}); *C–sulfate band* (1092 cm^{-1}); *D–hydrocarbon band* (2919 cm^{-1})

To distinguish between linseed, walnut, and poppy seed oils, mixed with pigments and aged for several hundred years, is not an easy task. Almost all of the unsaturated fatty acids which are responsible for oil drying are gone due to oxidation and polymerization into insoluble linoxin. Fortunately, there are two stable, fully saturated fatty acids (palmitic and stearic) present in all drying oils and their concentration ratios differ enough in all three types of oils to facilitate analytical differentiation. Gas chromatography (GC) and gas chromatography–mass spectrometry (GC–MS) methods for determination of palmitic/stearic (P/S) ratios, developed at the GCI, are based on the transesterification of linoxin which yields methylesters of both fatty acids which are needed for GC or GC–MS analysis.

Egg, animal glue, and casein are proteinaceous binding media used in so-called tempera paintings. Ground layers of tempera paintings were usually prepared by mixing fine chalk or gypsum with a dispersion of animal glue. Egg yolk or whole egg was used for tempera painting and egg white was sometimes used as a temporary varnish on oil paintings and in the form of glair in illuminated manuscripts. Both casein and animal glue temperas were used in mediaeval times as an alternative to the more common egg tempera.

All these types of binding media contain protein and their overall chemical composition can be rather simple and limited only to the presence of protein, as in the case of animal glue, or very complex containing a whole range of other chemical compounds, as it is in the case of casein and egg.

High performance liquid chromatography (HPLC) analysis of amino acids of proteinaceous binding media hydrolysates shows that the high concentration of hydroxyproline is specific to animal glue and that the concentration ratio between several other amino acids can be used to differentiate between casein, egg white, egg yolk or whole egg binding media. Using precolumn derivatization of protein hydrolysates with 9-fluorenylmethyl chloroformate followed by reverse-phase liquid chromatography using binary gradient elution and fluorescence detection allows for sub-picogram analysis of individual amino acids. The whole analytical procedure can be performed using paint samples as small as a few micrograms.[12]

Throughout the history of painting, artists have experimented with multiple-component binding media to achieve desired optical effects or working properties of paint. Analysis of multiple-component binding media is very challenging and usually a combination of several analytical methods is needed to provide the answers to all of the important analytical questions. Developments in the application of GC and GC–MS techniques in analysis of oils, waxes, resins, proteins, and gums achieved by a research team at The Getty Conservation Institute has lead to the development of a single-sample, one-technique, multiple-step procedure for the quantitative analysis of paint samples containing multiple-component binding media. With the new methodology it has been possible to detect cholesterol in a small sample of 600 year old paint (Figure 6). The presence of cholesterol indicates the use of egg tempera or egg-oil emulsion in paintings.

The knowledge of quantitative amounts of individual binding media in the

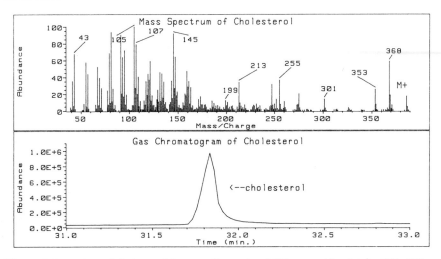

Figure 6 *Detection of cholesterol in a small sample of 600 year old paint by GC–MS*

sample allows the analyst to find out (a) what exact type of binding medium was used and (b) if its composition corresponds to binding media used by the same artist in different paintings or during a specific time period.

5 Dating of Paintings

Dating, provenancing, and authentication of paintings are procedures usually conducted by art historians; they require a trained eye and detailed knowledge of all other works by a given artist. This is supplemented by a great amount of archival work. The cases which are still inconclusive or cases in which art historians cannot form a uniform opinion about the painting are best suited for the application of scientific and analytical methods. Scientific dating of paintings usually starts with the search for the presence of pigments which would be inconsistent with known pigment chronology. If these studies do not provide enough supporting data for an opinion, it is necessary to use absolute dating methods to determine the date of creation of the artifact. There are pronounced similarities between the application of radiocarbon dating in archaeology and art research, but there are also major differences which make it difficult to simply take a methodology already developed for archaelogy and apply it directly to radiocarbon dating of art objects. Fine art studies deal with relatively recent time periods (~ 2000 BC to the present compared with 100 000 BC to ~ 500 AD for archaeology) and because the majority of art objects are unique, irreplaceable, and relatively small, the ability to date a minimum amount of sample is extremely critical. Typical sample sizes needed for conventional β-counting radiocarbon dating are equivalent to 10–100 g of pure carbon. Such an amount of sample is too large to take from a wooden panel or canvas of a painting. A new era of application of radiocarbon dating in art research came with the development of accelerator mass spectrometry (AMS)

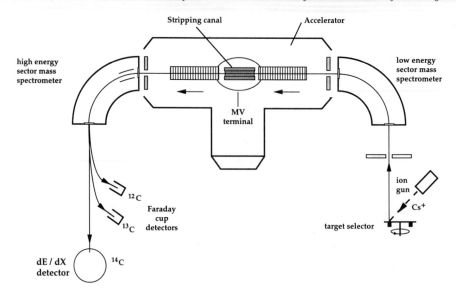

Figure 7 *Accelerator mass spectrometer as used to measure the concentration of radiocarbon*
^{14}C in binding media of paintings and for absolute dating of paint layer

which directly determines the concentration ratio of ^{14}C and stable carbon
isotopes in the analysed sample (Figure 7). A sample as small as 100 µg of pure
carbon can be dated with a precision of ± 30 years, in favorable cases, depending
on sample type and shape of the calibration curve in the region around a
targeted date.[14] However, to radiocarbon date, a wooden or canvas support
of a painting still might leave considerable doubts in the mind of art historians
who need more convincing evidence that the painting was or was not created
in the proposed time period. They are well aware that a knowledgeable forger
would use all the right pigments and that he or she would do everything to
obtain wooden or canvas support from the proper time period. If suspicion of
possible forgery still exists, only radiocarbon dating of the paint layer can
provide the necessary data needed for final art historical interpretation.

Direct AMS dating of a paint sample is not possible because some pigments
containing 'dead carbon' (for example, $CaCO_3$) might be present in the sample,
and the organic binding medium might be contaminated by more 'contemporary'
organic materials which could have been incorporated into the paint layer from
numerous past revarnishings using natural resins or possible canvas relining,
which uses hot wax or carbohydrate paste. One of the major goals of the Getty
Binding Media Project has been to develop an experimental strategy which
would allow chemical separation of complex binding media mixtures and for
isolation of a critical datable material as needed for AMS experiments. The
successful development of a separation strategy in combination with AMS
radiocarbon dating has opened a whole new chapter in the application of
scientific methods to art research.

Acknowledgements

The author would like to thank all members of The Getty Conservation Institute Binding Media Project team for their contribution: Michele Derrick, Luiz Souza, Dr Mary Striegel, Andrew Parker, Cecily Grzywacz, and Michael Schilling for providing graphic material and to Dr Arie Wallert for detailed information on the history of titanium white pigment.

References

1. A. Gweynne-Jones, 'The Life and Works of Sir Joshua Reynolds', *J. R. Soc. Arts*, CIV, 1956.
2. R. J. Gettens and G. L. Stout, 'Painting Materials', Dover, New York, 1966.
3. W. C. McCrone, L. B. McCrone, and J. G. Delly, 'Polarized Light Microscopy', Ann Arbor Science Publication, Ann Arbor, 1978.
4. Arie Wallert, private communication.
5. J. S. Mills and R. White, 'The Organic Chemistry of Museum Objects', Butterworths, London, 1987.
6. L. Masschelein-Kleiner, *Stud. Conserv.*, 1968, **13**, 105.
7. D. C. Stulik (ed.), 'Identification and Analysis of Binding Media in Paintings', to be published.
8. A. E. Parker and D. C. Stulik, to be published.
9. R. P. Haugland, 'Molecular Probes', Molecular Probes Inc. Publ., Eugene, USA, 1992.
10. M. R. Derrick, J. M. Landry, and D. C. Stulik, 'Methods in Scientific Examination of Works of Art: Infrared Microspectroscopy', The Getty Conservation Institute, California, 1991.
11. M. R. Derrick, D. C. Stulik, J. M. Landry, and S. P. Bouffard, *J. Am. Inst. Conserv.*, 1992, **31**, 225.
12. C. M. Grzywacz, Master Thesis, California State University, Northridge, 1992.
13. M. R. Schilling, to be published.
14. D. C. Stulik and D. J. Donahue, *MRS Bull.*, 1992, **17**, 53.

Flow Analysis in the Nineties

W. E. van der Linden

LABORATORY OF CHEMICAL ANALYSIS, DEPARTMENT OF CHEMICAL TECHNOLOGY, MESA RESEARCH INSTITUTE, UNIVERSITY OF TWENTE, PO BOX 217, NL-7500 AE ENSCHEDE, THE NETHERLANDS

Summary

After a short retrospect on the development of flow analytical methods over the past decades, some subjects which will need attention in the coming years are discussed.

- Flow analytical methods will increasingly be applied for process control and environmental purposes. This requires systems that can operate unattended and can be supervised by personnel that are not analytically trained. An autodiagnosis system which will give appropriate information on the functioning of the analytical system and which provides instructions for the operator about actions that should be taken has to be developed.
- To increase robustness and to improve portability, integration, and miniaturization of the analytical systems has to be pursued. Micromachining techniques offer the possibility to construct and integrate microdosing systems/pumps, micromixing devices, and flow meters on single chips. Combinations with detectors based on IC technology will eventually lead to completely integrated microanalysis systems.
- In the next few years attention will need to be given to the development of sample preparation methods that are compatible with (miniaturized) flow analytical methods.

1 Introduction

It is always challenging, but risky, to make any predictions for the future with regard to direction or developments in research. If, however, some forecast for the coming years has to be given, at first a short reflection on the past and the present state-of-the-art is required.

Flow analysis, which comprises both segmented and unsegmented flow systems, has come of age and is at present a widely accepted technique for handling samples in combination with a great variety of analytical detection methods. Starting with simple optical and electrochemical methods, flow injection analysis (FIA) is now also used in combination with more sophisticated methods, *e.g.* nuclear magnetic resonance (NMR) and inductively coupled plasma-mass spectrometry (ICP-MS).

There are a number of reasons why FIA in particular has become so popular. Firstly its simplicity and the ease with which systems can be assembled from cheap separate parts have to be mentioned. Related to this modular character, the flexibility of FIA systems has to be emphasized. Secondly, many classical chemical methods of analysis, which are sometimes rather capricious and often critically time-dependent, can be carried out relatively easily in a flow system. A well-known example is the determination of phosphate. The main advantage is the large number of samples that can be dealt with per unit time and the high reproducibility of the performance of the systems as long as no changes are made in the set-up or operating conditions such as the flow rates. This allows the analysis of a series of unknown samples and calibration with standard solutions within a very short period of time.

Discussing flow analysis in general and FIA in particular, it will not do justice to the technique and its 'missionaries', if emphasis is only given to improvements in the performance of existing methods of analysis. It has been said that the scope of application of FIA is only limited by the ingenuity of its users. Indeed, the scope of FIA analysis includes the performance of reproducible stopped-flow experiments, the use of membranes for sample clean-up and enhanced selectivity, the use of concentration gradients formed during the transportation of the sample zone, the possibility to cover an extended range of concentrations depending on which point at the tailing peak detection is done, the injection of reagent into a continuous sample stream to reduce reagent consumpion, *etc.* Ruzicka and Hansen,[1] as well as the Córdoba group headed by Valcárcel and Luque de Castro,[2] have in particular contributed a lot in this respect.

Recently, a new alternative was suggested by Ruzicka[3] called Sequential Injection (SI). In this technique, zones of different solutions including the sample, and, *e.g.* reagent, buffer, *etc.* are created in a single tube by suction in combination with a selection valve, which sequentially connects the containers filled with the respective solutions to the tube. If the leading zone consists of pure water, the pump, *e.g.* a piston-type pump, comes into contact with this water only. After formation of the desired sequence of zones the flow rate is reversed and the stream is directed to a detector. During transport the zones can partly interpenetrate allowing reactions to proceed, and the reaction products can be detected.

2 Theory

In the past ten years quite a lot of work has been done on the theoretical description of the spreading (dispersion) of the sample zone during its transportation from the point of introduction in the system to the detector, with or without a chemical reaction taking place.[4-10] Most of the models refer to straight tubes (the equations pertaining are summarized in Table 1) but also some results for coiled tubes are available.[11] Even the transportation of material across

Table 1 *Set of equations important for discussing miniaturization of Flow Injection Systems*

$$\Delta P = 8\eta \langle v \rangle L R^{-2} \qquad\qquad\qquad V_R = \pi R^2 L$$
$$t_v = L \langle v \rangle^{-1} \qquad\qquad\qquad f_v = \langle v \rangle \pi R^2$$
$$\tau = D t_v R^{-2} \qquad\qquad\qquad \sigma_v = f_v \sigma_t$$
$$Re \cdot Sc = 2 \langle v \rangle R D^{-1} \qquad\qquad f_{max} = 600 \sigma_t^{-1}$$
$$\sigma_t = \{ t_v R^2 (24D)^{-1} \}^{1/2}$$

ΔP is pressure drop; η is dynamic viscosity; L is length of tube; R is radius of tube; t_v is residence time; $\langle v \rangle$ is mean linear flow velocity; τ is 'reduced time'; D is diffusion constant; V_R is tube volume; f_v is volumetric flow rate; σ_v and σ_t are the standard deviations of the peak in terms of volume and time, respectively; f_{max} is the maximum sample frequency; Re is the Reynolds number; and Sc is the Schmidt number.

membranes in flow injection systems has been modelled.[12] These studies are partly numerical in nature and partly based on exact solutions of the relevant differential equations. They have provided more qualitative insight into the actual processes going on but, in spite of all mathematical efforts, exact predictions of the behaviour of all types of modules with respect to dispersion are virtually impossible. Hence an experimental procedure has been proposed for evaluating modules for their contribution to dispersion. This approach is based on concepts of convolution and deconvolution.[13] It can be mathematically described using the operator $*$ for the convolution process and $\#$ for the deconvolution process:

$$\text{convolution: } O(t) = h(t) * I(t)$$
$$\text{deconvolution: } h(t) = O(t) \, \# \, I(t)$$

where $I(t)$ is the input function, $O(t)$ is the output function, and $h(t)$ is the transfer function or impulse/response function. This latter function gives the response of a so-called delta injection, *i.e.* an injection within an infinitely short time element or an infinitely small width of the sample zone. The concept of convolution is illustrated in Figure 1.

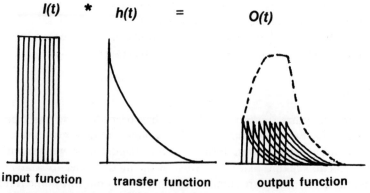

Figure 1 *Illustration of the principle of convolution*

For each element in the transport line between the point of injection and the point of detection a separate impulse/response function can be defined and determined, which describes its contribution to the dispersion of the sample. The final detector response $O(\text{det})$ is obtained by the successive convolution of the output function of the preceding element with the impulse/response function of the following element:

$$O(\text{det}) = \{[\{[\{[I(t) * h(t1)] * h(t2)\} * h(t3)] * \ldots * h(tn - 1)\} * h(tn)] * h(\text{det})\}$$

where $h(\text{det})$ is the impulse/response function of the detector itself.

In practice, the situation is even more complex because not only the axial component (*e.g.* time or length) has to be taken into account, but also the radial variation which accounts for the spatial inhomogeneity of the sample zone over the cross-section of the tubes and modules.

Due to the dispersing characteristics of most modules, the final response curves of complex flow systems have a much more Gaussian shape than the strongly asymmetric tailing peaks found with simple straightforward flow injection systems. Brooks *et al.*[14] have shown that most response curves of total flow injection systems can be described by a Gaussian function convoluted with an exponential decay function. The more smooth response profiles resulting from complex systems make them more suitable for standard data evaluation procedures like the ones used in chromatography.

2 Future Developments

Autodiagnosis

In the eighties an ever increasing interest for (total) quality control was observed. This interest has and will continue in this decade and will be accompanied by a growing need for continuous monitoring of all kinds of processes not only in the chemical process industry but also in the food industry, agriculture, *etc.*, and last, but not least, in medical care. Flow analysis can provide a major contribution to these areas provided that systems can be made to be reliable and robust. Several manufacturers have brought such systems to the market already, but as far as is known, none have autodiagnostic facilities which will automatically warn the user or operator of malfunctioning. This means that these systems cannot really operate unattended if, for instance, the control system of a process crucially depends on these measurements. So what is needed is the development of diagnosis systems that indicate if the equipment operates well and, if not, which part is the cause of the trouble.

Because all component parts contribute to the final response curve, in principle, all information about the operation of the system is contained in the shape of this curve. Thus the first step of an autodiagnosis system could be the detection of deviations from the exponentially modified Gaussian (EMG) function, via the calculation of the second statistical moment, which should be independent of concentration and peak height, and the location of the peak maximum. In simple systems where only a limited number of possibilities may lead to malfunctioning it should be possible to directly relate the kind and degree of deviations to particular causes. In more complex systems, the situation may be less clear and if any deviation is observed, flow rates in the individual transport lines should be checked first. Therefore flow meters have to be installed in the various lines. Information about flows, and characteristic figures describing the shape of peaks, can be incorporated in an expert system which may come up with possible causes of malfunctioning and suggest the actions that the user/operator should take to restore the correct operation.

Such systems are not yet available, however initial studies have been made.[15]

Integrated Systems

Some time ago Ruzicka and Hansen[16,17] suggested that microconduit systems could be developed by engraving channels in blocks of Perspex. Integration and miniaturization lead to a greater reliability and robustness. The present development of micromachining techniques offer many more possibilities nowadays. It is possible, for instance, to implement microdosing systems/pumps, micromixing devices, and flow meters on one single chip. When combined with detectors based on IC technology this should lead to totally integrated microanalysis systems. It may not be too long before complete neural networks can be implemented on a single chip; as well as allowing the integration of the analytical system with a highly sophisticated multivariate calibration method.

The scaling down of flow systems produces some problems with regard to, *e.g.* the injection volume, the effective volume of the detection system, and pressure drops in the system. Previously,[18] it was shown that with diminishing tube diameters, below 0.2 mm, the detector volume soon becomes a limitation if a maximum allowable pressure drop in the system of 5×10^4 Pa (0.5 bar) is not to be exceeded and if a sample frequency of at least 60 samples per hour is required. When miniaturizing a flow system to the size of (or part of) a Si-wafer, the length of the channels will become a limiting factor as well. Although it is possible to etch long spiralized channels, as shown by Angell *et al.*[19] for a miniaturized gas chromatograph, it is less attractive for large scale production of complex analytical systems. Where channels in flow analysis systems based on chemical reactions are also used for accomplishing mixing of a sample and reagent(s) and for creating a certain delay time to let the reaction proceed to some degree of completion, other means have to be found to achieve these ends in miniaturized systems.

The reaction time can be easily controlled by using stopped-flow procedures; the mixing requires a small scale mixing device. Fortunately, such devices have been reported recently.[20] They consist of a flat chamber of 2.0×2.0 mm, through which the sample stream flows, on top of which a flat plate is fixed with 400 micro-nozzles through which the reagent can merge with the sample stream (Figure 2).

Nowadays, it is also possible to manufacture integrated micro-liquid dosing systems which can supply flow rates in the order of 0.1 μl s^{-1}. The thermo-pneumatic principle of one such device is visualized in Figure 3. The air-chamber is periodically heated by the electrical heater by means of a square wave generator,

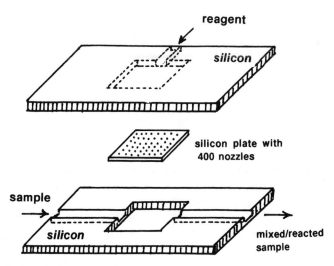

Figure 2 *Scheme of micromixing device*

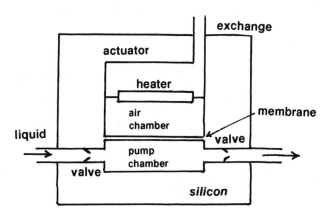

Figure 3 *Principle of thermo-pneumatic pumping device*

resulting in a periodical pressure change in the air-chamber. This causes a regular deflection of a membrane which separates the air- and pump-chamber. The valves ensure that the pressurized fluid can be sucked and expelled in one direction only. By adapting the electrical current and/or the frequency, the flow rate can be regulated. A flow sensor is desirable to measure the flow rate. Such a micro flow sensor can be simply made based on a resistor (heater) with at both sides, at a fixed distance, two temperature sensitive resistors. The temperature difference between the two thermistors, at constant heating power, is a measure for calculation of the flow rate (constant power mode). It is also possible to measure the temperature of the heater directly and keep it constant by varying the electrical current. The electrical power needed to keep the temperature constant is a measure of the flow rate (constant temperature mode). This latter type of sensor is very fast.

Sample Preparation

Micromachining and microelectronics open up challenging possibilities to develop integrated micro-sized sensing devices. However, analytical chemists should always keep in mind that most kinds of untreated samples are not compatible with analysing systems with internal diameters of transport lines and modules of less than 0.5 mm, simply due to the risk of clogging. Hence it is necessary to filter the test solutions to get rid of any particulate matter such as, *e.g.* debris from process streams or biological material in environmental samples. Because in several cases the analyte may be (partly) adsorbed on the particles, or is even a constituent of the solid material, stripping or extraction of the analyte from these particles has to be performed. It is most desirable that these sample pretreatment procedures be carried out in such a way that they are compatible with subsequent flow analytical systems. It may be necessary, for instance, to investigate the use of ultrasonic treatment to assist the process of dissolution, extraction, or desorption.

Coupling of the flow sample pretreatment system to the flow analytical system can probably best be accomplished by cross-flow filtration using various types of membranes, depending on the analyte and matrix concerned.

4 Concluding Remarks

Flow analytical methods will certainly continue to attract interest from those who are dealing with large numbers of samples, but above all, from those involved in continuous monitoring of processes, be it from the process control or the environmental point of view. With respect to continuous operation and handling of sometimes aggressive or corrosive media, the further development and evaluation of methods of transporting samples without contact with vital parts in the system that are easily vulnerable to deterioration, such as pump tubing, valves, *etc.*, has to be pursued. Sequential injection analysis mentioned earlier may be useful in this respect.

References

1. J. Ruzicka and E. H. Hansen, 'Flow Injection Analysis', 2nd edn., Wiley-Interscience, New York, 1988.
2. M. Valcárcel and M. D. Luque de Castro, 'Flow Injection Analysis, Principles and Applications', Ellis Horwood, Chichester, 1987.
3. N. Lacy, G. D. Christian, and J. Ruzicka, *Anal. Chem.*, 1990, **62**, 1482; *Anal. Chim. Acta*, 1992, **261**, 11.
4. J. M. Reijn, W. E. van der Linden, and H. Poppe, *Anal. Chim. Acta*, 1980, **114**, 105.
5. J. T. Vanderslice, K. K. Stewart, A. G. Rosenfeld, and D. J. Higgs, *Talanta*. 1981, **28**, 11.
6. J. M. Reijn, W. E. van der Linden, and H. Poppe, *Anal. Chim. Acta*, 1981, **123**, 229.
7. J. M. Reijn, W. E. van der Linden, and H. Poppe, *Anal. Chim. Acta*, 1981, **126**, 59.
8. C. C. Painton and H. A. Mottola, *Anal. Chim. Acta*, 1983, **154**, 1.
9. H. Wada, S. Hiraoka, A. Yuchi, and G. Nakagawa, *Anal. Chim. Acta*, 1986, **179**, 181.
10. S. D. Kolev and E. Pungor, *Anal. Chim. Acta*, 1986, **185**, 315.
11. R. Tijssen, *Anal. Chim. Acta*, 1980, **114**, 91.

12. S. D. Kolev and W. E. van der Linden, *Anal. Chim. Acta*, 1992, **268**, 7.
13. I. C. van Nugteren-Osinga, M. Bos, and W. E. van der Linden, *Anal. Chim. Acta*, 1988, **214**, 77.
14. S. H. Brooks, D. V. Leff, M. A. Herandez Torres, and J. G. Dorsey, *Anal. Chem.*, 1988, **60**, 2737.
15. I. C. van Nugteren-Osinga, Thesis, University of Twente, Enschede, The Netherlands, 1991.
16. J. Ruzicka, *Anal. Chem.*, 1983, **55**, 1040A.
17. J. Ruzicka and E. H. Hansen, *Anal. Chim. Acta*, 1984, **161**, 1.
18. W. E. van der Linden, *Trends Anal. Chem.*, 1987, **6**, 37.
19. J. B. Angell, S. C. Terry, and P. W. Barth, *Sci. Am.*, 1983, **248**, 36.
20. R. Miyake, T. S. J. Lammerink, M. Elwenspoek, and J. H. J. Fluitman, Proc. MEMS '93 Conference, 7–10 February 1993, Fort Lauderdale, FA, USA.
21. T. S. J. Lammerink, M. Elwenspoek, and J. H. J. Fluitman, Proc. MEMS '93 Conference, 7–10 February 1993, Fort Lauderdale, FA, USA.
22. T. S. J. Lammerink, M. Elwenspoek, and J. H. J. Fluitman, Proc. Eurosensor VI Conference, 4–10 October 1992, San Sebastian, Spain.

Some Ways of Improving the Sensitivity and Selectivity of Flow Electroanalysis

K. Štulík

DEPARTMENT OF ANALYTICAL CHEMISTRY, CHARLES UNIVERSITY, PRAGUE, THE CZECH REPUBLIC

Summary

Several examples are given of the present trends leading to improvements in the sensitivity and/or selectivity of electrochemical detection in flowing liquids. The advantages of voltammetric microelectrodes, use of rapid potential scan techniques, multichannel detection, and physico- and biochemical modification of electrodes are briefly discussed.

1 Introduction

Analysis in flowing liquids has been very important for some time and its practical importance is still rapidly increasing for several reasons: industrial and environmental control require rugged and reliable monitoring systems and continuous monitoring of substances in physiological fluids is often needed in medicine. Flow techniques—such as continuous-flow analysis (CFA) and flow-injection analysis (FIA)—greatly facilitate automation of laboratory analyses and are less expensive and more versatile than laboratory robots. Also, procedures based on high-performance liquid chromatography (HPLC) belong among flow-through methods. Practicing analysts sometimes do not realize that all the above methods have a common hydrodynamic basis and that the boundaries among them are diffuse; consequently, experience obtained in one kind of flow method can often be applied to advantage in another kind. Moreover, modern electrophoretic methods have numerous features in common with liquid flow methods.

Like any other analytical method, flow analysis involves three principal steps, namely, sampling, a literal or figurative separation of the analyte(s) from the sample matrix, and measurement of a suitable analytical property. The flow system can then be characterized by the number of theoretical plates.

In order to avoid distortion of the analytical signal and deterioration in the resolution of analyte peaks in HPLC or sample zones in FIA and CFA, several conditions must be satisfied: all the sample interactions taking place within the analytical system and all the signal handling procedures must be sufficiently rapid; the hydrodynamics of the system must be well defined and reproducible; the effective volumes of the injector and the detector cell must not substantially exceed the volume of a single theoretical plate—hence the demands become progressively more stringent on transfer from industrial monitors characterized by a few theoretical plates, though FIA and CFA systems (tens to hundreds of plates), to HPLC instruments (thousands and more plates). For more details, see references 1 and 2 and the references therein.

Any measuring method can be used for flow detection provided that it meets the above requirements. Spectral methods, especially absorption spectrophotometry, are most common. Of the various electrochemical methods, conductometry has found routine use in certain industrial monitors, *e.g.* total electrolyte content, and in ion chromatography. Ion-selective electrode potentiometry is quite popular in FIA. Voltammetry and coulometry have been used in all flow techniques for specialized applications where their inherent advantages of high sensitivity for certain analytes and controllable selectivity, can be fully utilized. They are also suitable for use in capillary separations, as voltammetric cells are readily miniaturized. The problems of electrode fouling, which often deter analysts from using electrochemical methods, are less serious in flow analyses than in batch measurements, as the electrodes are continuously washed with the pure carrier liquid and only briefly exposed to sample zones. On the other hand, electrode interactions with the test solution have made it possible to develop important techniques such as stripping voltammetry, where analytes are accumulated through (electro)chemical reactions and physical adsorption, and permitted the use of physico-chemically or biochemically modified electrodes, in which systems selectively interacting with analytes in solution or catalysing their electrochemical reactions are attached to the electrode.

The field of electrochemical flow detection is now well established; its general treatment and a survey of applications up to 1984 can be found in reference 1 and a more recent review in reference 2.

This paper aims to point out some of the most important directions for further development of flow electroanalysis. In the author's opinion, these directions involve the following:

(i) use of micro- and ultramicroelectrodes and their arrays;
(ii) physico-chemical and biochemical modification of the electrode surface and utilization of selective chemical reactions between analytes and the electrode material;
(iii) rapid potential-scan techniques, or dual and multichannel detection;
(iv) combination of several detection methods, spectrochemistry, and spectrochemical derivatization;
(v) use of membrane electrodes and solid-polymer electrolyte cells; and
(vi) charge transfer between two immiscible electrolyte solutions.

Only a few selected examples are discussed below.

2 Microelectrodes

The advantages of voltammetric microelectrodes and their arrays, namely, rapid mass transport and consequent rapid steady-state establishment and a very low ohmic drop in microelectrode measurements, are now well known and extensively used.[3] For flow detection, microelectrodes offer further attractive features: lateral diffusion (edge effect) enhances the signal; the signal dependence on the flow rate is suppressed; and with microelectrode arrays, the diffusion layer is replenished with the analyte during passage of the liquid over the insulator separating the microelectrodes which again results in signal enhancement. Hence, measurements with microelectrodes readily permit (a) rapid potential scanning, (b) the possibility of two-domain recordings (time and potential), and (c) operation in poorly conductive liquids without a base electrolyte.

Also, the accessible potential range is extended. The small dimensions of microelectrodes make them readily applicable as detectors in micropacked and capillary column separations.

Examples of single microelectrode detection in HPLC employ a carbon fibre inserted into the end of a fused-silica capillary column[4] or a microdisk wall-jet system with a carbon composite electrode[5] also used with a fused-silica capillary column. A single fibre electrode has been placed across the flow,[6,7] while an array of carbon fibres has been used either parallel with the flow[8] or perpendicular to it.[9] The properties of microcylindrical electrodes,[10] microelectrode arrays,[11,12] and a microdisk electrode[13] in FIA and HPLC systems have been studied in detail.

It can be expected that voltammetric microelectrode detection will soon find practical use in supercritical fluid chromatography and in capillary electrophoretic techniques, primarily thanks to extremely low ohmic voltage drop values, as pointed out above. Another very promising field in which microelectrodes should play an important role is discussed in the following section.

3 Rapid Potential-scan Techniques and Multichannel Detection

Rapid potential-scan techniques have been used for some time. The two principal problems of this approach, encountered when working with normal-size electrodes, namely, very large charging currents and distortion of the voltammograms owing to sluggish electrode response, have been partially overcome by using square-wave,[14] differential pulse,[15] phase-sensitive a.c.,[16] or coulostatic staircase[17] voltammetry.

However, microelectrodes are much better suited for the purpose, as already pointed out in the pioneering work on rapid-scan flow detection,[14] and it is possible to work successfully even with very fast linear potential scans. Good results were obtained using a derivative measurement[18] with a single carbon fibre microelectrode inserted into the end of a fused-silica capillary column even under LC conditions of gradient elution.[19] Also, well developed chromatovol-

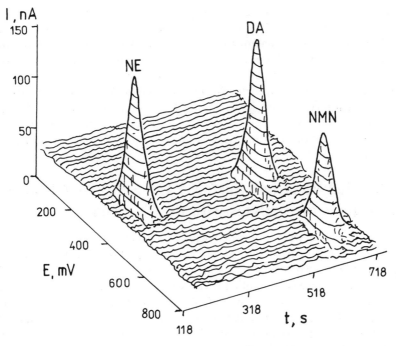

Figure 1 *A square-wave chromatovoltammogram of 10^{-5} M norepinephrine (NE), dopamine (DA), and normetanephrine (NMN). Flow rate, 1 ml min^{-1}; potential-scan range, 0.0 to +0.8 V (Ag/AgCl); scan frequency, 2s; pulse amplitude, 25 mV; pulse frequency, 100 Hz*

tammograms were obtained by employing rapid-scan square-wave voltammetry to a carbon fibre microelectrode (Figure 1).[20]

Nevertheless, the amount of information obtainable from rapid-scan voltammograms should not be over estimated; unknown substances can rarely be directly identified and the method is primarily useful for checking whether a particular zone contains one or more analytes. The selectivity of quantitative HPLC analyses is definitely improved, but the precision and accuracy of measurement may be somewhat poorer compared with constant-potential amperometry or coulometry.

Even greater possibilities are offered by multichannel detection employing an array of microelectrodes, each working at a different set of electrochemical conditions (*e.g.* at a different constant potential), in direct analogy to diode-array spectrometric detectors. Multichannel detection should not only yield chromatovoltammograms of a better quality than rapid-scan techniques, but should also be amenable to extensive application of pattern recognition and other chemometric procedures.

Of course, the technique is at the very beginning and many technical problems must still be solved to obtain reliable, sufficiently small detection cells with satisfactory hydrodynamic properties, containing a sufficiently large number of independently operated electrodes whose signals do not influence one another.

So far, a thin-layer array of 32 gold strip electrodes, 400 µm wide with gaps of 200 µm, has been described,[21] but without any experimental results. Two more recent papers discuss 16-channel cells, one with 50 µm gaps,[22] and the other with a circular array of disk electrodes of glassy carbon, gold, or carbon paste.[23] These two papers also deal with the multipotentiostats required and the computer control of the systems. Current signals for 80 potential values were obtained[22] within 0.27 s and typical limits of detection were 20 pmol of ferrocene derivatives[22] or 0.1 pg of epinephrine.[23] Selectivity can further be improved by using electrodes of different materials or different pretreatment of electrodes in the array.

4 Modified Electrode Surfaces and Selective Chemical Reactions of Analytes with the Electrode

This is a very active, wide, and varied field (for a general survey see, *e.g.* reference 24, reviews of polymer-modified electrodes[25] and modified carbon paste electrodes[26] are also available). Application of modified electrodes to flow measurements is discussed in other reviews.[2,27,28] It should be pointed out that only a few of the great number of physico-chemically modified electrodes described in the literature find real practical use, as the electrodes are very often insufficiently stable, both physico-chemically and mechanically. In flow measurements, especially in chromatographic detection, there is an additional requirement of a sufficiently rapid response of the modified electrode. A theoretical model has been described[29] for amperometric electrodes covered with a permselective layer which deals with the response rate and optimization of the response curve height.

Solid electrodes actually become modified during procedures commonly used for their pretreatment and activation (polishing, heat treatment, chemical and electrochemical processes—the field was reviewed recently[30]). Electrochemical techniques are especially attractive because they can readily be included in the overall measurement programme.

For example, the surface of carbon electrodes can be electrochemically oxidized, either at high positive potentials,[31-33] or at high anodic currents.[34] A great variety of oxygen-containing functional groups are formed, primarily phenols, carbonyls, and carboxyls. With very high electric charges passed, graphite oxide is produced on the electrode surface which no longer exhibits redox properties, but behaves as an ion-selective membrane.[33,35] Generally, with increasing charge passed, the concentration of surface phenols increases and that of carbonyls decreases and there is a certain optimal charge at which the electrode exhibits the highest activity.

Many electrode reactions are enhanced in this way, through catalysis and/or reactant adsorption. The sensitivity of batch determinations may be substantially increased, *e.g.* see the limits of detection[34] given in Table 1. The measuring selectivity can also be improved, as demonstrated on separation of the overlapping DPP peaks of dopamine and ascorbic acid after anodic treatment of the electrode.[32,34] The effect on the sensitivity of flow detection is less

Table 1 *Limits of detection of DPP determination of epinephrine, norepinephrine, and dopamine before and after anodic treatment of a glassy carbon working electrode*[34]

| | *Limit of detection*/mol l^{-1} | | |
	Epinephrine	Norepinephrine	Dopamine
Before treatment	6.3×10^{-6}	2.3×10^{-6}	1.3×10^{-6}
After treatment	3.6×10^{-8}	3.7×10^{-8}	4.3×10^{-9}

0.1 M NaClO$_4$, 0.8 s pulse of 20 mA

pronounced, because the electrode reaction enhancement, as pointed out above, stems partially or completely from adsorption accumulation of the analyte at the electrode surface and the response rate is thus decreased. However, a great improvement in selectivity of detection is attained when using a dual carbon electrode detection system with one electrode untreated and the other anodically pretreated,[36] as both the half-wave potentials and the current magnitudes are greatly affected by the pretreatment (Table 2).

Another possibility is laser irradiation of the electrode surface, either in solution[37,38] or in air.[39] In contrast to anodic polarization, the laser treatment removes a thin film of adsorbed particles and the electrode material from the surface, resulting in a decrease in the amount of chemi-sorbed oxygen. This is actually an effect of heat produced, which increases with increasing intensity of the laser pulse and increasing pulse length. Many electrode reactions are accelerated by this treatment. The use of a laser complicates the instrumentation, nevertheless it has been proposed for periodic *in-situ* electrode activation, both in batch DPP measurements,[40] and in HPLC detection.[41]

Both anodic polarization and laser irradiation lead to a certain increase in the electrode surface roughness, so that the electrodes must be repolished from

Table 2 *The half-wave potentials of several analytes at anodically pretreated and untreated glassy carbon electrodes in a dual-electrode thin-layer cell and the peak-current ratios for the pretreated and untreated electrode*[36]

| Analyte | Half-wave potential (V vs. Ag/AgCl) | | Peak current ratio (pretreated/untreated electrode), at a potential of +0.6 V (Ag/AgCl) |
	Pretreated electrode	Untreated electrode	
Epinephrine	+0.59	+0.73	20.0
(3,4-Dihydroxyphenyl) alanine	+0.47	+0.71	11.1
5-Hydroxytryptophan	+0.58	+0.70	5.9
4-Hydroxytryptophan	+0.52	+0.66	1.7
L-Tryptophan	+0.86	+0.92	100.0
Homovanillic acid	+0.69	+0.79	27.8

A 20 mA galvanostatic pulse, 0.8 s long, applied to one of the working electrodes in 0.1 M HClO$_4$

time to time. The electrode activity deteriorates during measurement and the anodic and laser treatment must be periodically repeated. Both the procedures have potential use in preparing surfaces with a defined and reproducible surface concentration of surface groups for further chemical modification.

The field of electrode modification with enzymes and similar highly selective biochemical systems develops extremely rapidly (for a brief survey of applications to flow detection see, *e.g.* reference 2). In practical use, the greatest problems lie in the relative instability of such systems, so that the lifetime of biochemically modified sensors is usually rather short, and often exhibit a sluggish response which may prevent application to flow measurements, especially in HPLC.

Direct immobilization of a biochemical system on the sensor surface is advantageous from the point of view of the response rate: therefore, with rapid response, the analyte zone is less dispersed and thus the sensitivity of measurement is improved. On the other hand, its immobilization in a preceding reactor results in greater sample capacity and prolonged lifetime, as a much larger amount of enzyme is available strongly bound to a suitable support.[42] A recent example is the very rapid, simple, and reliable determination of ethanol in serum, using a short reversed-phase HPLC column for separation of the blood components, followed by an alcohol oxidase reactor and a carbon fibre electrode to detect the hydrogen peroxide formed (Figure 2).[43]

Another selective detection technique involves complexation reactions of the analytes with copper(II) ions present in a porous film on a passive copper electrode. This method has been used by a number of authors for potentiometric

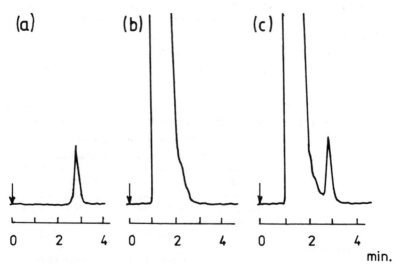

Figure 2 *Liquid chromatographic determination of ethanol in serum. Amperometric detection of hydrogen peroxide produced in a hydroxyethylmethacrylate reactor with immobilized alcohol oxidase. Flow rate, 0.2 ml min^{-1}; carbon fibre electrode potential, +0.9 V (Ag/AgCl); 20 μl sample. (a) Aqueous ethanol (1.72 × 10^{-3} mol l^{-1}); (b) Blank serum; (c) Serum spiked with 1.72 × 10^{-3} mol l^{-1} ethanol*

Table 3 *Calibration curve parameters for some analytes in distilled water as the mobile phase, using a solid polymer electrolyte cell*[46]

Analyte	Slope nA ng^{-1}	Intercept nA	Correlation coefficient	Limit of detection for a signal equal to twice the peak-to-peak noise/ng
Hydroquinone	62.6 ± 3.2	0.2 ± 0.8	0.998	2.5
4-Methylcatechol	115.5 ± 6.7	0.4 ± 0.4	0.998	1.4
$K_4[Fe(CN)_6]$	13.6 ± 1.8	0.3 ± 0.5	0.994	11.8

Sample volume, 50 µl; flow-rate, 0.3 ml min^{-1}; platinum electrode potential, +0.9 V (Ag/AgCl)

and amperometric detection of many inorganic and organic compounds that are capable of rapid complexation with copper(II) ions (primarily amino acids, short peptides, and various organic and inorganic anions). For a survey, see *e.g.* references 2 and 44.

A useful alternative to the use of ultramicroelectrodes in extending the

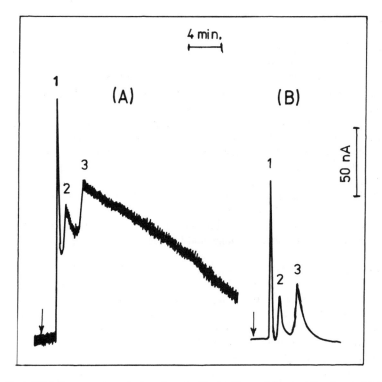

Figure 3 *A HPLC separation of phenols on a C_{18} column with a mobile phase of 4 %vol. dioxan in water. Flow rate, 0.3 ml min^{-1}; electrode potential, +0.9 V (Ag/AgCl). (A) Detector cell with a carbon fibre array electrode. (B) The solid polymer electrolyte cell with a platinum semi-microelectrode. 1—1.03 × 10^{-3} M hydroquinone; 2—1.15 × 10^{-3} M pyrocatechol; 3—2.59 × 10^{-3} M methylcatechol*

applicability of voltammetric detection to poorly electrically conductive liquids, are cells with a solid polymer electrolyte which separates the test solution from an auxiliary compartment containing a reference and counter electrode immersed in an electrolyte solution. So far, two such cells have been described, both containing the fluorinated copolymer Nafion (du Pont) as the solid electrolyte and employing either a large area working electrode obtained by chemical deposition of platinum on the Nafion membrane,[45] or a platinum wire semi-microelectrode piercing through the membrane.[46] Good operational parameters have been obtained (Table 3), even in HPLC detection (Figure 3).[46]

5 Conclusion

The few examples given above illustrate the present trends in electrochemical detection. It should be pointed out that these techniques will never become general detection methods. Their importance lies in certain specialized applications, in which their inherent advantages of high sensitivity and selectivity can be fully utilized, and when operated in combination with other detection methods.

References

1. K. Štulík and V. Pacáková, 'Electroanalytical Measurements in Flowing Liquids', Ellis Horwood, Chichester, 1987.
2. K. Štulík and V. Pacáková, *Sel. Electrode Rev.*, 1992, **14**, 87.
3. M. Fleischmann, S. Pons, D. R. Rolison, and P. P. Schmidt (eds.) 'Ultramicroelectrodes'. Datatech Systems Inc., Morgantown, NC, 1987.
4. L. A. Knecht, E. J. Guthrie, and J. W. Jorgenson, *Anal. Chem.*, 1984, **56**, 479.
5. M. J. J. Hetem, H. A. Claessens, P. A. Leclercq, C. A. Cramers, V. Pacáková, and K. Štulík, in 'Proceedings of the VIIIth Int. Symp. on Capillary Chromatography, Vol. II', ed. P. Sandra, Hüthig Verlag, Heidelberg, 1987, p. 1122.
6. H. Huiliang, Chi-Hua, D. Jagner, and L. Renman, *Anal. Chim. Acta*, 1987, **193**, 61.
7. Chi-Hua, K. A. Sagar, K. McLaughlin, M. Jorge, M. P. Meaney, and M. R. Smyth, *Analyst (London)*, 1991, **116**, 1117.
8. K. Štulík, V. Pacáková, and M. Podolák, *J. Chromatogr.*, 1984, **298**, 225.
9. Huamin Ji, Jun He, Shaojun Dong, and Erkang Wang, *J. Electroanal. Chem.*, 1990, **290**, 93.
10. J. E. Baur and R. M. Wightman, *J. Chromatogr.*, 1990, **482**, 65.
11. A. Aoki, T. Matsue, and I. Uchida, *Anal. Chem.*, 1990, **62**, 2206.
12. Tse-Yuan Ou and J. L. Anderson, *Anal. Chem.*, 1991, **63**, 1651.
13. D. L. Luscombe, A. M. Bond, D. E. Davey, and J. W. Bixler, *Anal. Chem.*, 1990, **62**, 27.
14. R. Samuelsson, J. O'Dea, and J. Osteryoung, *Anal. Chem.*, 1980, **52**, 2215.
15. J. J. Scanlon, P. A. Flaquer, G. W. Robinson, G. E. O'Brien, and P. E. Sturrock, *Anal. Chim. Acta*, 1984, **158**, 169.
16. A. Trojánek and H. G. de Jong, *Anal. Chim. Acta*, 1982, **141**, 115.
17. T. A. Last, *Anal. Chem.*, 1983, **55**, 1509.
18. J. G. White and J. W. Jorgenson, *Anal. Chem.*, 1986, **58**, 2992.
19. M. D. Oates and J. W. Jorgenson, *Anal. Chem.*, 1989, **61**, 1977.
20. S. P. Kounaves and J. B. Young, *Anal. Chem.*, 1989, **61**, 1469.
21. M. DeAbreen and W. C. Purdy, *Anal. Chem.*, 1987, **59**, 204.

22. T. Matsue, A. Aoki, E. Ando, and I. Uchida, *Anal. Chem.*, 1990, **62**, 407.

23. J. C. Hoogvlied, J. M. Reijn, and W. P. van Bennekom, *Anal. Chem.*, 1991, **63**, 2418.

24. R. W. Murray, 'Chemically Modified Electrodes', in 'Electroanalytical Chemistry', Vol. 13, ed. A. J. Bard, Marcel Dekker, New York, 1983, p. 191.

25. A. R. Hillman, 'Polymer Modified Electrodes: Preparation and Characterization', in 'Electrochemical Science and Technology of Polymers—1', ed. R. G. Linford, Elsevier Applied Science, London, New York, 1987, pp. 103; 291.

26. K. Kalcher, *Electroanalysis*, 1990, **2**, 419.

27. J. Wang, *Anal. Chim. Acta*, 1990, **234**, 41.

28. Erkang Wang, Huamin Ji, and Weying Hou, *Electroanalysis*, 1991, **3**, 1.

29. W. Breen, F. Cassidy, and M. E. G. Lyons, *Anal. Chem.*, 191, **63**, 2263.

30. K. Štulík, *Electroanalysis*, 1992, **4**, 829.

31. J.-X. Feng, M. Brazell, K. Renner, R. Kasser, and R. N. Adams, *Anal. Chem.*, 1987, **59**, 1863.

32. J. Wang and M. S. Lin, *Anal. Chem.*, 1988, **60**, 1459.

33. L. J. Kepley and A. J. Bard, *Anal. Chem.*, 1988, **60**, 1459.

34. J. Mattusch, K.-H. Hallmeier, K. Štulík, and V. Pacáková, *Electroanalysis*, 1989, **1**, 405.

35. V. Majer, J. Veselý, and K. Štulík, *J. Electroanal. Chem.*, 1973, **45**, 113.

36. J. Mattusch, G. Werner, K. Štulík, and V. Pacáková, *Electroanalysis*, 1990, **2**, 443.

37. E. Hershenhart, R. L. McCreery, and R. D. Knight, *Anal. Chem.*, 1984, **56**, 2256.

38. M. Poon and R. L. McCreery, *Anal. Chem.*, 1986, **58**, 2745.

39. K. Štulík, D. Brabcová, and L. Kavan, *J. Electroanal. Chem.*, 1988, **250**, 173.

40. M. Poon and R. L. McCreery, *Anal. Chem.*, 1987, **59**, 1615.

41. K. Sternitzke, R. McCreery, C. S. Bruntlett, and P. T. Kissinger, *Anal. Chem.*, 1989, **61**, 1989.

42. V. Pacáková, K. Štulík, D. Brabcová, and J. Barthová, *Anal. Chim. Acta*, 1984, **159**, 71.

43. V. Pacáková, K. Štulík, Kang Le, and J. Hladík, *Anal. Chim. Acta*, 1992, **257**, 73.

44. K. Štulík, V. Pacáková, Kang Le, and B. Hennissen, *Talanta*, 1988, **35**, 455.

45. T. W. Kaaret and D. H. Evans, *Anal. Chem.*, 1988, **60**, 657.

46. L. Loub, F. Opekar, V. Pacáková, and K. Štulík, *Electroanalysis*, 1992, **4**, 447.

Flow Injection Analysis with Ion-selective Electrodes: Recent Developments and Applications

L. K. Shpigun

KURNAKOV INSTITUTE OF GENERAL AND INORGANIC CHEMISTRY, RUSSIAN ACADEMY OF SCIENCES, MOSCOW 117907, RUSSIA

Summary

The field of Flow Injection Analysis with Ion-Selective Electrodes (Flow Injection Potentiometry, FIP) is reviewed and its current status assessed. Some recent results concerning developments and applications of conventional ion-selective electrodes (ISEs) for potentiometric detection in flow-injection systems are presented. The main attention is focused on the role of the dynamic response characteristics of potentiometric devices and FI manifold parameters on the form of resulting signal and the development of optimized FIP systems. The limitations of ISEs and the use of flow-injection systems to extend the capabilities of ISEs, particularly as far as detection limits and matrix interferences, are discussed. These topics are illustrated by results from a detailed study of the solid-state membrane Ag-, Cu-, Pb-, and Cd-ISEs. Different approaches are illustrated for the FIP determinations of transition and heavy metal ions. Special attention is paid to the development of new flow-through detectors with poly(vinyl chloride)-matrix membrane electrodes based on synthetic macrocyclic complexing reagents. FIP systems for routine analysis of water are proposed. Some areas of current intensive activity and future potential for FIP techniques are indicated.

1 Introduction

The use of potentiometric detectors in flow analysis is not new, but has recently achieved special importance due to the increased needs for automatic analytical techniques and for miniaturization of measuring devices.[1-3] Pungor *et al.*[4-6] and Ruzicka and Hansen *et al.*[7-9] were the first to employ potentiometric ion-selective sensors for solving analytical problems by means of flow injection analysis (FIA) techniques. Nowadays, detectors based on classical membrane ion-selective electrodes (ISEs), coated wire electrodes, or field effective transistors hold very important positions in conjunction with FIA.[10-12] This hydrodynamic modern electroanalytical method called Flow Injection Potentiometry (FIP) offers great possibilities when designing new analytical systems for large numbers of determinations of many inorganic and organic compounds.

Generally, FIP techniques allow numerous problems to be solved in a rather simple and convenient way in the field of clinical chemistry and in pharmaceutical product control.[13-15] A variety of electrodes have been employed as sensors in agricultural and environmental analyses.[16-18] Some examples of FIP applications in basic electrochemical research have been also reported.[19-21] Finally, the modified reversed FIP technique with ISEs was shown to be of use for process analysis and monitoring purposes.[22]

ISEs are typical surface detectors; the surface concentration which is sensed by the electrode in a FIA system is examined under highly controlled hydrodynamic conditions during the fixed measuring time. The FIA methodology is thus a tool to achieve improved sensor capabilities. Compared to manual ISE potentiometry, the use of ISE-detectors in FI systems has several undisputed advantages: (i) the reproducibility of the analytical signal; (ii) the reproducible use of small sample volumes; (iii) faster electrode response and the ability to achieve a rapid sample throughput; and (iv) the improved selectivity due to kinetic discrimination against interfering and membrane-inactive species, as well as to on-line preliminary separation of the analyte from an interfering matrix. Moreover, in flow-injection systems, ISEs can be used for potentiometric detection not only in the Nerstian range, but also in the sub- or super-Nernstian ranges of electrode functioning. Further key features among the capabilities of FIP are: high detection sensitivity; wide measuring concentration range; and simple and low cost experimental set-up. The theoretical and practical aspects of FIP, as well as the advantages of ISEs when used in FIA, have been reviewed extensively by Pungor et al.,[23] Frensel,[24-26] and others.[27-30]

In general, the formation of the potentiometric signal produced by any ISE-detector is based on the dynamic local equilibrium reactions at the sensor interface where the membrane potential is measured. The most important parameters affecting sensitivity and selectivity of the signal in the combination of FIA with ISEs, are the electrochemical properties of an ion-sensing membrane and the dynamic response characteristics of a potentiometric measuring cell. Thus the development of ISE-detectors for FIA is one of the key problem areas in hydrodynamic electroanalytical chemistry. Other factors, which should be taken into account when using ISE-detectors in flow-injection systems, are mainly due to the effects of the FIA operating parameters on the dynamic electrode response.

During the last decade, a number of solid-state membrane ISE-detectors have been proposed for FIP.[31-34] However, until now, little work has been carried out on the application of FIP for metal determinations, especially for transition and heavy metal cations. J. F. van Staden has demonstrated the suitability of tubular solid-state membrane cation-selective electrodes for FIA.[35-39] For example, a coated tubular solid-state membrane copper(II)-ISE has been employed as sensor for analysis of effluents, tap water, and copper sulfate plating bath solutions.[35] In our laboratory, the use of commercially available solid-state membrane ISEs in flow-injection systems incorporating potentiometric cells of the cascade or wall-jet type was systematically investigated in order to improve their performance in analysis.[39,40]

A few liquid or poly(vinyl chloride) (PVC) membrane-based potentiometric sensors have been incorporated in FIP manifolds, mainly for sodium-, potassium-, and lithium-ion determinations.[1,7,41-44] However calcium ISEs for FIA have been also described.[45] We have developed a flow-through potentiometric sensor with a PVC-matrix membrane containing dibenzo-18-crown-6 or dibenzo-diaza-18-crown-6 as ionophores.[46]

Recently, FIP systems incorporating various types of flow cells with ISEs have been investigated for the determination of heavy metal ions and other ionic species in environmental and geochemical analysis. The first part of this review will outline the common methodological aspects of FIP with ISEs, as well as experience with conventional solid-state membrane heavy metal cation-selective electrodes (Ag-, Cu-, Pb-, or Cd-ISEs) used in cascade and wall-jet model flow cells. The second part will describe the development and analytical performance of new metal ISE-detectors based on tubular PVC-matrix membranes containing macrocyclic complexing components as ionophores. The third part will demonstrate the utility of the FIP method with ISEs for water analysis, with special emphasis on trace inorganic analysis of sea water.

2 Experimental

Reagents and Solutions

All the solutions were prepared from analytical-reagent grade chemicals and re-distilled water. The diluted heavy metal solutions were prepared daily from 1000 p.p.m. aqueous standard solutions for AAS (BDH Chemicals).

The new synthetic macrocyclic compounds (Figure 1) used as membrane ionophores were received from the Department of Organic Chemistry of the Moscow State University (Russia).

High relative molecular mass poly(vinyl chloride) (PVC), 2-nitrophenyl octyl ether (NFOE), potassium tetrakis-*p*-chlorophenylborate (KT*p*ClPB) and other organic reagents were supplied by Fluka (Buchs, Switzerland).

Detector Design and Electrodes

All potentiometric measurements with solid-state membrane ISEs were carried out with laboratory-made cascade or wall-jet types of flow cells containing a membrane electrode and AgCl/Ag-reference electrode similar to those described previously.[7,11] Commercially available metal sulfide-based membrane electrodes, sensitive to silver-, copper-, lead-, or cadmium-ions, an iodide-ISE ('Critur') and an Orion 90-02-00 reference electrode were used.

A laboratory-made tubular membrane measuring flow cell was used for potentiometric detection with PVC-matrix membrane ISEs.[46] Plasticized

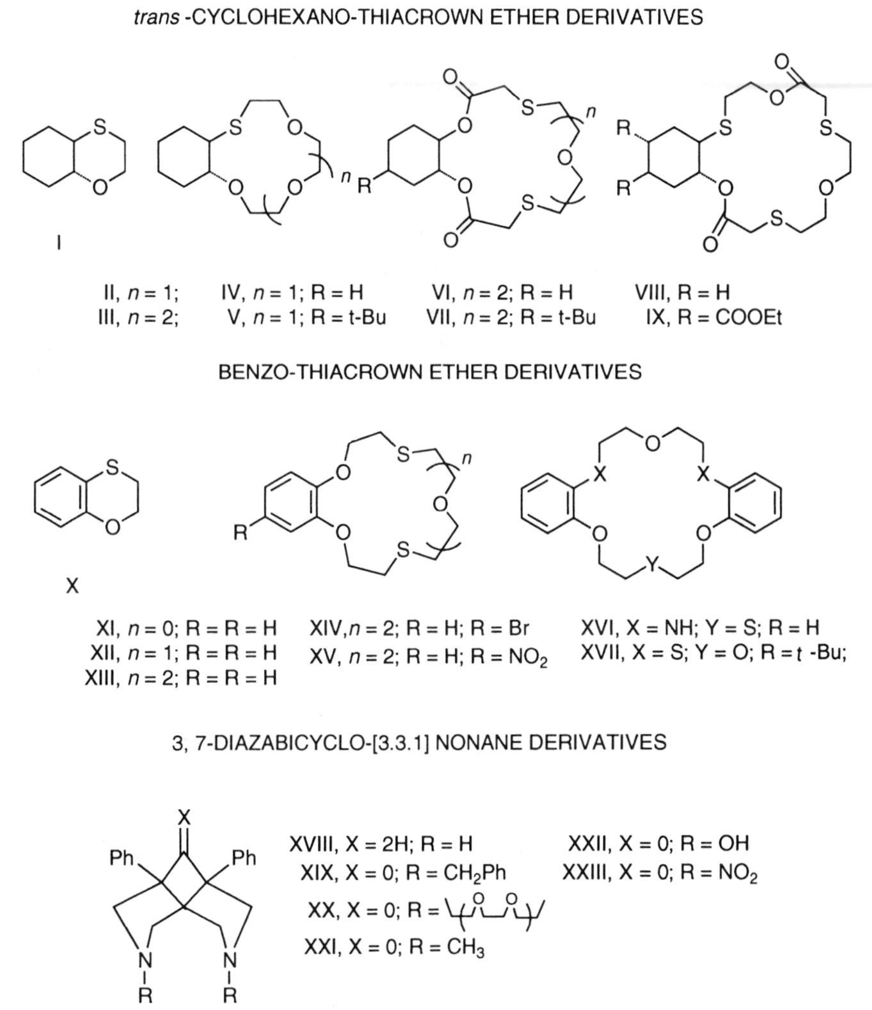

trans-CYCLOHEXANO-THIACROWN ETHER DERIVATIVES

II, $n = 1$;	IV, $n = 1$; R = H	VI, $n = 2$; R = H	VIII, R = H
III, $n = 2$;	V, $n = 1$; R = t-Bu	VII, $n = 2$; R = t-Bu	IX, R = COOEt

BENZO-THIACROWN ETHER DERIVATIVES

XI, $n = 0$; R = R = H	XIV, $n = 2$; R = H; R = Br	XVI, X = NH; Y = S; R = H
XII, $n = 1$; R = R = H	XV, $n = 2$; R = H; R = NO$_2$	XVII, X = S; Y = O; R = t-Bu;
XIII, $n = 2$; R = R = H		

3, 7-DIAZABICYCLO-[3.3.1] NONANE DERIVATIVES

XVIII, X = 2H; R = H	XXII, X = 0; R = OH
XIX, X = 0; R = CH$_2$Ph	XXIII, X = 0; R = NO$_2$
XX, X = 0; R =	
XXI, X = 0; R = CH$_3$	

Figure 1 *The synthetic macrocyclic compounds tested as PVC–matrix membrane ionophores*

PVC-matrix membranes were prepared by the procedure of Moody and Thomas[47]; the membranes contained (mass ratio in percentages): the sensing compound, 2.0; PVC, 32–33; *o*-NFOE, 64–65; KT*p*ClPB, 0–0.2.

Flow-Injection Manifolds and Apparatus

A FIAStar-5020 flow-injection analyser ('Tecator') equipped with a digital pH-meter ('Orion' 811) was used connected to a pen-recorder ('Linear Instrument Corporation').

The potentiometric measurements were made using different single- or multi-line FIP manifolds in which all the connecting tubes were of 0.5 mm i.d. The volumetric flow-rates, v; the injection sample volumes, S; and the lengths, L of the mixing or reaction coils and the carrier compositions were as given in the figures or in the text.

Evaluation Parameters of Potentiometric Selectivity

The potentiometric selectivities of ISE-detectors were evaluated using two different parameters: the selectivity coefficient, $K_{M,N}^{Pot}$ [48] and the relative selectivity factor, $F_{M,N}^{Pot}$. Both parameters were determined using the separate solution method based on the following flow-injection procedure. Equal concentrations, C of primary M^{z+} or foreign N^{n+} metal ions were separately injected into the carrier stream. The potential changes ΔE as peak height, H with respect to the base-line were recorded and used for calculation of potentiometric selectivity parameters.

The selectivity coefficients $K_{M,N}^{Pot}$ were calculated from the following equation:

$$\log K_{M/N}^{Pot} = [-(H_M - H_N)/\alpha] + [(1 - \bar{\bar{Z}}_M/\bar{\bar{Z}}_N)\log C/D \tag{1}$$

where H_M and H_N are the peak heights measured for the primary M^{z+} and foreign N^{n+} cations, respectively; z_M and z_N are the charge numbers of these cations; C_M is the primary cation concentration in the injected solution (mol l^{-1}); D is the total dispersion and α is the slope of primary ion response function.

The relative selectivity factor $F_{M,N}^{Pot}$ was calculated from the equation:

$$F_{M,N}^{Pot} = (H_N - H_M)/H_M \tag{2}$$

The major and important difference between these two parameters is that the relative selectivity factor $F_{i,j}^{Pot}$ can be used in the cases of non-Nernstian dynamic electrode behaviour. From this point of view, this characteristic seems to be more suitable than the commonly used selectivity coefficient $\log K_{i,j}^{Pot}$ for evaluation of the selectivity of the dynamic electrode response in non-equilibrium FIA conditions.

In cases of Nernstian electrode response, the two selectivity parameters can be connected by the following relationship ($z_M = z_N$):

$$F_{M,N}^{Pot} = (\alpha/H_M)\log K_{M,N}^{Pot} \tag{3}$$

If the electrode response to primary ion M^{z+} is Nernstian with slope α, then the peak height can be given as $H_M = k\alpha$. Hence, relationship (3) becomes

$$F_{M,N}^{Pot} = \log K_{M,N}^{Pot}/n \tag{4}$$

where n is equal to the number of decades of the primary ion concentration range corresponding to the linear electrode response with the slope α.

3 Results and Discussion

The height H of the recorded signal is measured as the maximal potential change caused by the injected solution of the primary ion M^{z+} with respect to the base-line, the potential, E, of which is caused by the carrier solution composition and can be expressed by the equation:

$$E_{\text{base-line}} = E_0 + \alpha \log(a_{\text{M,carrier}} + K_{\text{M,B}}^{\text{Pot}} a_B{}^{Z_M/Z_B}) \tag{5}$$

where α is the slope of the electrode response, mV decade^{-1}; $K_{\text{M,B}}^{\text{Pot}}$ is the selectivity coefficient value against ion B^{z+} in the base-line supporting solution (BLSS); $a_{\text{M,carrier}}$ and a_B are the activities of the primary ion M^{z+} and foreign ion B^{z+} in the carrier stream, respectively.

For dynamic equilibrium FIP measurements with ISEs (Nernstian range), the peak height H measured on injection of the solution of primary ion activity in a carrier solution containing trace amounts of primary ions is given by the following expression (if $D = 1$):

$$H = \Delta E(t) = \alpha \log[a_M/(a_{\text{M,carrier}} + K_{\text{M,B}}^{\text{Pot}} a_B{}^{Z_M/Z_B})] \tag{6}$$

This means that the composition of the carrier solution together with the total dispersion D of the injected sample zone in a flow manifold are also of importance. For these reasons, the use ISEs in FIA requires an optimization of the flow manifold, the measuring cell design, and the chemical conditions in the FIP system.

It is known that the FIA signal is the result of a complex interaction of a number of factors but depends largely on the dynamic response characteristics of the detectors.[49]

The dynamic response of an ISE-detector can be characterized with the help of the time parameter τ' which is the period of time taken for the potential to change with the maximal slope $\Delta E/\Delta t$ on the dynamic potential–time curve (Figure 2). Hence, the analytical signal detected with an ISE in a FIP system is influenced by the ratio of the contact time Δt of the injected sample zone with the membrane surface and the dynamic response time τ' of the detector.

When $\Delta t > \tau'$, the resulting signal is caused by the difference in the steady-state values of the electrode potential and is practically independent of the FIP parameters. When $\Delta t < \tau'$, the potentiometric signal is transient and decreases with increasing flow-rate v and decreasing injected sample volume. Under the latter conditions, some anomalies in the sensor response and deviations from the Nernst equation are expected. Often a super-Nernstian slope of the dynamic response can be observed.

The selectivity properties of a sensor are also contact time-dependent and usually the potentiometric selectivity increases with decreasing contact time. The dependence of $K_{\text{M,N}}^{\text{Pot}}$ values on the hydrodynamic conditions, as well as, the sample and carrier solution composition, has been reported by other authors.[25–28]

It should be pointed out that when an analytical procedure based on flow-injection potentiometry principles is being developed, the selection of the

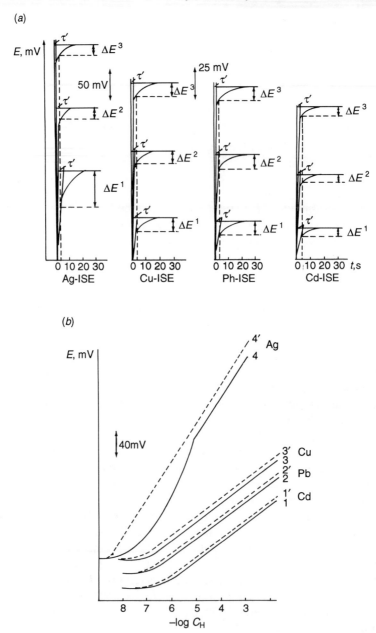

Figure 2 (a) *Dynamic response–time curves recorded for an ISE in a dispersion FIP system by injecting 3 ml of solutions containing $1 \times 10^{-5}\,mol\,l^{-1}(1)$, $1 \times 10^{-3}\,mol\,l^{-1}$ (2), and $1 \times 10^{-1}\,mol\,l^{-1}(3)$ primary ion; (b) Experimental calibration curves obtained for $t = 5\,s(—)$ and under the dynamic equilibrium conditions $(---)$ for potentiometric measurements with Cd-ISE, Pb-ISE, Cu-ISE, and Ag-ISE. The carrier is $0.5\,mol\,l^{-1}$ potassium nitrate solution with $5 \times 10^{-7}\,mol\,l^{-1}$ primary ion added*

Δt is most important and should be evaluated from the dynamic response–time curves recorded under the chosen flow conditions.

In order to optimize the dynamic measurement range and obtain a Nernstian response for the ISE-detector, the following relationship must be fulfilled:

$$\tau' \leqslant \Delta t = 60SD/V \, (s) \qquad (7)$$

On the basis of this relationship, it is possible not only to choose the FIP manifold variables but also to predict the behaviour of a given electrode in a flow-injection system with the selected variables.

Below, some results are given of an evaluation of the proposed approach for optimizing the potentiometric signal in FI systems with conventional ISEs, for the determination of transition and heavy metal ions with good sensitivity and selectivity.

Performance of Solid-state Membrane Cation-selective Electrodes for Flow-injection Potentiometry

A series of experiments were conducted to study the dynamic electrochemical behaviour of metal sulfide membrane-based Ag-, Cu-, Pb-, and Cd-ISEs in a low-dispersion flow-injection system. FIP signals were obtained by injection of 200 μl of the primary ion standard solutions in 0.5 mol l^{-1} potassium nitrate into a carrier stream (0.5 mol l^{-1} potassium nitrate solution containing 5×10^{-7} mol l^{-1} primary ion) with a flow rate of 2.8 ml min^{-1}. The data were compared with the corresponding dynamic response–time curves recorded for injections of large volumes of the primary ion (M^{z+}) solutions into the same flow manifold (Figure 2a). Table 1 shows the peak profile characteristics as well as the values of τ' obtained for the different primary ion concentrations for measurements with Ag-, Cu-, Pb-, and Cd-ISEs. Because the contact time Δt in the flow-injection potentiometric measurements was 5 s, the potential differences ΔE between the values detected at this time and the near equilibrium response time τ' were determined with respect to different activities (concentrations) of a primary ion and compared with each other. The results indicated that a fast and reversible electrode response was observed with Cu-, Pb-, and Cd-ISEs and for these electrodes, the differences ΔE were found to be practically

Table 1 *Dynamic Response Time and Peak Profile Charactertistics for the Studied Solid-state Metal ISEs ($n = 5$, $P = 0.95$)*

| Parameter | Concentration of primary ion/ mol l^{-1} | Ion-selective Electrode | | | |
		Ag-ISE	*Cu-ISE*	*Pb-ISE*	*Cd-ISE*
τ'	$\begin{cases} 1 \times 10^{-5} \\ 1 \times 10^{-3} \\ 0.1 \end{cases}$	15.0 3.0 2.0	5.0 2.5 2.0	5.0 2.5 1.0	4.5 2.0 1.5
$H = \Delta E/\text{mV}$	1×10^{-3}	255 ± 2	88 ± 1	82 ± 1	70 ± 1
t_b/s	1×10^{-3}	1140 ± 4	270 ± 2	240 ± 2	210 ± 2
t_w/s	1×10^{-3}	1135 ± 4	265 ± 2	235 ± 2	205 ± 2

independent of the primary ion activity (concentration). Thus, an almost Nernstian relationship between the recorded peak heights H and the primary ion concentrations C_M in the injected solution was obtained over a wide concentration range (Figure 2b).

In contrast, the value of ΔE measured with the Ag-ISE increased with a decrease in the primary ion activity. These results correspond well to the super-Nernstian slope on the calibration curve observed in the flow-injection measurements with an Ag-ISE. This electrode only shows Nernstian behaviour for silver ion concentrations above $1 \times 10^{-4}\,\mathrm{mol\,l^{-1}}$. At the lower silver ion concentration range the super-Nernstian slope, α, of the electrode response is obtained (Figure 2b). An increase of the silver ion concentration in the carrier solution from 5×10^{-7} to $5 \times 10^{-5}\,\mathrm{mol\,l^{-1}}$ causes the decrease in the slope, α. The peak heights $H = E(\Delta t)$ obtained by the Cu-, Pb-, and Cd-ISEs slowly decreased with increases in the carrier flow rate, v. In the case of the Ag-ISE, the effect of variation of the carrier flow rate v was more pronounced, especially with low silver ion concentrations. The peak width (t_b) decreased with increase in the flow-rate and silver ion concentration in the carrier solution and with decrease in the injected volume. These results clearly demonstrate that the influence of the main FIP system variables on the transient potentiometric signals is set by the properties of the membrane material of an ISE and by the primary ion activity (concentration) in the carrier and in the injected solutions.

The potentiometric selectivity of the electrodes were also found to be dependent on the dynamic electrode response–time-characteristics of an ISE. Figure 3a shows the electrode response–time curves recorded with an Ag-ISE for injection of large volumes of $0.1\,\mathrm{mol\,l^{-1}}$ metal ion solutions into the same

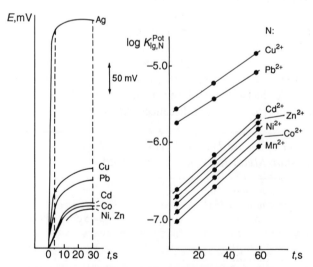

Figure 3 (a) *Dynamic response–time curves obtained in the FIP system with an Ag-ISE by injection of large volumes of $0.1\,mol\,l^{-1}$ primary or foreign ion solutions; and the dependence of potentiometric selectivity coefficients on the contact time t of the injected solution zone with the electrode surface*

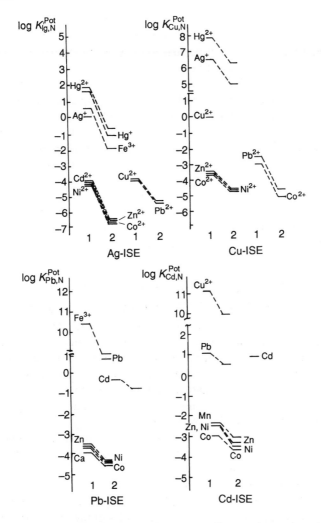

Figure 3 (*Continued*) (b) *Potentiometric selectivity coefficients,* log $K_{K,N}^{pot}$, *calculated from the potentiometric measurements with the solid-state cation-selective electrodes under continuous flow equilibrium conditions* (1) *and FIP conditions* (2)

FIP manifold. It can be seen that the values of the time parameter τ' for the primary and foreign ions are quite different and that the potential differences E between the electrode responses for primary and foreign ions depend on the contact time, Δt and mainly decrease with an increase of this time. These differences produce the effect of kinetic discrimination of foreign ions while making FIP measurements. Under the conditions $\Delta t < \tau'$, the selectivity coefficients calculated using equation (1) were found to be dependent on the contact time Δt and increased with decreasing Δt (Figure 3a). Hence, the selectivity data evaluated for $\Delta t = 5\,\mathrm{s}$ were smaller than those obtained by steady-state measurements (Figure 3b).

The utility of precipitate-based ISE-detectors for direct potentiometric measurements is often hampered by poor potentiometric selectivity, a limited range of determined ions, and gradual fouling of the electrode surface due to adsorption of large organic species or reaction products.

One approach that offers a possible solution to the above problems is the use of liquid or quasi-liquid membrane-based ISE-detectors. As mentioned earlier, only a few papers deal with such sensors in FIP. Therefore, our investigations also studied the design, performance characteristics, and applications of metal ion-selective detectors based on PVC–matrix membranes containing various classes of lipophilic organic complexing agents or their complexes with metal ions.

The following section describes our recent investigation of the characteristic behaviour of new synthetic *S*- and *N*-containing macrocyclic compounds, in order to choose the most selective ion-sensing material for preparing high-performance ISE-detectors for heavy metal ions.

Ion-selective Electrode Detectors for Heavy Metal Ions Based on PVC–Matrix Membranes with Macrocyclic Complexing Agents

A low-dispersion FI manifold was used to investigate the dynamic electrode response characteristics of series of polymeric membranes containing a number of synthetic macrocyclic compounds, namely cyclohexano-thiacrown ethers and their 4-alkyl derivatives, mono- and dibenzo-thia- or azacrown ether derivatives and 3,7-diazabicyclo-[3.3.1]nonane derivatives (Figure 1).

The significance of electrically neutral ionophores like macrocycles as metal cation-sensing materials in ISEs was first recognized by W. Simon *et al.*[50] and since then an increasing interest has been focused on the molecular design of the structures and their application in ISEs for flow injection determination of alkali and alkaline earth metal cations.[51]

The results demonstrate the influence of the structures of the macrocyclic compounds, and the kinetics of the complexing reactions at the solution–membrane interface, on the dynamic response characteristics of the membrane. It has been found that the potentiometric selectivity parameters for various metal ions were quite different and dependent upon the nature of the ionophores and the other membrane-forming compounds. When various metal ion salts were used as carrier solutions in a flow-injection system, changes in the selectivity parameters for the same membrane were also observed. These results are explained by the known effect of the influence of a carrier solution composition on the base-line electrode potential. Hence, for a comparative evaluation of the potentiometric selectivity of different ionophores, the dynamic selectivity parameters for metal cations, referred to the potassium cation, were calculated from the peak heights recorded for injection of 200 µl of pure solutions of metal ions ($0.01 \, \text{mol} \, l^{-1}$) into the $0.01 \, \text{mol} \, l^{-1}$ magnesium chloride carrier solution.

Figure 4 shows the relative selectivity factors $F_{K,N}^{Pot}$ obtained for membranes based on several *trans*-cyclohexano-thiacrown ethers I–IX. The data demonstrates the high selectivity for mercury and silver ions over other metal cations

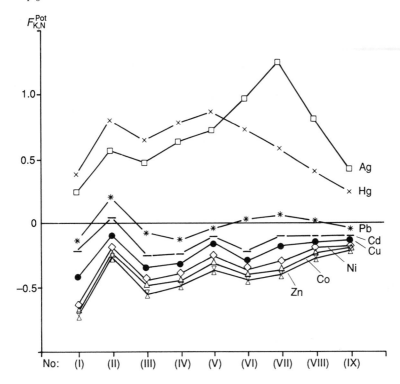

Figure 4 *The relative selectivity factors, $F_{K,N}^{Pot}$, obtained for membranescontaining trans-cyclohexano-thiacrown ethers I–IX*

for almost all of the compounds tested. A selectivity pattern of Hg > Ag > Pb > Cd ≫ Zn,Ni,Co,Mn was obtained for cyclohexano-thiacrown ethers I–III. The membranes based on cyclohexano-thia-12-crown-4(II) showed positive values of the relative selectivity factors $F_{K,N}^{Pot}$ not only for mercury and silver cations but also for lead and cadmium cations.

In the case of cyclohexano-dioxcodithiacrown ethers IV–VII, the $F_{K,N}^{Pot}$ values were significantly dependent on the size of the ring as well as on the nature of the 4-substituent R in the cyclohexane fragment. The selectivity for silver ions was increased by increasing the ring size from 15/5 to 18/6, whereas it decreased for mercury ions. Introducing a butyl substituent in the cyclohexane fragment of 15- and 18-member ring macrocycles (V, VII) improved the selectivity for mercury and silver over Group I and II ions. In contrast, increasing the number of sulfur atoms in the 18-member ring of compounds VIII and IX reduced the potentiometric selectivity for these metal ions.

In general, membranes containing t-butyl-cyclohexano-dioxcodithia-18-crown-6 (VIII) show good preference for silver ions over all of the other cations and can thus be recommended for preparing Ag-ISE.

Similar results were obtained for membranes containing benzo-thiacrown ether derivatives X–XV. Figure 5a shows a pictorial representation of selectivity factors and signals for several benzo-dithiacrown ethers recorded for injection

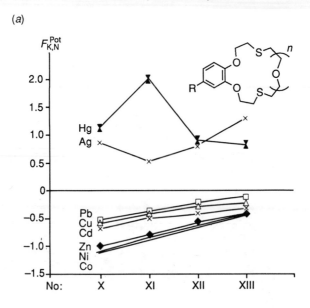

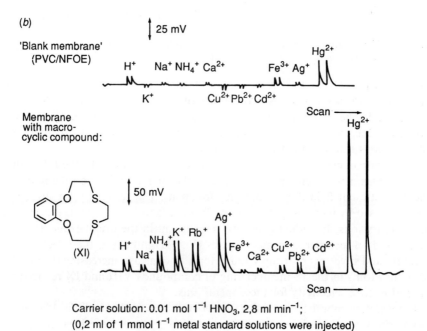

Carrier solution: 0.01 mol 1⁻¹ HNO₃, 2,8 ml min⁻¹;
(0,2 ml of 1 mmol 1⁻¹ metal standard solutions were injected)

Figure 5 (a) *The relative selectivity factors $F_{K,N}^{Pot}$ for PVC-matrix membranes with benzo-thiacrown ethers X–XII; (b) The dynamic electrode responses for the 'blank' membrane and the membrane containing compound XI*

of different metal ion solutions into an acidic carrier solution stream. All the macrocycles showed a remarkable response to mercury(II) and silver ions. High values of $F_{Hg,N}^{Pot}$ were obtained for membranes containing benzo-7,10-dithia-12-crown-4(XI). These data indicate the excellent potentiometric selectivity exhibited to mercury ions over the Group I and II ions as well as some transition and heavy metal ions (Figure 5b). The selectivity of the membrane with benzo-dithia-18-crown-6 derivatives (XII–XVI) decreased in the order: $Ag^+ > Hg^{2+} > Pb^{2+}, Cd^{2+} > Cu^{2+} > Zn^{2+}, Co^{2+}, Ni^{2+}$. It was found that the introduction of R-substituents ($R = NO_2$, Br) into the benzene fragment of the 18-member ring macrocycle improved both the sensitivity and selectivity of the membrane towards the heavy metal ions compared with Group I and II ions (Figure 6a). For example, the membrane containing ionophore (XV) responded rapidly not only to silver and mercury ions, but showed a near-Nerstian response to cadmium ions over the range $1 \times 10^{-4.5}$ – 1×10^{-2} mol l^{-1} for CdCl$_2$ solutions (Figure 6b).

Experiments were also carried out to compare the properties of two N- and S-containing dibenzo-18-crown-6 derivatives XVI and XVII. The membrane containing the dibenzo-dithia-18-crown-6 derivatives (XVI) produced fast and reproducible electrode responses to silver ions and that with dibenzo-diazathia-18-crown-6 (XVII) showed good dynamic responses to lead ions.

The data in Figure 7 demonstrates that manganese(II) ions can be sensed with surprisingly high selectivity using membranes containing a new class of ionophore—3,7-diazabicyclo-[3.3.1]nonane derivatives (XVIII–XXIII). These compounds were also found to exhibit some silver sensor activity, however the dynamic response to the other metal ions such as nickel(II), cobalt(II), zinc, cadmium, and lead ions was very small.

On the basis of the membrane systems studied, new potentiometric PVC–matrix membrane sensors for determining some heavy metal ions were prepared and incorporated into FI systems. The potential–time transients were recorded and the optimum working conditions of each electrode were selected for potentiometric detection in FI systems, based on the previously discussed optimization criteria.[7]

In optimizing the performances of the ISEs, the effect of other compounds in the membranes, such as plasticizing solvent mediators and lipophilic anions, was investigated. The use of NFOE as a solvent mediator, together with lipophilic salts like KTpClPB, improved the dynamic response time τ' and these components were therefore used in membrane preparation for all electrodes. It should be mentioned, however, that Pb-, Cd-, and Mn-ISEs with membranes containing the TpClPB$^-$ anion showed slightly poorer selectivity for primary cations over mercury, silver, caesium, potassium, ammonium, and hydrogen ions. The addition of KTpClPB to the membrane of Hg- and Ag-ISEs produced a stabilization of the signal and improved the selectivity to mercury and silver ions over the other heavy metal ions, but reduced the selectivity against hydrogen ions. It should also be noted that a dynamic flow system improves the apparent selectivity of all these electrodes, demonstrating one of the advantages of the FIP method compared to batch ISE potentiometry. The linear ranges, slopes,

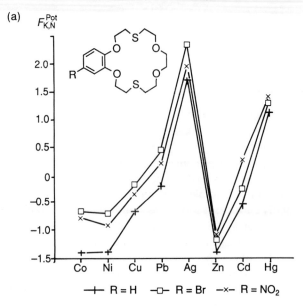

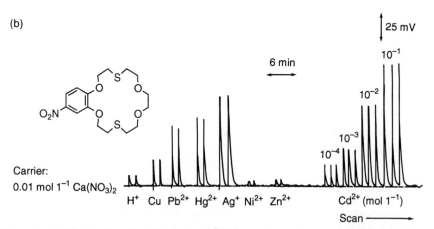

Figure 6 (a) *The relative selectivity factors $F_{K,N}^{Pot}$ for membranes containing benzo-dithia-18-crown-6 with different 4'-substituents in the benzolic fragment;* (b) *The dynamic electrode responses for membrane with the compound XII. The injected metal ion concentration was $0.01 \, mol \, l^{-1}$*

and detection limits of the proposed ISE-detectors together with the sampling frequency are summarized in Table 2. All of the new PVC–matrix membrane ISE-detectors described have comparable or advantageous characteristics compared to the alternative solid-state membrane detectors.

A discussion of the application of FIP to water analysis follows.

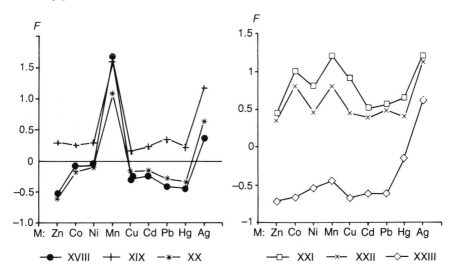

Figure 7 *Pictorial representation of the relative selectivity factors for PVC–matrix membranes, containing 3,7-diazabicyclo[3.3.1]nonane derivatives XVIII–XXIII*

Table 2 *Some Performance Characteristics of the Proposed Tubular PVC–Matrix Membrane Metal ISE-detectors*

ISE	Total measuring range/$mol\,l^{-1}$	Slope of electrode response/ $mV\,dec^{-1}$ (25 °C)	Interfering heavy metal cations	Sampling rate/ h^{-1}
Silver	1×10^{-6}–1×10^{-2}	55.7 ± 0.5	High level of Hg	30
Mercury	5×10^{-8}–1×10^{-4}	33.2 ± 0.7	High level of Ag	20
Lead	5×10^{-6}–1×10^{-2}	59.1 ± 0.5	Ag, Hg	40
Cadmium	5×10^{-5}–1×10^{-2}	27.8 ± 0.4	Ag, Hg, Pb	50
Manganese	1×10^{-5}–0.1	29.2 ± 0.3	Ag	60

Flow Injection Potentiometry Systems for Water Analysis

Although potentiometric detection does not have widespread use in FIA, there are some examples of its application to significant problems in water analysis. For such purposes, the initial concept of single-line FIP systems must be partly superseded by multi-line flow manifolds of different complexity, in order to recognize the different interfering effects by samples and processing steps and to provide appropriate chemical conditions for potentiometric measurements. There is a sustained interest in interfacing of FI systems for sample preconcentration and chemical modification with ISE-detectors.

Our attention has often focused on the development of effective FIP systems for trace inorganic analyses of sea water[52–54] and procedures for the determination

Table 3 *FIP Procedures for Sea Water Analysis*

Analyte	Species in sea water	Measuring range	Sampling frequency/ h^{-1}	FI Features
Chloride	Cl^-	$0.2-15.0 \, mmol \, l^{-1}$	120	two-line manifold
Fluoride	F^-, MgF^+	$0.5-5.0 \, mg \, l^{-1}$	60–80	two-line manifold
Sulfur	SO_4^{2-}, $NaSO_4^-$	$2-28 \, mmol \, l^{-1}$	40	on-line dilution, ind. with Pb-ISE
	HS^-	$0.05-5.0 \, mg \, l^{-1}$	30–60	on-line gas-diffusion–separation
Calcium	Ca^{2+}	$100-800 \, mg \, l^{-1}$	40–60	two-line manifold
Potassium	K^+	$200-800 \, mg \, l^{-1}$	70	single-line manifold
Copper	$CuCO_3$	$0.1-5.0 \, \mu g \, l^{-1}$	2–10	on-line ion-exchange preconcentration

of a number of these analytes have been developed (Table 3). All have been satisfactorily applied to routine sea water analysis aboard a research vessel working in oxic and anoxic marine basins. From our experiments it was concluded that the application of FIA makes possible analytical results which satisfy the quality demands for oceanographic investigations.

The determination of heavy metals in marine waters requires the development of FIA procedures with very low detection limits. One way to greatly improve the sensitivity and detection limits of FIP methods is the use of systems with on-line metal preconcentration steps followed by potentiometric detection of the analyte in an acidic eluate zone.[55] In these cases, however, some difference in pH along the eluate zone and the carrier stream can produce a blank response which has a negative effect on the resulting signal. Moreover, the peak height in such a system is dependent on the sample dispersion in the flow manifold and on the sample zone contact time with the ion-sensing membrane. Thus special attention must be paid to the chemical composition of the carrier solutions, in addition to the variables of the flow manifold, for optimal performance in trace level potentiometric determinations.

Optimal conditions were found for the direct potentiometric detection of Ag-, Cu-, Pb-, or Cd- ion concentrations (C_x) in $0.5 \, mol \, l^{-1}$ nitric acid solutions by means of solid-state membrane detectors in a two-line flow-injection manifold. The base-line supporting buffer solution stream containing trace amounts of the primary ions $(C_{M,carrier})$ was pumped through the second channel in order to ensure the optimal pH and ionic strength along the injected sample zone. It was also necessary to ensure that a reasonable change in concentration was caused by the injection and the background signal was in the linear range of the response function of the ISE.

The potentiometric signal (peak height H), caused by injection of the primary ion concentration C_x, was measured relative to the base-line and can be expressed as follows:

$$H = \alpha \log[(C_x/D + C_0/k)k/C_0] = \alpha \log[(C_x k/C_0 D) + 1] \qquad (8)$$

Table 4 *Analytical Characteristics of the Solid-state Heavy Metal ISEs Employed*
in a Two-line FI Manifold for Potentiometric Detection of Metals in
Acidic Eluate Zone ($0.5\,\mathrm{mol\,l^{-1}}$ HNO_3) ($n = 4$, $P = 0.95$)

ISE	Linear range/ $\mathrm{mol\,l^{-1}}$	Equation of calibration graph	Limit of detection/ $\mathrm{mol\,l^{-1}}$	R.S.D./ %	Sampling frequency/ $\mathrm{h^{-1}}$
Cu	1×10^{-4}–0.1	$(169 \pm 2) + (29.8 \pm 0.7)$ $(r = 0.9998)$	5×10^{-5}	0.4–2.0	40–60
Pb	5×10^{-5}–0.1	$(150 \pm 1) + (28.8 \pm 0.6)$ $(r = 0.9996)$	1×10^{-5}	0.9–4.0	60–90
Cd	1×10^{-5}–0.1	$(245 \pm 1) + (30.0 \pm 0.6)$ $(r = 0.9994)$	5×10^{-6}	0.6–3.8	40–60
Ag	1×10^{-6}–0.1	$(431 \pm 4) + (59.7 \pm 1.2)$ $(r = 0.992)$	5×10^{-7}	1.2–6.0	5–10
	1×10^{-6}–0.1	$(167 \pm 1) + (57.6 \pm 0.3)^*$ $(r = 0.9996)$	1×10^{-7}	0.3–4.4	20

*rFI system with injection of 0.1 ml of silver standards into the flowing sample solution.

where k is the dilution constant determined by $(v_1 + v_2)/v_2$.

Under selected experimental conditions, $C_x k/D C_0 \gg 1$, and thus a H vs. $\log C_{\hat{x}}$ plot was found to be linear over a wide concentration range for all ISEs examined (Table 4). As can be seen from the data in Table 4 the sampling rate of the proposed FIP procedure for silver ion determination with the Ag-ISE was much poorer than for the other cases because of the slow return to the base-line potential. The analytical characteristics of the method were improved by using a reverse approach when a known concentration of silver was injected.

A new flow-through Hg-ISE detector has been evaluated for use in a FIP system for direct potentiometric detection of mercury(II) ions in nitric acid solutions of various concentration.

The performance characteristics of the transient potentiometric responses of this detector were dependent on the acid concentration in the injected sample $(C_{a,inj.})$ as well as in the carrier stream $(C_{a,carrier})$. When 560 µl of $1 \times 10^{-5}\,\mathrm{mol\,l^{-1}}$ mercury (II) nitrate solutions, with the different acid concentration, were injected into the carrier solution of constant nitric acid concentration, the recorded peak height H increased with increasing acid concentration in the injected solution from 10^{-4} to $0.01\,\mathrm{mol\,l^{-1}}$ and then sharply decreased with subsequent increasing $C_{a,inj.}$. If the value of $C_{a,inj.}$ was constant, a steady increase in the peak height H was observed with increase in the acid concentration in the carrier solution from 10^{-4} to $0.01\,\mathrm{mol\,l^{-1}}$ HNO_3. Thus, optimal measurement conditions were obtained when $C_{a,inj.} = C_{a,carrier} = 0.01\,\mathrm{mol\,l^{-1}}$ HNO_3.

With operating conditions corresponding to a single-line FI manifold, the system showed super-Nernstian response. The calibration graph was linear over the mercury concentration range 5×10^{-7}–$1 \times 10^{-4}\,\mathrm{mol\,l^{-1}}$ with a regression coefficient of 0.9997 and a slope of $82.8 \pm 0.6\,\mathrm{mV\,pHg^{-1}}$. The detection limit $(S/N = 2)$ was $1.10^{-7}\,\mathrm{mol\,l^{-1}}$ $Hg(II)$.

The conditions corresponding to direct potentiometric detection of the mercury ion concentration in nitric acid solutions in a two-line FI manifold, incorporating the new Hg-ISE detector, were also studied. The peak height dependence on the pumping rates of the 0.01 mol l^{-1} nitric acid carrier solution, v_1, and of the 0.01 mol l^{-1} citric buffer solution (pH = 3.6), v_2, sample loop volume S and mixing coil length L were investigated. According to the data shown in Figure 8, both flow-rates and the injected sample volume have a great influence on the peak height. The v_1/v_2 ratio was a key factor in achieving a high electrode response. The optimum ratio of 1.5:2.0 was obtained by using a flow-rate of 1.5 ml min^{-1} for the carrier stream and 2.0 ml min^{-1} for the buffer solution stream. The potentiometric signal height increases with increase in the

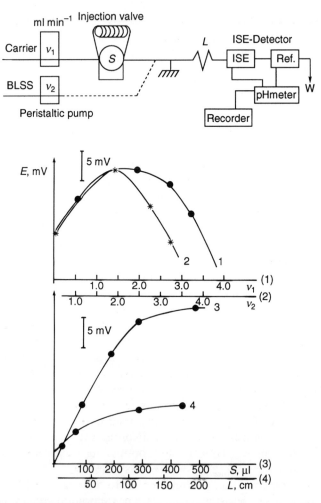

Figure 8 *Flow-injection manifold employed for the study of the dynamic behaviour of a Hg-ISE and effects of flow rates v_1 (1) and v_2 (2); injected volume S (3) and mixing coil length L (4) on peak height ΔE. Concentration of the injected mercury ion standard solution was 1×10^{-6} mol l^{-1}*

sample volume from 30 to 300 µl and with increase in the length of the mixing coil from 20 to 120 cm. With the manifold described, a plot of peak height H vs. concentration C_{Hg} of mercury (II) ions in the $1 \, mol \, l^{-1}$ nitric acid solution injected was linear over the range $200 \, \mu g \, l^{-1}$–$20 \, mg \, l^{-1}$ ($r = 0.9997$):

$$H = (62.9 \pm 0.2) + (39.9 \pm 0.3) \log C_{Hg}$$

The detection limit was $20 \, \mu g \, l^{-1}$ of mercury (II) ions (S/N = 2/1). The contact time was about 16 s and the time required for return to the base-line was about 100 s which indicates that *ca.* 30 samples can be processed per hour. The results obtained for analysis of some artificial effluent water samples are given in Table 5.

The proposed FIP procedures can be utilized for determination of trace amounts of the corresponding heavy metal ions in acidic eluate zones formed in FI systems with an on-line ion-exchange preconcentration step.[55]

Another interesting use of FIP methods for determining trace amounts of metals in waters is in the potentiometric monitoring of metal-catalysed reactions.[56] We have developed two-step catalytic-potentiometric FI procedures for the determination of trace molybdenum(VI) and manganese(II) in drinking waters, based on their catalytic effect on the oxidation reaction of iodide by hydrogen peroxide and periodate-ions, respectively. In order to optimize the sensitivity of detector response and the sampling rate, the sample and the two reagent solution streams were mixed and pumped through a temperature controlled reaction coil placed in the injector valve, and the stop-flow mode was used in the loading position of the valve (Figure 9a). The reaction mixture solution stream was stopped in the injector (reaction) coil for the fixed time T_{stop} to allow the oxidation reaction to proceed to a sufficient extent, and then 200 µl of this reaction mixture was injected into the carrier stream and pumped continuously through with a home-made wall-jet iodide ion-sensing device. Thus, the residual amounts of iodide in the injected zone were detected as the indicator substance. The characteristics of the proposed procedures are summarized

Table 5 *Recovery of Mercury Ions from Artificial Effluent Water Samples by Proposed FIP Method with Hg-ISE Detector* ($n = 6$, $P = 0.95$)

No.	Sample Composition	Mercury (II)/mg l^{-1} Taken	Found	R.S.D./ %	Recovery/ %
1.		0.20	0.19 ± 0.01	6.4	95.0
	$1 \, mol \, l^{-1}$ HNO$_3$	1.00	0.97 ± 0.03	2.6	97.0
		5.00	5.02 ± 0.04	0.5	101.0
2.	Zn (65 mg l^{-1}),				
	Cd (112 mg l^{-1}),	0.50	0.52 ± 0.02	4.9	95.2
	Ni (59 mg l^{-1}),	1.00	0.96 ± 0.04	3.7	96.0
	$1 \, mol \, l^{-1}$ HNO$_3$				
3.	Fe (56 mg l^{-1}),				
	Cu (64 mg l^{-1}),	0.20	0.20 ± 0.01	6.2	100.0
	$1 \, mol \, l^{-1}$ HNO$_3$	2.00	2.01 ± 0.03	1.3	100.2

(a)

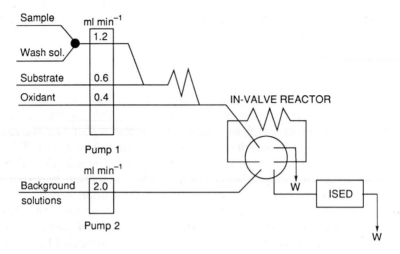

(b)

Analyte	Substrate	Oxidant	T, s stop	Measuring range μg l^{-1}
Mo (VI)	KI	H$_2$O$_2$(HCl)	40	25–200
			180	2–50
Mn (II)	KI	KIO (o-phen)	30	10–100
			120	1–25

Figure 9 (a) *The flow-injection manifold for the potentiometric monitoring of a metal-catalysed reaction with primary ion; (b) The analytical characteristics of the FIP procedures for determination of Mo(VI) and Mn(II) metal ions*

in Figure 9b. Such types of FI manifolds are very advantageous for slow chemical reactions.

The possibilities of using Ag-, Cu-, Pb-, and Cd-ISEs for indirect potentiometric determination of membrane-inactive Zn^{2+}, Co^{2+}, Ni^{2+}, and Mn^{2+} cations, which do not respond directly to the indicator electrode, were also investigated using chemical reactions between an inactive analyte and carrier (or reagent) solutions to produce detectable concentration changes of the primary ion. In this case, an appropriate complexing organic reagent solution was chosen to be pumped as a second carrier stream. The procedure was based on the response of a M^{z+}-ISE to an increase of the free primary cation M^{z+} concentration caused by the reaction of an analyte ion with the free ligand in the primary ion buffer solution. Such combination of a copper(II)-ISE and a copper-ion–buffer solution has already been described by Ishibashi et al.[57] One of the characteristics of the method is that all metal ions which form complexes with the buffer ligand can be quantified with equal sensitivity, as long as the electrode senses only the primary ion and its response is fast. Nevertheless, the sensitivity often differs for each metal ion, because the concentration of primary ion displaced is dependent on the difference in the stability constants of the complexes between the metal ions and the chosen buffer ligand. Hence, the composition and concentration of the ion buffer stream are expected to be very important for indirect determination of metal ions.

We have studied the effect of the complexing ligands on the solid-state cation-sensitive membrane electrode response[58] and have shown the electrode potential (E) depends on the nature and concentration (C_R^0) of these substances in a sample solution.

Under dynamic equilibrium conditions, the peak height H caused by injection of reagent solution R of concentration C_R^0 can be given by the following equation, which corresponds to the formation of a 1:1 complex:

$$H = E = -\alpha \log(1 + \beta'[R']) \tag{9}$$

where β is the mean complex stability constant and $[R']$ is the free equilibrium concentration of injected reagent R.

If $\beta[R'] > 1$ and $C_M^0 < C_R^0/D$, then $[R'] \approx C_R^0/D$. Thus, equation (9) can be simplified to:

$$H = -\alpha \log \beta' + \alpha \log D - \alpha \log C_R^0 = const. - \alpha \log C_R^0 \tag{10}$$

We investigated the dynamic potentiometric response of the Ag-, Cu-, Pb-, and Cd-ISEs to several metal ion complexing organic substances using a two-line FIA system. The organic reagents were chosen to give a wide variety of stabilities of complexes with all metal ions studied. It was found that the electrode response produced negative peaks with a near-Nernstian or non-Nernstian relationship to total injected reagent concentration, depending on the nature of both the membrane ISE and the organic reagent used. The experimental calibration plots for several compounds are shown in Figure 10. As can be seen in the case of the Ag-ISE, higher signals and a linear response down to $10^{-5}\,mol\,l^{-1}$ were observed for ammonium nitrilotriacetate (NTA). For the Cu-ISE, the peak height decreased in the order: thiourea (TM) > sodium citrate (Cit) > NTA > glycine (Gly). The potential changes obtained by the Pb- and Cd-ISEs decreased in the following subsequence: TM > NTA > Cit ≫ Gly. Comparison of the potentiometric data obtained here for different ISEs and reagents in the buffered media have shown that the best results are achieved in the case of fast complex formation reactions on the membrane surface and when $5 < \log \beta' < 20$. If $\log \beta' > 20$, a super-Nernstian slope is obtained, especially in the range of high reagent concentrations. In contrast, if $\log \beta' < 5$ a sub-Nernstian slope of the dynamic electrode response is observed. In flow conditions when $\Delta t < \tau'$ and in cases of slow complex formation reaction and significant adsorption of free organic reagent on the membrane surface, small peak heights are expected and the electrode response function will be non-linear

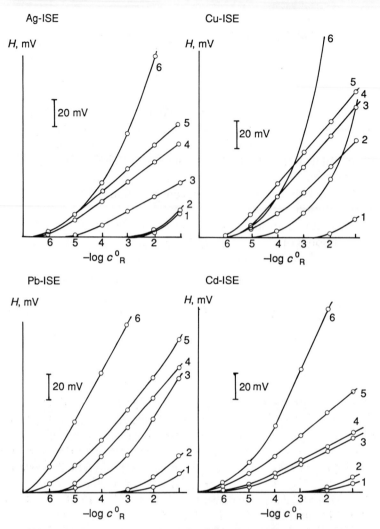

Figure 10 *The dynamic responses of solid-state membrane Ag-, Cu-, Pb-, and Cd-ISEs for injections of 0.2 ml of different organic compounds: ammonium acetate (1), glycine (2), sodium citrate (3), ammonium nitrilotriacetate (4), thiourea (5), and EDTA (6)*

in all concentration ranges. Overall, it can be concluded that the analytical signal parameters and analytical characteristics of the ISE dynamic function are dependent on the stability constant of the reagent-potential-forming metal ion complex as well as on the complex formation kinetic and the reagent adsorption and desorption rates on the membrane surface.

All the reagents studied were utilized as components of a metal ion buffer solution stream in the two-line FIP system for indirect determination of membrane inactive metal ions.[59]

The sample solution (0.2 ml) containing analyte N^{z+} (C_N^0) was injected into the water stream which then mixed with the metal ion buffer solution MR–R $(C_M^0 + C_R^0)$. In the mixing zone, N^{z+} reacts with the free ligand R to form the complex NL (charges of complexes are ignored) and the complex MR dissociates, so the concentration of the free M^{z+} increases. The change in the concentration of M^{z+} is monitored with a M-ISED and the concentration of N^{z+} can be determined indirectly.

The electrode response height was found to be a function of metal ion and reagent concentrations as well as the ratio of the conditional (mean) stability constants of the complexes formed. With the assumption that the potential response of the ISE obeys the Nernst equation, the recorded peak height can be approximately described by the following equation:

$$H \cong E_0' - \alpha \log(C_R^0 - C_M^0) + \alpha \log(C_N^0/D) \tag{11}$$

The best results were obtained in systems with a Cu-ISE and sodium citrate as the complexing reagent and with an Ag-ISE and ammonium nitrilotriacetate (Figure 11). The determination range for the metal ions was between 1×10^{-4} and $0.1 \, mol \, l^{-1}$. The FI methods for indirect potentiometric analysis based on complex formation reactions in the flowing solution or on the electrode surface have been shown to be advantageous as compared to the manual potentiometric complexing titration technique. Sixty samples per hour can be analysed.

4 Conclusion

It is concluded that potentiometric measurements will play an increasing role among the wide variety of detection methods applied in FIA. The number of publications on the FIA–ISE combination is constantly growing.[60-62]

The results obtained in our work, as well as that in the literature, have clearly demonstrated that the FIP method offers wide possibilities in designing new automated analytical systems and ISEs are promising to provide a reliable detection approach for analyses in even complex samples like sea water.

In the area of ISEs, the FIA approach assists in overcoming limitations in the practical application of these sensors in batch ISE potentiometric analysis via increased sensitivity, decreased chemical and mechanical interferences, faster response times, and very good reproducibility of results. It is suggested that the use of different FI configurations may allow the development of interesting procedures for trace determination of both membrane-active and membrane-inactive species following a sustained research effort. Possibly future support in the FIP area should be aimed at designing FI systems for multi-detection and series differential detection. In this respect the hybrid systems and flow-injection measurements without chemical elimination of interferences with data processing by means of factorial analysis appears promising.

FIA systems with PVC–matrix membrane electrodes in various modes are of broad applicability because of the selectivity for a large number of analytes which can be achieved by careful choice of the appropriate membrane-active material.

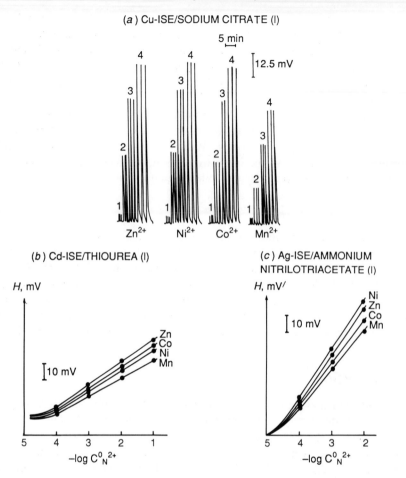

Figure 11 *The typical signals of a Cu-ISE (a) and calibration plots of a Cd-ISE (b) and Ag-ISE (c) obtained for indirect potentiometric determination of membrane inactive metal cations*

References

1. K. Toth, J. Fucsco, E. Lindner, Zs. Fecher, and E. Pungor, *Anal. Chim. Acta*, 1986, **179**, 359.
2. J. Koryta and K. Stulik, 'Ionselectivni Electody', in 'Mir', Moscow, 1989, p. 104 (in Russian).
3. P. Petak and K. Stulik, *Anal. Chim. Acta*, 1986, **185**, 171.
4. E. Pungor, K. Toth, and G. Nagy, *Hung. Sci. Instrum.*, 1975, **35**, 1.
5. E. Pungor, K. Toth, and G. Nagy, in 'Ion and Enzyme Electrodes in Biology and Medicine', ed. M. Kessler, Urban ans Schwarzenberg, Munchen, 1976.
6. Zs. Fecher, G. Nagy, K. Toth, and E. Pungor, *Anal. Chim. Acta*, 1978, **98**, 193.
7. J. Ruzicka, E. H. Hansen, and E. A. Zagatto, *Anal. Chim. Acta*, 1977, **88**, 1.

8. E. H. Hansen, F. J. Krug, A. K. Groser, and J. Ruzicka, *Analyst (London)*, 1977, **102**, 714.
9. J. Ruzicka and E. H. Hansen, *Anal. Chim. Acta*, 1978, **99**, 37.
10. M. Trojanowicz, in 'Modern Trends in Analytical Chemistry, Part A,' Akademia Kiado, Budapest, 1984, p. 189.
11. L. K. Shpigun, *Zh. Analyt. Khimii*, 1990, **45**, 1045 (in Russian).
12. L. K. Shpigun and O. V. Bazanova, in 'Ion-Selective Electrodes in Organic Analysis', Nauka, Moskva, 1989, pp. 30 (in Russian).
13. T. P. Tougas, *Int. Lab.*, 1988, **4**, 18.
14. E. Pungor and K. Toth, *Anal. Proc.*, 1983, **20**, 562.
15. P. D. van der Wal, E. J. R. Sudholter, and D. N. Reinhoudt, *Anal. Chim. Acta*, 1991, **245**, 159.
16. L. Ilcheva and K. Kammann, *Fresenius' Z Anal. Chem.*, 1985, **322**, 323.
17. J. F. van Staden, *Analyst, (London)*, 1988, **113**, 885.
18. T. J. Cardwell, R. W. Cattrall, P. C. Hauser, and L. C. Hamilton, *Anal. Chim. Acta*, 1988, **214**, 359.
19. R. Y. Xie, V. P. Y. Gadzekpo, and A. M. Kadry, *Anal. Chim. Acta*, 1986, **184**, 259.
20. S. D. Kolev, K. Toth, E. Lindner, and E. Pungor, *Anal. Chim. Acta*, 1990, **234**, 49.
21. K. Toth, E. Lindner, E. Pungor, and S. Kolev, *Anal. Chim. Acta*, 1990, **234**, 57.
22. W. Frenzel, *Fresenius' Z. Anal. Chem.*, 1988, **329**, 698.
23. E. Pungor, Z. Fecher, G. Nagy, K. Toth, and G. Horvai, *Anal. Chim. Acta*, 1979, **109**, 1.
24. W. Frenzel, *Analyst (London)*, 1988, **113**, 1039.
25. W. Fennel and P. Bratter, *Anal. Chim. Acta*, 1986, **185**, 127.
26. W. Frenzel and P. Brater, *Anal. Chim. Acta*, 1986, **187**, 1.
27. M. Trojanowicz and W. Matuszewaki, *Anal. Chim. Acta*, 1983, **151**, 77.
28. L. Ilcheva, M. Trojanovicz, and T. Kraczynski vel Krawczyk, *Fresenius' Z. Anal. Chem.*, 1987, **328**, 27.
29. H. Muller and J. Kramer, *Fresenius' Z. Anal. Chem.*, 1989, **335**, 205.
30. R. J. Forster and D. Diamond, *Anal. Chem.*, 1992, **64**, 1721.
31. J. Slanina, W. A. Lingerak, and F. Bakker, *Anal. Chim. Acta*, 1980, **117**, 91.
32. J. F. van Staden, *Anal. Chim. Acta*, 1986, **179**, 407.
33. W. Frenzel, *Fresenius' Z. Anal. Chem.*, 1989, **335**, 931.
34. O. Elsholz, W. Frenzel, C-Yu Lui, and J. Muller, *Fresenius' Z. Anal. Chem.*, 1990, **358**, 159.
35. J. F. van Staden and C. C. P. Wagener, *Anal. Chim. Acta*, 1987, **197**, 217.
36. J. V. van Staden, *Fresenius' Z. Anal. Chem.*, 1988, **332**, 157.
37. J. F. van Staden, *Fresenius' Z. Anal. Chem.*, 1989, **333**, 226.
38. J. F. van Staden, *Fresenius' Z. Anal. Chem.*, 1988, **331**, 594.
39. L. K. Shpigun and O. V. Bazanova, *Zh. Anal. Khim.*, 1989, **44**, 1640 (in Russian).
40. L. K. Shpigun, O. V. Bazanova, and Yu. A. Zolotov, *Sensors Actuators B*, 192, **10**, 15.
41. M. E. Meyerhoff and P. M. Kovach, *J. Chem. Educ.*, 1983, **60**, 767.
42. V. P. Y. Gadzekpo, G. J. Moody, and J. D. R. Thomas, *Anal. Proc.*, 1986, **23**, 62.
43. G. J. Moody, B. B. Saad, and J. D. R. Thomas, *Analyst (London)*, 1989, **114**, 15.
44. M. Telting-Diaz, D. Diamond, and M. R. Smyth, *Anal. Chim. Acta*, 1991, **251**, 149.
45. E. H. Hansen, J. Ruzicka, and A. K. Grose, *Anal. Chim. Acta*, 1978, **100**, 151.
46. E. A. Novikov, L. K. Shpigun, and Yu. A. Zolotov, *Zh. Anal. Khim.*, 1987, **42**, 1945 (in Russian).
47. A. Craggs, G. J. Moody, and J. D. R. Thomas, *J. Chem. Educ.*, 1974, **51**, 541.
48. K. Srinivasan and G. A. Rechnitz, *Anal. Chem.*, 1969, **41**, 1203.
49. E. Linden, K. Toth, and E. Pungor, 'Dynamic Characteristics of Ion-Selective Electrodes', CRC Press, Inc., Boca Raton, Florida, 1988, pp. 130.

50. W. Simon and U. E. Spichiger, *Int. Lab.*, 1991, **9**, 35.
51. L. K. Shpigun and R. D. Tsingarelli, in 'Macrocyclichecki Soedinenija v analiticheski Khimil', eds. Yu. A. Zolotov and N. M. Kuzmin, Nayka, Moskva, 1993, Ch. 5.
52. L. K. Shpigun, H. S. Zamokina, G. M. Varshal, and Yu. A. Zolotov, in 'Methods of the Environmental Analysis', Nauka, Moskva, 1983, p. 133 (in Russian).
53. L. K. Shpigun, I. D. Eremina, and Yu. A. Zolotov, *Zh. Anal. Khim.*, 1986, **41**, 1557 (in Russian).
54. Yu. A. Zolotov, L. K. Shpigun, I. Ya. Kolotyrkina, E. A. Novikov, and O. V. Bazanova, *Anal. Chim. Acta*, 1987, **200**, 21.
55. L. K. Shpigun, O. V. Bazanova, and N. M. Kuzmin, *Zh. Anal. Khim.*, 1988, **43**, 2200 (in Russian).
56. H. Muller and V. Muller, *Anal. Chim. Acta*, 1986, **180**, 30.
57. N. Ishibashi, T. Imato, and K. Trukiji, *Anal. Chim. Acta*, 1986, **190**, 185.
58. L. K. Shpigun and O. V. Bazanova, *Zh. Anal. Khim.*, 1992, **47**, 1581 (in Russian).
59. L. K. Shpigun and O. V. Bazanova, *Zh. Anal. Khim.*, 1992, **47**, 1588 (in Russian).
60. D. E. Davey, D. E. Mulcahy, and G. R. O'Connell, *Analyst (London)*, 1992, **117**, 761.
61. R.-M. Lui, D.-J. Lui, and Al-L. Sun, *Analyst (London)*, 1992, **117**, 1335.
62. B. H. van der Schoot, S. Jeanneret, A. van der Berg and N. F. de Rooij, *Anal. Methods Instrum.*, 1993, **1/1**, 38.

Microelectrodes: New Trends in their Design and Development of Analytical Applications

P. Tuñón-Blanco and A. Costa-García

DEPARTMENT OF PHYSICAL AND ANALYTICAL CHEMISTRY, UNIVERSITY OF OVIEDO, 33006 OVIEDO, SPAIN

Summary

The substantially different properties of microelectrodes in comparison to regular-area sized electrodes contribute to widening of the range of possibilities in electroanalytical applications. The most attractive features of microelectrodes, include their high rates of mass transport and greatly reduced ohmic and capacitance effects which allow studies under conditions not attainable with conventional electrodes.

Herein some recent trends and the preparation of mercury coated microelectrodes obtained by electrodeposition of mercury on carbon fibre electrodes (7.5 μm of diameter), are outlined. The critical aspects affecting the design of the microelectrode, such as geometry and size of the carbon support, pretreatment, activation procedure, etc., together with the reproducibility of the voltammetric signal, are discussed.

The possibility of carrying out adsorptive stripping voltammetry on mercury film microelectrodes, which also can be coupled to a FIA system or to HPLC, enlarges the range of analytical applications of this technique.

1 Introduction

Microelectrodes consisting of minute carbon fibres or noble metals wires (5 to 10 μm diameter, 0.1 to 0.3 cm length), either bare or mercury coated, find increased use in electroanalytical chemistry. Because many aspects of electrochemical and electroanalytical techniques can be improved using microelectrodes, their use has grown rapidly from the early 1970s. Detailed reviews are available[1-7] and a special issue of 'Electroanalysis' has appeared in 1990 (Vol. 2, No. 3, Verlag Chemie, Weinheim).

In comparison to conventional size electrodes the special properties of microelectrodes have widened the range of analytical possibilities in different techniques. As a consequence, some electroanalytical techniques can be performed in such a way that would not be possible using conventional electrodes: more electrochemical information can be obtained and the miniaturized devices contribute to solving real problems, *i.e.* microzone analysis, analysis '*in vivo*',[6] *etc.* The main advantages derived from their use may be summarized:[4] (i) Very small currents (rates of electrochemical reaction), can be relatively easily measured. (ii) The iR drop in solution is reduced at small electrodes. This means that in practical work, voltammetric measurements can be obtained in the absence of supporting electrolyte. (iii) Capacitative charging currents, the limiting factor of the sensitivity which can be reached in all the transient electrochemical techniques, are reduced to insignificant proportions allowing a better discrimination of the faradaic current (analytically useful). (iv) The rate of mass transport to and from the electrode surface increases as the electrode size decreases. It means that steady states of mass transfer are rapidly established. As a consequence of both reduced capacitative charging currents and increased mass transport rates, microelectrodes exhibit a signal to noise ratio very favourable for analytical purposes. (v) Other additional advantages include: low cost and simplicity.

In the following review, some of the main current research topics and relevant analytical applications are considered. Special emphasis is on the use of mercury film coated microfibres electrodes.

2 The Choice of the Material and Geometry of the Microelectrodes

From an analytical point of view the choice of the nature of the electrode, solid or liquid (mercury), depends on the electrochemical properties of the analyte, and the range of potentials to be explored both for microelectrodes and conventional electrodes.

Many materials have been used to make solid electrodes, including different forms of carbon. They are studied mainly due to the increasing interest in the determination of organic molecules based on their oxidation processes since the mercury electrode presents a shorter anodic range of potentials. These materials are also preferred in the design of new electrochemical detectors for coupling to HPLC because the devices are simpler and rugged. Finally solid microelectrodes show more possibilities when their surface is modified (functionalized) for electroanalytical applications. Electrochemical detection at positive potentials is usually done with solid electrodes constructed from platinum, gold, or various forms of carbon. When a noble metal is used, special analytical advantages (selectivity enhancement) can be obtained by generating a surface oxide phase.[8] Nowadays, an increasing use of carbon fibre as an electrode material is noted. Much of this interest is a consequence of their application for voltammetry *in vivo* in clinical applications. Carbon fibres (6–10 μm diameter) are produced by pyrolysis at high temperatures of polymeric materials, and they exhibit interesting surface phenomena that can influence significantly the reversibility of the redox processes and the resulting voltammetric response. The main ractical and theoretical aspects of the carbon fibre electrodes has been reviewed by Edmonds.[2]

Three kinds of arrangements have been mainly reported: the exposed single microelectrode with cylindrical geometry, single microelectrodes or arrays of microelectrodes embedded in an insulated material with disk geometry, and brushes or arrays of exposed microelectrodes.

Some important characteristics influence the choice of the type of microelectrode for analytical purposes. Simple design and handling coupled with easy surface renewing or cleaning are considered to be the main requirements. Cylindrical electrodes are easier to prepare although they can be easily damaged during their handling and always require some kind of chemical or electrochemical cleaning procedure, which takes time. On the other hand, disk electrodes are sometimes easier to clean (by polishing) but more difficult to prepare. Other geometric forms are also possible, *i.e.* band and ring electrodes. They are more robust and can be cleaned by polishing but their analytical use is more limited.

The most critical step in the preparation of a single microelectrode is the sealing of fibres, foils, and metallized glass rods either in glass or in a low density epoxy holder. The risk always exists of an incompleted sealing or the introduction of air bubbles, which makes the microelectrode useless. For this reason, several refinements to the construction of microelectrodes have been introduced, new types of sealant have been developed, and new methods of finally sealing the microfibre have been devised.[9]

Several groups have reported the construction of carbon fibre of different types.[10–15] The most general procedure is based on sealing a fibre into a glass capillary using epoxy resin. The capillary is pulled to a controlled tip diameter by a vertical puller, and the fibre is inserted or floated into the pulled capillary and then sealed with epoxy or nonconducting wax. Electrical contact from the fibre to a stainless steel wire is made by filling the pipette with mercury or conducting epoxy. Other procedures involve heat sealing the fibre in glass and making electrical contact via mercury to a silver or platinum wire[16,17] or threading the fibre into capillary through a microbore Tygon tube which has been placed inside a heat-shrinkable tube.[18]

All the procedures summarized above are laborious and time-consuming. We have developed a novel method[19] which is both simple and rapid since it avoids the use of heat sealing (only drying in an oven is required) and the necessity for a vacuum. Furthermore, the procedure is cheap, the design is more robust and the success rate of manufacture is high (above 95%), therefore offering savings in time. Figures 1a and 1b illustrate the two arrangements of the fibre which can be obtained allowing cylindrical or disk surface of the exposed fibre. The optional use of a plastic sheath protects the fibre from damage. The most time consuming step, once the assembly is ready, is making sure of the complete polymerization of the resin (A.R. Spurr, California) at 70 °C for 8 hours.

This kind of microelectrode has been successfully tested. Single or mercury coated fibre microelectrodes have proven to be effective when carrying out stripping voltammetric measurements using static or flow cells as discussed below.

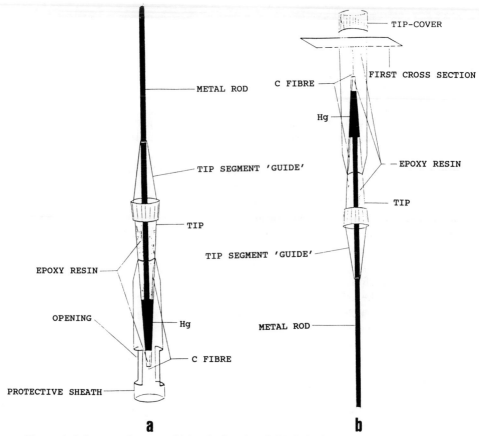

Figure 1 *Schematic drawing of* (a) *cylindrical and* (b) *disk ultramicroelectrode*

3 The Choice of the Pretreatment of the Microelectrode

There is no doubt that when microelectrodes are used as analytical tools the main disadvantage is the lack of reproducible measurements. The surface of a solid electrode changes with time due to adsorption of species from solution or due to chemical changes to the surface itself. These changes often result in variations in sensitivity or reversibility and in extreme cases may lead to complete inhibition of charge transfer. It is widely recognized that pretreatment of solid electrodes has a marked effect on their response to many species and it constitutes a critical and essential step in order to get well-defined and reproducible voltammetric signals. Several activation procedures have been devised for pretreating solid electrodes; these include polishing, chemical treatment, flaming, vacuum heat-treatments, radio frequency plasma treatment, and laser treatment.[20-23]

Several groups have also pointed out the convenience of submitting carbon electrodes or carbon fibre electrodes, to an electrochemical pretreatment prior to analytical use.[24-27] In our experience, a combination of two different kinds of pretreatment, *i.e.* chemical and electrochemical, and the activation procedure must be followed carefully. The choice of the activation procedure is not a simple and easily answered question since firstly, it always depends on the analyte assayed and secondly, electrochemical activation procedures have been known to actually damage the microelectrode. This dependence is related to the electrochemical behaviour of the compound (electrode process mechanism) and also to the functionalized groups created on the microelectrode surface when an electrochemical activation procedure is followed.

From an analytical point of view, a pretreatment procedure that can provide a sensitive and reproducible active surface is desired. For this reason, when adsorption products are involved at a microelectrode surface, an electrochemical activation procedure is recommended. In such cases the electrochemical activation should be evaluated systematically with respect to the duration of applied potential, frequency of activation regime, potential range, and solution conditions.

Figures 2a–c show some of the results obtained when an alternating potential programme using a triangular waveform was applied to a carbon fibre microelectrode and the analytical response was evaluated by measuring the DPV response of folic acid at 1.0×10^{-6} M. The parameters examined were potential range, duration, and frequency. Due to the damage of the fibre when a positive potential is held for long periods, the pretreatment selected as optimum was a potential cycle from 0 to 2 V at a frequency of 10 Hz for a period of 1 min in pH 2 Britton–Robinson (BR) buffer. This kind of pretreatment is also necessary before beginning and between each measurement. Using the optimum activation procedure between each Differential Pulse Voltammetric (DPV) scan, the response was constant with a relative standard deviation better than $\pm 2.5\%$ ($n = 10$). As a consequence of this electrochemical pretreatment of the carbon fibre, acceptable folic acid determinations are possible.[28]

4 Analytical Applications

Apart from the well established clinical applications using ultramicroelectrodic voltammetry, two main analytical research areas stand out, namely the use of coated mercury microelectrodes and the use of ultramicroelectrodes as sensors and detectors. Besides these, research for the improvement of previous electroanalytical methods using microelectrodes in the place of other conventional size electrodes is being carried out. Recently M. Reviaj *et al.*[29] have demonstrated the application of a gold fibre microelectrode in trace analysis of mercury. The Differential Pulse Adsorptive Stripping Voltammetry (DPASV) method described introduced improvements such as the use of small sample volumes, accumulation without stirring, and also enables a considerable reduction in the amount of Au used for construction of the working electrode. The method was successfully applied to the analysis of some environmental water samples and fertilizers. Many other examples could be given where trace metal analysis by Anodic Stripping Voltammetry (ASV) is performed in better conditions than when using

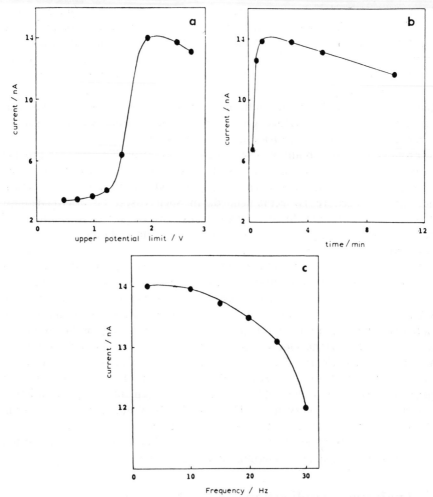

Figure 2 *Effect of pretreatment parameters on the DPV response of* 1.0×10^{-6} M *folic acid.* (a) *Effect of upper potential limit;* (b) *effect of duration of potential cycle;* (c) *effect of frequency. Duration* (a, c), 1 min, *frequency* (a, b), 10 Hz; *upper potential limit* (b, c) 2 V. *Supporting electrolyte:* 0.1 M *perchloric acid*

other conventional electrodes, *i.e.* Hanging Mercury Drop Electrode (HMDE).[30–32]

The enhanced mass transfer characteristics of microelectrodes reduce the preconcentration times and eliminates the need for convective hydrodynamics during the deposition step; furthermore, because the ohmic drops are negligible, the measurements can be obtained even in poorly conductive media without addition of any supporting electrolyte.[33–41] For these reasons, the combination of DPASV with ultramicroelectrodes should prove to be extremely useful for analytical applications involving ultratrace levels of heavy metals in various matrices such as rain water, sea water, or wine.[42,43]

5 Mercury-coated Carbon Fibre Electrodes

It has been shown previously that mercury film electrodes give increased resolution compared to the HMDE and that carbon fibre electrodes produce a very small charge current, hence it would be extremely desirable to combine both properties. Furthermore, this combination may provide additional advantages such as easy handling, low cost, and other well known analytical advantages associated with the use of the ultramicroelectrode itself or with the mercury properties. The stability and size of these electrodes suggest a wide variety of uses. However in the following, only the main analytical findings will be discussed.

J. Osteryoung et al. have prepared such Mercury-coated Carbon Fibre Microelectrodes (MCCFE) and examined their properties using Square Wave Voltammetry (SWV),[16] an enhanced stripping response to lead(II) at a mercury-coated fibre was observed.[17]

Low-modulus carbon fibres were satisfactorily used as the support for a mercury film electrode in differential pulse anodic stripping voltammetry mode (DPSAV) with the additional advantages mentioned above. The results confirmed a better sensitivity than when no coating mercury film was used for the DPSAV analysis of heavy metals.[12] Peaks showed better resolution than at HMDE as expected, but as very small volumes of mercury are present on the electrode surface, intermetallic compound formation occurs to a greater extent than on conventional mercury drop electrodes. The effect of intermetallic compound formation results in splitting of the analyte peak and rapid loss of linearity. Many other examples could be given when heavy metals are examined but it is generally accepted that the possibility of intermetallic interferences is high in contrast to organic interferences. The absence of organic interferences allows direct measurement of some samples without pretreatment.[42,43]

Other relevant aspects of the performance of these electrodes are the use of very small cells (microlitres capacity),[44] the possibility to carry out rapid-multicomponent-trace determination of heavy metals by using linear sweep voltammetry at high scan rates $(1-100 \, V \, s^{-1})$[45] and that no time-consuming deoxygenation of samples was necessary. Some analytical applications in non-aqueous media have also been reported.[39]

Special analytical advantages were found using the cathodic stripping voltammetry of Se(IV) employing a coated mercury carbon fibre microelectrode where the mercury film was electrogenerated in situ.[46] The in situ formation of mercury-coated fibre microelectrodes was achieved by applying a potential of $-1.2 \, V$ for $150 \, s$ in $5 \, M$ hydrochloric acid containing $1.0 \times 10^{-4} \, M$ Hg(II). Selenium(IV) deposition and stripping analysis were performed in a quiescent solution, subsequent to film formation, by scanning the potential in differential pulse mode. It has been demonstrated that the time taken for the overall stripping procedure could be reduced if after the first stripping signal is recorded, the solution is stirred briefly, allowed to quiesce, and the Se(IV) deposition and stripping steps are then repeated. In this way, up to nine stripping measurements could be carried out on the same mercury film with a RSD of 2.7% at $50 \, ng \, ml^{-1}$ level $(n = 7)$. Removal of the mercury film was best achieved by transferring the microelectrode to 0.1 M sodium acetate electrolyte, and applying a potential of 0.5 V for 30 s. Normally a single fibre can be used for two to three days before the sensitivity of the stripping signal declines. Due to the excellent mass transport conditions exhibited by the microelectrode, substantial Se(IV) deposition occurs during the scan period and by this reasoning the deposition potential is not a specific potential, as is the case at conventional size electrodes, but is rather a range of potentials over which continuing deposition of Se(IV) occurs. This peculiar behaviour is also observed when the deposition time is studied (Figure 3). As the deposition time enhancement of the peak current was confined to relatively short time values it is not necessary to choose a specific deposition time. In the same way, any plating potential can be chosen except one in which the oxidation of mercury occurred (in order to avoid calomel formation). The result is the simplification and speeding up of the voltammetric procedure avoiding, also, other potential sources of error such as the control over the time parameter or the necessity to maintain reproducible convective transport conditions. The analytical characteristics of the method are similar to those based on the accumulation of mercury(II) selenide[47,48] with the additional advantages mentioned above.

Finally, an important research field is devoted to the type of electrochemical technique applied. Square Wave Anodic Stripping Voltammetry (SWASV) has been shown to be very effective in order to enhance the stripping current produced by lead and cadmium.[49] Also Short-pulse Rapid Scan Stripping Voltammetry (SPRSSV) produced well resolved peaks of lead and cadmium with substantial increase in signal to background ratio compared to DPSAV due to the higher replating efficiency reached between pulses.[50] Potentiometric stripping multianalysis performed with mercury coated carbon fibre have been also studied by various authors and important applications in organic or inorganic fields has been shown.[13,51,52] The small background signals obtained lead to excellent signal

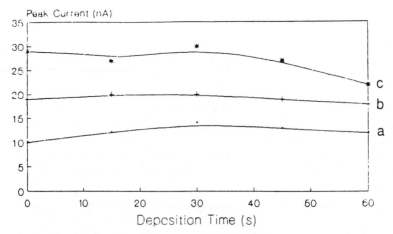

Figure 3 *Deposition time dependence of selenium DPCS peak current: (a) 1 ng ml⁻¹; (b) 2 ng ml⁻¹; (c) 3 ng ml⁻¹ Se(IV)*

to background ratios. The cut-fibre electrodes are advantageous in the analysis of small sample volumes, due mainly to their small size. Their application in flow systems seems very promising, especially if computerized flow potentiometric analysis is used. Here the sensitivity is normally so high that the time for sample electrolysis is less than ten seconds.[53]

In the following sections a few illustrative examples of adsorptive stripping voltammetry of selected pteridines or related antitumour drugs will be shown.

6 Adsorptive Stripping Voltammetry on Mercury-coated Carbon Microelectrodes

Adsorptive Stripping Voltammetry (AdSV) is an electrochemical technique which exploits the adsorptive characteristics of the analyte as a way of preconcentration.

Nowadays AdSV is a very interesting technique for trace and ultratrace analysis[54] due to its excellent sensitivity, accuracy, precision, and the low cost of instrumentation. In an effort to combine the adsorptive features of mercury electrodes and the electroanalytical advantages of microelectrodes we carried out several studies to show the capability of this techique when it is performed using mercury thin film microelectrodes (MTFMs). As indicated above, the majority of the work in this area is related to the use of MTFMs for anodic stripping of metals and cathodic stripping of organic compounds has not been previously reported.

Prior work in our laboratories on the adsorptive stripping voltammetric behaviour of some pteridines with biological significance[55-58] or related antitumour drugs[59] has made possible the study of mercury film deposition on carbon fibre microelectrodes and comparison of results with those obtained using classical mercury electrodes. The ultramicroelectrode illustrated in Figure 1 was used throughout as support for the mercury film.

7 Mercury Thin Film Formation

Mercury deposition on carbon fibres may be in the form of a thin film or may be increased so as to form an almost spherical shape. Deposition of larger amounts of mercury facilitates adsorption of greater amounts of the analyte but the electrode would then not possess the superior stability of a MTFE. Thus the mercury plating conditions on the fibre are very critical and need to be studied carefully. The film can be generated *in situ* or *ex-situ* depending on the kind of compound which is studied. When the analyte precipitates in the presence of mercury salts, an *ex-situ* formation of the mercury film is required and therefore the stability of the film during medium exchange is another additional critical factor to be considered.[16] When the analyte does not precipitate with mercury(II) salts, further additional advantages can be obtained, *e.g.* allowing the use of the analyte and the salt as ingredients of the carrier in FIA. Under optimum preplating conditions the film can be characterized simply by anodic stripping of the mercury thin film from the carbon fibre surface.

It has been found that chemical pretreatment of the fibre is necessary in order to obtain consistent films. The fibre was activated by dipping in concentrated chromic acid for 5 min. After washing with distilled water, the fibre was dipped in concentrated nitric acid for 2 min and then washed again with distilled water. The electrode was then ready for the mercury film deposition.

Two mercury preplating parameters have to be studied: the applied preplating potential E_{film} and the preplating time t_{film}. Moreover, the background supporting electrolyte composition exerts a strong influence on the stripping signal of the analyte. Apart from the mercury salt, the choice of the supporting electrolyte is important. Hydrochloric acid allows good film deposition and, consequently, good stripping responses when pteridines are assayed while, perchloric acid was more suitable when mitoxantrone was studied.[60] Figure 4 shows a typical study of the influence of the supporting electrolyte composition on the a.c. stripping signal of 4.0×10^{-8} M folic acid using the first reduction process common to many pteridines, in which the pteridine ring is reversibly reduced to the 5,8-dihydro corresponding derivative. Following previous studies carried out in these laboratories[61] a solution of 10^{-3} M Hg(NO$_3$)$_2$ in 5 M HCl was chosen for the *ex-situ* formation of the mercury film. Films formed using these conditions are thought to exist as mercury microdroplets of high surface area facilitating efficient adsorption of the analyte compound. When using higher salt concentrations the anodic stripping peak of the mercury film became broader and shorter due to the formation of large mercury droplets of lower surface area and inferior stability. Conversely, at lower concentrations the

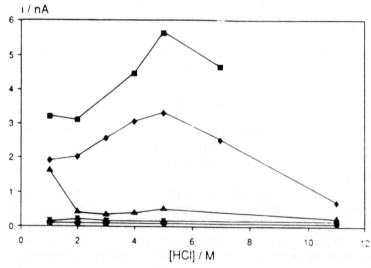

Figure 4 *Optimization of thin mercury film conditions: influence of the supporting electrolyte composition on the a.c. stripping voltammetric response of 4.0×10^{-8} M folic acid. Hg(NO$_3$)$_2$ concentration: ● = 1.0×10^{-5}; ✗ = 1.0×10^{-4}; ▲ = 1.0×10^{-3}; ♦ = 0.01; ■ = 0.1 M, t_{film} = 30 s; E_{film} = −0.2 V. Stripping peak in acetate buffer (pH 5.0)*

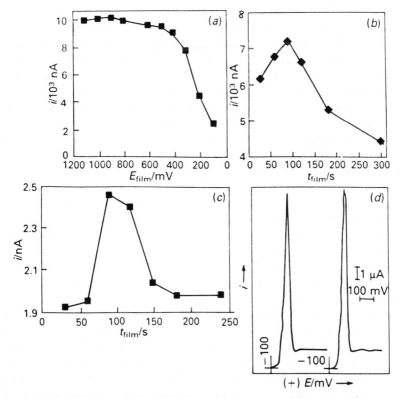

Figure 5 (a) *Optimization of the MTF formation in terms of E_{film} using direct current adsorptive stripping voltammetry of 5×10^{-8} M aminopterin in pH 5 acetate buffer. Mercury salt solution = 1×10^{-3} M in 0.5 M HCl; $t_{film} = 60$ s; $E_{acc.} = 0.00$ V, $t_{acc.} = 60$ s; film was stripped anodically at 100 mV s^{-1}. (b) Optimization of the MTF formation in terms of t_{film} using direct current adsorptive stripping voltammetry of aminopterin. $E_{film} = -800$ mV. Rest of conditions as in (a). (c) Optimization of the MTF formation in terms of t_{film} versus a.c. stripping voltammetry of aminopterin 5×10^{-9} M. Rest of conditions as above. (d) Two typical anodic stripping peaks of the MTF under optimum deposition conditions; $t_{film} = 90$ s; $E_{film} = -800$ mV; scan rate = 100 mV s^{-1}*

surface coverage to the fibre decreased producing a less efficient electrode which was more prone to surface fouling.

Figures 5a–d show the typical behaviour observed in the optimization of the Mercury Thin Film (MTF) *ex-situ* formation for the different electroplating parameters affecting the adsorptive stripping voltammetry of 5×10^{-9} M aminopterin solution.[61] Similar behaviour has been shown for other pteridines studied.[60,62] Because two separate cells had to be employed, once the mercury film was generated, the MTFM was quickly transferred to the analytical cell containing the buffer of pH 5. The transfer procedure lasted aproximately 10 s in total. Throughout the procedure a closed circuit was maintained with the potential set at the optimum film deposition potential (-0.800 V). Once the electrodes had been placed in the analytical cell the potential was moved to

− 1.400 V for 30 s to ensure that a clean film was present for the first analysis. The mercury preplating was optimized by employing a 60 s preplating time at various potentials. Each film was subsequently stripped anodically and the peak current was measured. In close agreement with a previous report,[16] the film characteristics depend strongly on the deposition potential applied. As the applied potential becomes increasingly more negative the number of active sites on the carbon fibre becomes increasingly larger and hence the number of mercury droplets increase. At potentials above − 800 mV hydrogen gas is generated causing droplet detachment. Therefore a potential of − 800 mV was chosen to ensure the stability of the film. The film preplating time was optimized in a similar way by holding a preplating potential and increasing the preplating time. As shown in Figures 5b and 5c the stripping current began to decrease when higher deposition times were employed. It seems that at deposition times of greater than 90 s a growth in droplet size rather than in droplet number occurs, thus causing an overall decrease in the surface area of mercury available for analyte adsorption, and also producing a less stable film. Therefore, an optimum deposition time of 90 s was chosen.

Between each film formation the fibre was regenerated producing a clean surface of mercury. The optimum conditions for this regeneration were found to be the application of a potential of + 790 mV for 40 seconds, which allowed the total oxidation of the mercury on the fibre surface. Figure 5d shows two typical anodic stripping signals of the mercury film deposited under optimum conditions. The reproducibility of the formed film was excellent (RSD of 1.11%, $n = 6$). But the transfer reproducibility yielded only a RSD of 13.3%. Therefore for analytical purposes, this suggested the necessity to use the same film throughout the whole analysis run and regeneration of the film surface between each measurement. This is not necessary when the analyte does not precipitate with mercury(II) for example Mitoxantrone (MXT), a well known antitumour drug, can be analysed by generating the mercury film *in situ* and different films can be used for recording different voltammograms. In any case it was noticed that after each stripping recorder signal, electrochemical activation of mercury film is necessary in order to remove the adsorbed reduction products. It was normally carried out by holding the potential at a rather negative value for a short time if the same mercury film had to be used throughout the analysis run or more positive if the mercury film was renewed in each measurement.

Once the mercury thin film microelectrode had been transferred to the analytical cell it is ready for any analytical purposes and accumulation studies of the selected molecules could be made.[60-62] Utilizing the special features of ultramicroelectrodes coupled with adsorptive preconcentration and a.c. voltammetry a very sensitive method was obtained allowing detection of extremely low concentrations of the compounds with good accuracy and precision. By careful selection of the accumulation time, different concentration ranges could be studied. Table 1 summarizes some of the main analytical characteristics of the compounds under examination.

When the performance of these electrodes was compared to those previously reported using classical mercury electrodes several advantages of the ultramicro-

Table 1 *Analytical characteristics of selected pteridines/antitumour drugs using a.c. stripping voltammetry on thin layer film ultramicroelectrodes*

Analyte	Ref.	Calibration range	t_{acc}	L.D. $(S/N) = 3$	Reproducibility (RSD)
Folic acid	60	1×10^{-9}–2.5×10^{-8} M	60	9×10^{-10}	1.44% (at 10^{-8} M, $n = 10$)
Mitoxantrone	60	5×10^{-10}–2×10^{-8} M	300	5×10^{-10}	5.05% (at 10^{-9} M, $n = 10$)
Aminopterin	61	2×10^{-10}–8×10^{-9} M	180	1×10^{-10}	3.57% (at 10^{-10} M, $n = 10$)
Edatrexate	62	1×10^{-10}–5×10^{-9} M	240	5×10^{-14}*	1.39% (at 10^{-8} M, $n = 10$)*
Metotrexate	62	1×10^{-10}–3×10^{-9} M	180	3×10^{-13}*	—
				*$t_{acc} = 350$ s.	*$t_{acc} = 10$ s.

electrodes was evident. Primarily, the charging current related to the ultramicroelectrode is minimal compared to that of the classical electrodes. This fact combined with the greater mass transport characteristics of ultramicroelectrodes allowed the employment of much lower accumulation times and accumulation from quiescent solution. These factors are particularly advantageous for analysis of biological fluids since longer accumulation times and stirring of the solution enhance the diffusion of interferring large compounds which normally diffuse very slowly to the electrode surface in quiescent solutions. Much lower limits of detection were achieved using the ultramicroelectrodes and the precision of the signal compared favourably with that obtained using classical electrodes. The same fibre could be used for a period of at least 8 weeks with no significant diminution of performance.

8 Analytical Approach for Analysis of Real Samples

The analytical approach to real (complex) samples was carried out by first evaluating the adsorptive preconcentration/stripping responses of various analytes in the presence of surfactants. Once it was shown that this interference was lower than when other conventional electrodes were used, the next step involved assessing the performance of the mercury thin film ultramicroelectrodes for direct analysis of real biological samples. The results obtained were quite satisfactory depending on the sample matrices used. From urine samples, compounds such as mitoxantrone,[60] aminopterin,[61] or edatrexate[62] could be determined directly. In all cases a dilution of the spiked urine sample was necessary in order to minimize interferences arising from the medium and better limits of detection could be reached after sample clean-up using solid phase extraction methodology. Again, due to the greater mass transport properties of the ultramicroelectrode very short accumulation times were employed. After introduction of the sample into the cell, a standard addition method was followed and the initial concentration of the urine was evaluated by extrapolation. Table 2 shows some of the results obtained when Edatrexate is determined in urine. Although the results are acceptable for normal therapeutic levels, as expected, the relative error increases at lower analyte concentration. It was due to modification of the mercury film or slight electrode passivation by compounds naturally present in urine. For when the same samples were submitted to the

Table 2 *Direct analysis of Edatrexate in urine*

	$n = 3$		
True Concentration/M	Concentration determined/M	Relative Error/%	RSD/%
5×10^{-6}	5.26×10^{-6}	5.09	1.48
1×10^{-6}	1.04×10^{-6}	7.24	7.07
5×10^{-7}	5.65×10^{-7}	12.87	4.17

above mentioned previous extraction clean up procedure a limit of detection of 1.0×10^{-8} M Edatrexate in urine was recorded.

The main problem was the lifetime of the microelectrode, which was lowered by gradual passivation of the fibre by compounds present in urine. However, the use of an optimized mercury film and employment of high urine dilutions and short accumulation times minimized this effect. Surface fouling was also alleviated by regenerating the carbon fibre for 1 min in concentrated chromic acid. The same fibre could normally be used for five different urine analysis runs involving approximately 35 measurements before it had to be replaced. This problem had not been noticed when the mercury film was generated *in situ*.[60]

Finally, improvements in the analytical performance of mercury thin film ultramicroelectrodes have been discovered[63] using a coated mercury film microelectrode, described above, in flow systems. The T shape arrangement (flow cell) as shown in Figure 6, a.c. was used for adsorptive stripping analysis of MXT in flowing systems and gave better reproducibility of the measurements. The method is more versatile and lower detection limits (9.0×10^{-10} M MXT)

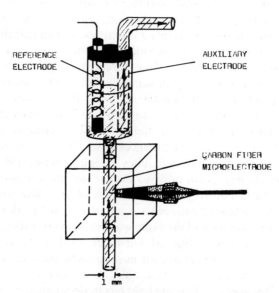

REFERENCE ELECTRODE

AUXILIARY ELECTRODE

CARBON FIBER MICROELECTRODE

1 mm

Figure 6 *Schematic illustration of the carbon fibre electrode in the flow cell*

were obtained. Due to the use of a carrier, the mercury film seemed to be more resistant to interferences arising from the matrix when real samples (serum) were analysed. Again, the main inconvenience was the impossibility of using the same fibre for making more than one calibration graph when serum samples are analysed hence the need to search for new materials for mercury film support.

9 Ultramicroelectrode Sensors and Detectors

Ultramicroelectrodes have found new and interesting applications as sensitive electrochemical sensors and detectors in a variety of media. The main research topics are devoted to their behaviour in gas phase measurements, particularly in gas chromatographic applications,[64] and in HPLC.[65-71] Several flow cells have been reported using ultramicroelectrodes in HPLC applications. Most of them are based on the positioning of a single fibre in the outlet of a chromatographic column.[72,73] The main advantages of these devices are smaller dead volume of the device, a more convenient signal to noise ratio, and a reduced requirement of the supporting electrolyte in the solution. The theory describing the amperometric performance of an ultramicroband electrode in flowing solution has been recently developed.[74] The elimination of supporting electrolyte constitutes an important advantage for chromatographic systems using organic solvents as eluents. Here the potential window for the electrochemical detection can actually be widened significantly. With the higher current density, better sensitivity per unit area is expected for the ultramicroelectrode compared to conventional size electrodes.

Further advantages were obtained when microelectrode array arrangements are used[65,68,69,75,76] since the dependence of the signal on the mobile phase flow-rate is severely suppressed and the ratio S/N is enhanced as the radial diffusion towards each microelectrode produces a signal greater than that obtained at a continuous electrode with the same active area. Moreover, the diffusion layer is partly depicted with the analyte during passage of the solution from one microelectrode to another. It seems that current measurements on multiple carbon fibres in a flow-through cell are easy to measure when compared to the microdisk and batch of carbon fibres.

Important biomedical and environmental analytical applications using different chromatographic techniques or FIA coupled to different scanning potential modes on ultramicroelectrodes have been recently reported. These applications include the electrochemical degradation of trichloroethylene,[77] the determination of biogenic amines using SWV,[78] furenes by gel permeation-HPLC using amperometric and fast scan-rate cyclic voltammetry[79] and tocopherols using HPLC-ED at a surface-oxide modified platinum microelectrode.[80]

Finally, important findings are related to the use of microelectrodes in the design of new biosensors. Quite recently, the design and applications of neutral-carrier-based ion-selective microelectrodes have been reviewed.[81] But a considerable body of literature is devoted to make practical micro-enzyme electrodes for rapid analysis of clinically important substances. Field effect transistor (FET)-based enzyme electrodes sensors have been developed and present a promising approach for the fabrication of micro-enzyme sensors, whose working principle is potentiometry.[82,83] More recently, different micro-enzyme electrodes prepared on platinized platinum and based on amperometric measurements have shown excellent performance as biosensors of clinically important metabolites, *i.e.* glucose.[84-87]

10 Future Trends

The microfabrication of enzyme electrodes will continue to show that they constitute one of the most important approaches, not only for the realization of high performance autoanalysing systems based on electroanalysis of clinically important substances, but also for *in vivo* multifunctional biosensors which are usable in artifical organs. Effective strategies will also be demanding attention in the field of biosensor technology, including the miniaturization of immunosensors.

Besides new analytical applications of these devices coupled to different chromatographic techniques, other additional and ideal features of the microelectrodes will be continually exploited as recently demonstrated.[88] Fast-scan cyclic voltammetry (up to $11\,000\,V\,s^{-1}$) can be used as an analytical tool in trace detection of biocompounds due to the advantages obtained from redox reactions which involve weak adsorption and diffusion. This approach should enlarge the analytical applications for compounds which are not suitable for adsorptive stripping voltammetry.

Acknowledgements

The authors would like to thank Mr Damien Boyd for his valuable contribution to the grammatical corrections of the original manuscript and to Dr. Mª. J. García Calzón for providing bibliographic aid.

The work has been financially supported by DGYCIT (Spain). Project No PB-87-1031.

References

1. R. M. Wightman, *Anal. Chem.*, 1981, **53**, 1125A.
2. T. E. Edmonds, *Anal. Chim. Acta*, 1985, **175**, 1.
3. S. Pons and M. Fleischman, *Anal. Chem.*, 1987, **59**, 1391A.
4. M. Fleichsman, S. Pons, D. Rolinson, and P. P. Schmidt (eds.), 'Ultramicroelectrodes', Datatech Science, Morgantown, NC, 1987.
5. R. M. Wightman and D. O. Wipf, in 'Electroanalytical Chemistry', A. J. Bard (ed.), Marcel Dekker, NY, 1988, Vol. 15, p. 267.
6. P. A. Broderick, *Electroanalysis*, 1990, **2**, 241.
7. Z. Tojek, *Mikrochim. Acta*, 1991, **11**, 353.
8. D. L. Lumscobe and A. L. Bond, *Talanta*, 1991, **38**, 65.
9. T. E. Edmonds, E. M. Palshis, and P. Rushton, *Analyst (London)*, 1988, **113**, 705.

10. J. L. Ponchon, R. Cespuglio, F. Gonon, M. Jouvet, and J. F. Pujol, *Anal. Chem.*, 1979, **51**, 1483.
11. M. A. Dayton, J. C. Brown, K. J. Stutts, and R. M. Wightman, *Anal. Chem.*, 1980, **52**, 946.
12. M. R. Cushman, B. G. Bennet, and C. W. Anderson, *Anal. Chim. Acta*, 1981, **130**, 323.
13. G. Schulze and W. Frenzel, *Anal. Chim. Acta*, 1984, **159**, 95.
14. J. Jennigs and J. E. Morgan, *Analyst (London)*, 1985, **110**, 121.
15. J. X. Feng, M. Brazell, K. Renner, R. Kasset, and R. N. Adams, *Anal. Chem.*, 1987, **59**, 1863.
16. J. Golas and J. Osteryoung, *Anal. Chim. Acta*, 1986, **181**, 211.
17. J. Golas and J. Osteryoung, *Anal. Chem. Acta*, 1986, **186**, 1.
18. M. J. Neuwer and J. Osteryoung, *Anal. Chem.*, 1989, **61**, 1954.
19. A. J. Suárez-Fernández, J. A. García-Calzón, A. Costa-Garciá, and P. Tuñón-Blanco, *Electroanalysis*, 1991, **3**, 413.
20. G. N. Kamau, W. S. Willis, and J. F. Rusling, *Anal. Chem.*, 1985, **57**, 451.
21. W. E. van der Linden and J. W. Dieker, *Anal. Chim. Acta*, 1980, **119**, 1.
22. J. F. Evans and T. Kuwana, *Anal. Chem.*, 1977, **49**, 1632.
23. K. Stutts, R. Kovach, W. Khur, and R. M. Wightman, *Anal. Chem.*, 1983, **54**, 1632.
24. R. C. Engstrom, *Anal. Chem.*, 1982, **54**, 2310.
25. H. Gunasingham and B. Fleet, *Analyst (London)*, 1982, **107**, 896.
26. F. G. Gonon, C. M. Fombarlet, M. J. Buda, and J. F. Pujol, *Anal. Chem.*, 1981, **53**, 1386.
27. J. Wang, P. Tuzhi, and V. Villa, *J. Electroanal. Chem.*, 1987, **234**, 119.
28. T. J. O'Shea, A. Costa García, P. Tuñón Blanco, and M. R. Smyth, *J. Electroanal. Chem.*, 1991, **307**, 63.
29. M. Rievaj, S. Mesáros, and D. Bustin, *Anal. Quim.*, 1993, **89**, 347.
30. J. Wang, *Anal. Chem.*, 1982, **54**, 221.
31. K. R. Wehmeyer and R. M. Wightman, *Anal. Chem.*, 1985, **57**, 1989.
32. E. B.-Tan Tay, S.-Beng Khoo, and S.-Wai Loh, *Anal. Chem.*, 1989, **114**, 1039.
33. J. O. Howell and R. M. Wightman, *Anal. Chem.*, 1984, **56**, 524.
34. A. M. Bond, M. Fleischmann, and J. J. Robinson, *J. Electroanal. Chem.*, 1984, **168**, 299.
35. A. M. Bond, M. Fleischmann, and J. J. Robinson, *J. Electroanal. Chem.*, 1984, **172**, 11.
36. M. Ciszkowska and Z. Stojek, *J. Electroanal. Chem.*, 1986, **213**, 189.
37. Z. Stojek and J. Osteryoung, *Anal. Chem.*, 1988, **60**, 131.
38. A. M. Bond and P. A. Lay, *J. Electroanal. Chem.*, 1986, **199**, 285.
39. J. Wang and P. Tuzhi, *Anal. Chim. Acta*, 1987, **197**, 367.
40. M. Ciszkowska, Z. Stojek, and J. Osteryoung, *Anal. Chem.*, 1990, **62**, 349.
41. S. Daniele and G. A. Mazzocchin, *Anal. Chim. Acta*, 1993, **273**, 3.
42. S. Daniele, Mᵃ. A. Baldo, P. Ugo, and G. A. Mazzocchin, *Anal. Chim. Acta*, 1989, **219**, 9.
43. S. Daniele, Mᵃ. A. Baldo, P. Ugo, and G. A. Mazzocchin, *Anal. Chim. Acta*, 1989, **219**, 19.
44. W. Frenzel, *Anal. Chim. Acta*, 1987, **196**, 141.
45. A. S. Baranski, *Anal. Chem.*, 1987, **59**, 662.
46. K. MacLaughin, J. R. Barreira-Rodriguez, A. Costa-García, P. Tuñón-Blanco, and M. R. Smyth, *Electroanalysis*, 1993, **5**, 455.
47. G. E. Batley, *Anal. Chem. Acta*, 1986, **187**, 109.
48. S. B. Adeljou, A. M. Bond, M. H. Briggs, and H. C. Hughes, *Anal. Chem.*, 1983, **55**, 2076.
49. M. Wojciechowski and J. Balcerzak, *Anal. Chim. Acta*, 1991, **249**, 433.
50. J. P. Sottery and C. W. Anderson, *Anal. Chem.*, 1987, **59**, 140.
51. V. J. Jennings and J. E. Morgan, *Analyst (London)*, 1985, **110**, 121.
52. Chi Hua, D. Jagner, and L. Renman, *Talanta*, 1988, **35**, 525.
53. H. Huillang, Chi Hua, D. Jagner, and L. Renman, *Anal. Chim. Acta*, 1987, **193**, 61.
54. M. G. Paneli and A. Voulgaropoulos, *Electroanalysis*, 1993, **5**, 355.

55. A. J. Miranda Ordieres, PhD Thesis, University of Oviedo, 1985.

56. J. Mª. Fernández-Alvárez, A. Costa-García, A. J. Miranda-Ordieres, and P. Tuñón-Blanco, *J. Electroanal. Chem.*, 1987, **225**, 241.

57. J. Mª. Fernández-Alvárez, A. Costa-García, and P. Tuñón-Blanco, in 'Contemporary Electroanalytical Chemistry', Ivaska-Lewenestam-Sara (eds.), Plenum Press, NY, 1990, p. 329.

58. M. A. Malone, A. Costa-García, P. Tuñón-Blanco, and M. R. Smyth, *J. Pharm. Biomed. Anal.*, 1993, **11**, 939.

59. J. Cortina Villar, A. Costa-García, and P. Tuñón-Blanco, *Talanta*, 1993, **40**, 333.

60. J. Amez del Pozo, A. Costa-García, and P. Tuñón-Blanco, *Anal. Chim. Acta*, 1993, **273**, 101.

61. M. A. Malone, A. Costa-García, P. Tuñón-Blanco, and M. R. Smyth, *Analyst (London)*, 1993, **118**, 649.

62. M. A. Malone, A. Costa-García, P. Tuñón-Blanco, and M. R. Smyth, *Anal. Meth. Instr.*, 1993, **1**, 164.

63. J. Amez del Pozo, Agustín Costa-García, and P. Tuñón-Blanco, *Anal. Chim. Acta*. 1994, **289**, 169.

64. R. Brina, D. Pons, and M. Fleischman, *J. Electroanal. Chem.*, 1988, **244**, 81.

65. W. L. Caudill, J. O. Howell, and R. M. Wightman, *Anal. Chem.*, 1982, **54**, 2532.

66. L. A. Knetch, E. J. Guthrie, and J. W. Jogersen, *Anal. Chem.*, 1984, **56**, 479.

67. J. G. White, R. L. St. Claire III, and J. W. Jogersen, *Anal. Chem.*, 1986, **58**, 293.

68. K. Stulík, *Analyst (London)*, 1989, **114**, 1519.

69. K. Stulík, V. Pacakova, and M. Podolak, *J. Chromatogr*, 1984, **298**, 225.

70. S. P. Kounaves and J. B. Young, *Anal. Chem.*, 1989, **61**, 1469.

71. M. Ghoto and K. Shimada, *Chromatographia*, 1986, **21**, 631.

72. C. Hua, Y. Wang, and T. Zhou, *Anal. Chim. Acta*, 1990, **235**, 273.

73. C. Hua, K. A. Sagar, K. McLaughlin, M. Jorge, M. Meaney, and M. R. Smyth, *Analyst (London)*, 1991, **116**, 1117.

74. P. Pastore, I. Lavagnini, C. Amatore, and F. Magno, *J. Electroanal. Chem.*, 1991, **301**, 1.

75. S. B. Khoo, H. Gunansingham, and B. Tay, *J. Electroanal. Chem.*, 1987, **216**, 115.

76. H. Ji, J. He, S. Dong, and E. Wang, *J. Electroanal. Chem.*, 1990, **290**, 93.

77. T. Nagaoka, J. Yamashita, M. Kaneda, and K. Ogura, *J. Electroanal. Chem.*, 1992, **335**, 187.

78. S. P. Kounaves and J. B. Young, *Anal. Chem.*, 1989, **61**, 1469.

79. B. Soucaze-Guillous, W. Kutner, and K. M. Kadish, *Anal. Chem.*, 1993, **65**, 669.

80. D. L. Luscombe and A. M. Bond, *Talanta*, 1991, **38**, 65.

81. T. Bührer, P. Gehrig, and W. Simon, *Anal. Sci.*, 1988, **4**, 547.

82. C. Caras and J. Janata, *Anal. Chem.*, 1980, **52**, 1935.

83. Y. Miyahara, T. Moriizumi, and K. Ichimura, *Sensors Actuators*, 1985, **7**, 1.

84. Y. Ikariyama, S. Yamauchi, T. Yukiashi, and H. Ushioda, *Anal. Lett.*, 1987, **20**, 1791.

85. Y. Ikaryama, S. Yamauchi, T. Yukiashi, and H. Ushioda, *Anal. Lett.*, 1987, **20**, 1407.

86. Y. Ikariyama, N. Shimada, S. Yamauchi, T. Yukiashi, and H. Ushioda, *Anal. Lett.*, 1988, **21**, 953.

87. Y. Ikariyama, S. Yamauchi, T. Yukiashi, and H. Ushioda, *J. Electroanal. Chem.*, 1988, **251**, 267.

88. H. Hsueh and AZ. Brajter-Toth, *Anal. Chem.*, 1993, **65**, 1570.

Electrochemical Sensors Based on Bulk-modified Carbon Composite Electrodes

J. Wang

DEPARTMENT OF CHEMISTRY AND BIOCHEMISTRY, NEW MEXICO STATE UNIVERSITY, LAS CRUCES, NM 88003, USA

1 Introduction

Sensors, which combine a recognition layer with a transduction element, are capable of real-time, on-site, analysis of complex environments. Amperometric devices hold a very promising position among the sensor systems, due to their remarkable sensitivity, fast response, and wide linear range, and the miniaturization capability of their solid-electrode transducers. The deliberate modification of solid electrodes can lead to a variety of improvements, including the preferential collection or biorecognition of target analytes, acceleration of electrochemical reaction rates, and exclusion of potential interferences. Such analytical applications of modified electrodes have been reviewed.[1-3]

Traditionally, the recognition element (the modifier) has been attached to the electrode transducer at the electrode–solution interface via polymeric coatings, covalent bonding, or chemisorption (Figure 1A). This review focuses on an alternative, and more practical route for preparing modified-electrode amperometric sensors, based on integration of the recognition element (the modifier) within the bulk of carbon electrodes (Figure 1B). No such integration of recognition and transducing elements has been reported previously in the field of chemical sensors. The deliberate manipulation of the bulk properties can lead to a variety of desirable effects, and can meet the needs of many sensing problems. In particular, the bulk modification strategy allows the controlled and versatile 'loading' of several modifiers (and hence leads to reagentless devices), results in renewable probes (via polishing, cutting, or extrusion, when the bulk of the sensor serves as a continuous 'reservoir' of the modifier), and ensures a fast response due to the immediate proximity of the recognition and transducing elements. Carbon composites, based on the dispersion of carbon powder within an insulator,[4] are particularly suitable to serve as hosts for the chemical or biological modifier, and will thus be the focus of the following

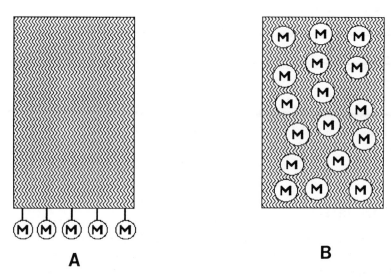

Figure 1 *Amperometric sensors based on tailoring of the electrode-solution interface* (A) *or manipulating the electrode interior* (B). *M is the modifying entity*

discussion. The fabrication of such bulk-modified carbon composite electrodes, and the sensing advantages accrued from such deliberate tailoring of the electrode **interior** will be discussed.

2 Chemical Sensors Based on Modified Carbon Composites

Modified Carbon Paste Electrode

Mixed modifier–carbon paste electrodes are extremely versatile in that numerous modifiers can be easily incorporated into the paste matrix with no need to devise individual attachment schemes for each modifier.[5] This is usually accomplished by hand-mixing appropriate portions of the modifier, graphite powder, and the organic pasting liquid. The resulting mixture is then packed into the electrode assembly, with fresh surfaces being generated by extrusion or cutting. Such modified electrodes maintain the advantages (of low residual current and array character) inherent in carbon paste electrodes.

Ravichandran and Baldwin[6] first demonstrated the utility of mixed modifier–carbon pastes for electrocatalytic detection of target analytes. The same group has developed powerful flow detectors based on cobalt phthalocyanine containing carbon paste matrices.[2] Ruthenium dioxide containing carbon paste is an extremely useful electrocatalytic sensor for carbohydrates.[7] Sensor arrays, comprising carbon paste doped with different metal oxides, used with statistical regression techniques, have been employed recently for the detection of amino acids and carbohydrates.[8] The recent introduction of metallized (Pd-, Pt-, Ru-) graphite powders has led to powerful electrocatalytic carbon paste electrodes with built-in catalytic activities.[9]

The use of modified electrodes as a means of preferentially accumulating inorganic analytes has also attracted intense interest. Selective chemistry can be utilized with molecularly tailored carbon pastes, offering exciting prospects for trace electroanalysis. For example, the collection and detection of trace nickel or gold has been accomplished with carbon pastes modified with dimethylglyoxime[10] or dithizone.[11] Useful preconcentrating carbon paste sensors have been developed based on the ability of crown ether or cryptand modifiers to recognize various ions.[12] Ion exchanger modified carbon pastes (based on appropriate resins or liquid ion exchangers) have been shown to be useful for electrostatic collection of copper[13] or nitrite.[14]

Modification of Robust Carbon Composites

To enhance the day-to-day practicality of sensors based on modified electrodes, it is desirable to renew their surface by simple polishing procedures as is common with conventional solid electrodes. Mechanically rugged and polishable probes can be obtained by replacing a soft carbon paste with a robust carbon composite surface. Shaw and Creasy[15] were first to describe a useful approach for preparing rigid modified electrodes via the copolymerization of styrene and divinylbenzene, in the presence of the modifier and carbon black. A simpler route to achieve the same goal is to employ commercially available kits for the formation of epoxy-bonded graphites.[16] Such modification of graphite epoxy is easily accomplished by mixing the desired quantity of the modifier with the two individual (graphite resin and accelerator) components of the epoxy. The successful incorporation of various chemical modifiers (preconcentrating agents, electrocatalysts) into robust carbon composite materials has been reported.[15-18] For example, the data in Figure 2 illustrates the electrocatalytic detection of the antibiotics streptomycin and novobiocin at a ruthenium dioxide modified graphite-epoxy composite electrode.

3 Biosensors Based on Biocomposites

Amperometric biosensors, based on the incorporation of biological entities within composite carbon matrices are gaining considerable attention and have reached the commercial stage.[19] The bulk material in these biocomposites serves as a source for the biological activity, 'fresh' biosurfaces can be easily obtained by renewing the surface. Very short response times accrue from the absence of supporting membranes and the close proximity of the biocatalytic and graphite sites. Such strategy allows the coimmobilization of an enzyme, its cofactor, a mediator, or another enzyme (as needed). For example, reagentless glucose

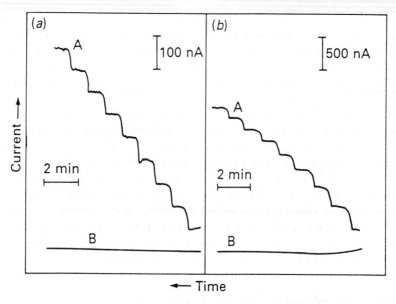

Figure 2 *Amperometric response to addition of* 5×10^{-5} M *novobiocin* (a) *and streptomycin* (b) *at* RuO_2-*modified* (A) *and unmodified* (B) *graphite epoxy electrode. (Reproduced with permission from reference 15)*

sensors have been developed by Gorton *et al.* based on the coimmobilization of glucose dehydrogenase, its NAD^+ cofactor, and Meldola blue as mediator within carbon pastes.[20] The complex reaction mechanism of such biocomposite electrodes has been elucidated by Kulys *et al.*[21]

The surprising fact that enzymes retain their biocatalytic activity within the hostile paste environment is related to the activity of enzymes in organic media. In various situations, the hydrophobic paste environment has beneficial (protective) effects on the enzyme stability.[22] Factors affecting the long-term stability of carbon paste enzyme electrodes were explored by Amine *et al.*[23] Various enzyme stabilizers (*e.g.* fumed silica, polyethyleneimine) can be coimmobilized to extend the operational stability. Other additives can be incorporated to eliminate potential interferences. For example, an *in situ* elimination of metal inhibitory effects can be accomplished using EDTA-containing carbon paste bioelectrodes[24] (Figure 3). Similarly, coimmobilization of protease enzymes can be used to 'destroy' coexisting proteins. Hydrophobic mediators or polymer relay systems are often used to facilitate the electron shuttling between an enzyme redox center and the graphite particles.[25,26] Alternately, metallized (Ru-)graphite can be used to enhance the selectivity by preferentially facilitating the detection of liberated peroxide or NADH species.[27] For example, the biomonitoring of glucose can be accomplished with no interference from acetaminophen or ascorbic and uric acids. Polishable and robust enzyme electrodes can be achieved using mechanically stable graphite–epoxy[28] or graphite–Teflon[29] matrices. For example, a renewable sensor for bilirubin can be fabricated by immobilizing bilirubin oxidase within graphite epoxy.[30]

Figure 3 *Ligand-containing carbon paste biosensor for in-situ elimination of metal inhibitory effects*

The biocomposite sensor route is not limited to enzyme electrodes. Novel and useful biosensors have been based on the incorporation of other biocomponents, *e.g.* algae,[31] plant tissues,[32] or yeasts[33] within carbon pastes.

4 Conclusions

While for many years the emphasis in sensor production was on modification via surface coatings, it has become clear that tailoring of the electrode interior offers a more practical approach for amperometric sensing. Representative examples of practical devices resulting from the bulk modification strategy are given in Table 1. The ability to embed all the necessary modifiers within the solid electrode transducer is particularly attractive, as is the convenient regeneration of fresh sensing layers. No other chemical sensor offers such unique integration of its recognition and transducing components. Despite their great

Table 1 *Representative Examples of Amperometric Sensors Based on Embedding the Modifiers Within the Carbon Transducer*

Modifier	Recognition Mechanism	Analyte	Ref.
Dimethylglyoxime	Coordination	Nickel	10
Dithizone	Coordination	Gold	11
Dowex resin	Ion exchange	Copper	13
Liquid anion exchanger	Ion exchange	Nitrite	14
Ruthenium dioxide	Electrocatalysis	Streptomycin	15
Phthalocyanine	Electrocatalysis	Glutathione	16
Lactate oxidase	Biocatalysis	Lactate	19
Bilirubin oxidase	Biocatalysis	Bilirubin	18
Algae	Bioaccumulation	Copper	29
Yeast	Biocatalysis	Alcohol	31

practicality, much work remains to be done in order to fully understand the complex response mechanism of bulk modified electrodes. New high-resolution techniques, such as STM or SECM, hold great promise for elucidating the structural-reactivity correlations, and hence for optimizing the composition of these devices for best performance. The introduction of new forms of carbon (*e.g.* carbon foams or aerogels) and advances in composite materials should lead to many exciting developments in the near future.

References

1. R. W. Murray, A. G. Ewing, and R. A. Durst, *Anal. Chem.*, 1987, **59**, 379A.
2. R. Baldwin and K. Thomsen, *Talanta*, 1991, **38**, 1.
3. J. Wang, *Electroanalysis*, 1991, **1**, 255.
4. D. Tallman and S. Petersen, *Electroanalysis*, 1990, **2**, 499.
5. K. Kalcher, *Electroanalysis*, 1990, **2**, 419.
6. K. Ravichandran and R. P. Baldwin, *J. Electroanal. Chem.*, 1981, **126**, 293.
7. J. Wang and Z. Taha, *Anal. Chem.*, 1991, **63**, 1714.
8. Q. Chen, J. Wang, G. Rayson, B. Tian, and Y. Lin, *Anal. Chem.*, 1993, **65**, 251.
9. J. Wang, N. Naser, L. Angnes, H. Wu, and L. Chen, *Anal. Chem.*, 1992, **64**, 1285.
10. R. P. Baldwin, J. K. Christensen, and L. Kryger, *Anal. Chem.*, 1986, **58**, 1790.
11. K. Kalcher, *Fresenius' Z. Anal. Chem.*, 1986, **325**, 181.
12. S. Prabhu, R. Baldwin, and L. Kryger, *Electroanalysis*, 1989, **1**, 13.
13. J. Wang, B. Green, and C. Morgan, *Anal. Chim. Acta*, 1984, **158**, 15.
14. K. Kalcher, *Talanta*, 1986, **33**, 489.
15. B. Shaw and K. Creasy, *Anal. Chem.*, 1988, **60**, 1241.
16. J. Wang, T. Golden, K. Varughese, and I. El-Rayes, *Anal. Chem.*, 1989, **61**, 508.
17. D. Leech, J. Wang, and M. Smyth, *Analyst (London)*, 1990, **115**, 1447.
18. S. Wring, J. Hart, and B. Birch, *Analyst (London)*, 1989, **114**, 1563.
19. 'Bioselective Electrodes', Orion Research, Boston, 1993.
20. G. Bremle, B. Persson, and L. Gorton, *Electroanalysis*, 1991, **3**, 77.
21. J. Kulys, W. Schuhmann, and H. L. Schmidt, *Anal. Lett.*, 1992,. **25**, 1011.
22. A. Amine and J. M. Kauffmann, *Bioelectrochem. Bioenerg.*, 1992, **28**, 117.
23. A. Amine, J. M. Kauffmann, and G. J. Patriarche, *Anal. Lett.*, 1991, **24**, 1293.

24. J. Wang and Q. Chen, *Anal. Chem.*, 1993, **65**, 2698.
25. J. Wang, L. Wu, Z. Lu, R. Li, and J. Sanchez, *Anal. Chim. Acta*, 1990, **228**, 251.
26. P. Hale, L. Boguslavsky, T. Inagaki, H. Karan, S. Lee, T. Skotheim, and Y. Okamoto, *Anal. Chem.*, 1991, **63**, 677.
27. J. Wang, L. Fang, D. Lopez, and H. Tobias, *Anal. Lett.*, 1993, **26**, 1819.
28. J. Wang and K. Varughese, *Anal. Chem.*, 1990, **62**, 318.
29. J. Wang, A. J. Reviejo, and L. Angnes, *Electroanalysis*, 1993, **5**, 575.
30. J. Wang and M. Ozsoz, *Electroanalysis*, 1990, **2**, 647.
31. J. Gardea-Torresday, D. Darnall, and J. Wang, *Anal. Chem.*, 1988, **60**, 72.
32. J. Wang and M. Lin, *Anal. Chem.*, 1988, **60**, 1545.
33. W. Kubiak and J. Wang, *Anal. Chim. Acta*, 1989, **221**, 43.

Biosensors for *In Vivo* Monitoring

M. Mascini,[1] D. Moscone,[2] and M. Anichini[3]

[1] DIPARTIMENTO DI SANITÁ PUBBLICA, EPIDEMIOLOGIA E CHIMICA ANALITICA AMBIENTALE, SEZIONE DI CHIMICA ANALITICA VIA GINO CAPPONI 9, 50121 FIRENZE
[2] DIP DI SCIENZE E TECNOLOGIE CHIMICHE, II UNIVERSITÁ DEGLI STUDI DI ROMA, VIA ORAZIO RAIMONDO, 00173 ROMA, ITALY
[3] OSPEDALE POGGIOSECCO, INRCA, VIA INCONTRI 30, FIRENZE

Summary

Biosensors have been developed for *in vivo* monitoring of glucose in animals and humans by implanting, in the subcutaneous fluid, a hollow fibre and by perfusing it with a physiological buffer.

The hollow fibre mimics the function of a blood vessel; substances in higher concentration (like glucose) in the extracellular fluid outside the hollow fibre diffuse in.

A flowing cell assembled with a glucose biosensor has been connected in series to a microdialysis probe. Glucose can be analysed in the range $0.1–20\,mmol\,l^{-1}$. Experiments with rabbits and humans submitted to a 'glucose load' have been successfully carried out.

1 Introduction

Electrochemical biosensors have found wide interest in clinical chemistry and medicine. Physiologists, cardiologists, and diabetologists dreamt for years about the possibility of continuously monitoring chemical parameters to feed back appropriate action to restore the values to normal levels. In the last 15 years a large number of publications, reviews, books, and workshops have been devoted to this topic. No operating, completely implantable, biosensor is presently commercially available, but many approaches have been reported and great advances have been achieved.[1-4]

One major challenge has been *in vivo* monitoring and several laboratories in Europe, USA, and Japan have been very active in proposing different solutions.

Two approaches should be mentioned: the first is the assembly of a needle glucose electrode for subcutaneous implant and the second is the microdialysis probe for sampling subcutaneous fluid. The first approach was pioneered by

Shichiri *et al.*[5] and then several other groups developed similar strategies. This approach found a limitation in the body reaction to the implanted needle. Biocompatible strategies have to be studied to overcome this problem.

A new technique for sampling *in vivo* has been recently applied in our laboratory for the purpose of developing a glucose continuous monitor; it is called microdialysis. This technique is a complementary approach to the implantable biosensors. The idea is to mimic the function of a blood vessel by implanting a 'microdialysis probe' into the tissue.[6] The essential component of the probe is a thin dialysis tube perfused with a physiological solution much like the blood perfuses a blood vessel (Figure 1). Substances in higher concentration in the extracellular fluid outside the probe diffuse in. Once substances are carried out of the body by the perfusion liquid their concentration can be determined by appropriate analytical techniques. Biosensors can be easily coupled to microdialysis devices and can monitor the appropriate metabolite without a pre-separation step. A flowing cell assembled with a glucose biosensor has been connected in series to microdialysis probes. Optimization of glucose determination under the above conditions has been performed.[7] Glucose has been analysed in the range $0.1-20\,mmol\,l^{-1}$. Experimental results using a single sterile fibre inserted subcutaneously by a small intravenous needle in animals and humans have been successfully obtained and are presented in this paper.

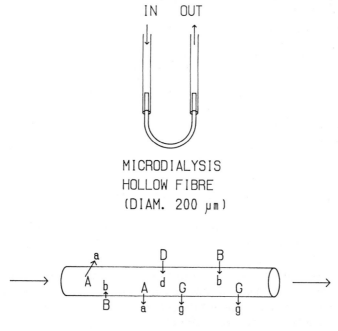

Figure 1 *Microdialysis Probe*

2 Materials and Methods

Glucose oxidase (GOD EC 1.1.3.4, from *Aspergillus niger*, type VII, 132 000 U g^{-1}) was obtained from Sigma. (A GOD-immobilized-nylon net membrane was prepared as previously reported.[7])

The glucose solution was prepared with β-D(+)glucose from Farmitalia Carlo Erba (Milan, Italy), allowed to equilibrate overnight, and suitably diluted. The buffer solution, Dulbecco's physiological buffer (pH 7.4), was prepared in doubly distilled water. All chemicals were of analytical grade.

Cellulose acetate (53% acetyl) and polyvinyl acetate of high molecular weight were obtained from Farmitalia Carlo Erba (Milan, Italy). For casting the cellulose membrane a precision gauge tool (from Precision Gauge and Tool Co, Dayton, OH) was used. This membrane, with about a 100 Da Molecular Weight Cut-off (MWCO), was prepared as follows.

Cellulose acetate (1.98 g) and polyvinyl acetate (20 mg) were dissolved in 20 ml of cyclohexanone and 30 ml of acetone. The solution was cast on a glass plate 200 µm thick, with the aid of a casting tool. After complete evaporation (10 h) the membrane was immersed in water and peeled off the glass surface. The dried membrane has a thickness of about 20 µm.

The peristaltic pump, Minipuls 3, for flow analysis was from Gilson (France).

A CMA/160 On-Line Injector automatically activated by a CMA/100 Microinjection Pump (CMA/Microdialysis, Stockholm, Sweden) was used.

A wall-jet flow cell, was obtained from Metrohm (Mod 656 Electrochemical Detector, Herisau, Switzerland). However, the working electrode was substituted with a platinum electrode, with a diameter of 1.6 mm, obtained from BAS (Model MF 2013). This cell was connected to a CV-37 Voltammograph (also from BAS) and the amperometric detector was connected to a L-6512 recorder (Linseis, Selb, Germany).

To assemble the microdialysis probes we used Spectra/Por '*in vivo*' microdialysis hollow fibre, regenerated cellulose (i.d. 150 µ, wall thickness 15 µ) obtained from Spectrum Medical Industries Inc. (Los Angeles, CA). We inserted into the fibres a 50 µm diameter tungsten wire obtained from Goodfellow (Cambridge, UK). Nylon tubings (i.d. 0.250 mm, o.d. 0.750 mm, wall thickness 0.250 mm) from Firie (Genova, Italy) were used to connect the hollow fibre to the flow-system.

3 Assembling the Sensor

The wall-jet flow cell includes three separate electrodes: the working electrode (platinum disk with diameter of 1.6 mm), the reference (Ag/AgCl), and the auxiliary electrode.

A thin (20 µm) membrane of cellulose acetate is stretched over the platinum electrode surface: it removes the electrochemical interferences (uric acid, ascorbate, *etc.*) with its nominal MWCO of 100 Da;[8-10] this figure is obtained as a rough number and mainly means that ascorbic and uric acid do not reach the electrode surface while hydrogen peroxide passes through easily. A nylon net (thickness 100 μm), with the immobilized glucose oxidase enzyme, is placed over the electrode area.

A sleeve of suitable diameter is used to stretch both the membranes on the Pt electrode surface.[11]

4 Procedures

The flow system for *in vivo* experiments is shown in Figure 2. For *in vitro* experiments the microdialysis probe was immersed in a beaker containing a physiological solution.

The peristaltic pump drives the carrier solution through the microdialysis probe, immersed in glucose standard solutions, at a constant rate. A steady-state current is obtained. The standard solutions, where the microdialysis probe was immersed, were manually changed.

The hollow fibre probe (200 μm) was connected to nylon tubes of the flow system.

To place the microdialysis hollow fibre subcutaneously a sterilized needle was inserted transcutaneously through the skin for about 1 cm and the needle tip was pulled out. Then the sterilized fibre was inserted from the needle tip and the needle was taken out leaving the hollow fibre under the skin. The fibre was connected to nylon tubes and fixed with cyanoacrylic glue. A flow injection system was used to check variations in sensitivity during the experiments. An injection valve with a 20 μl sample loop was introduced in the flow system just after the microdialysis probe (Figure 2). A glucose standard buffered solution filled the loop of the injection valve and flowed through the glucose biosensor. We obtained a current profile similar to a peak. The dispersion coefficient of the apparatus is 1.1 defined as $D = C^0/C_{max}$ where C^0 is the concentration of the solution and C_{max} is the concentration profile.[11]

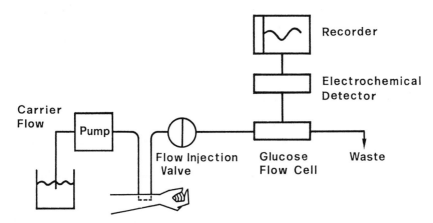

Figure 2 *Diagram of the flow system*

5 Results and Discussion

The wall jet used in this work shows a current signal more stable and less sensitive to external devices, such as pump pulsations, in comparison with a thin layer flow cell used previously.[7,11] The geometry of the flow cell and the MWCO of the dialysis fibre is very important in *in vivo* experiments. We have demonstrated how sensitivity can be stabilized by suitable choice of such parameters.[11]

In such conditions glucose concentrations in the range 1–20 mM gave a linear calibration curve. This is the range useful for measuring glucose in the interstitial liquid.

Figure 3 shows the results of an experiment with a rabbit (3 kg) where a glucose load of 2 g (30 ml of glucose solution, 7%) was infused in the ear vein for about 8 min. The hollow fibre has a MWCO of 9000 and was inserted subcutaneously in the back of the animal. After about 2–3 min from the start of the infusion procedure we could notice a variation of current measured by the cell. Such an increase, due to the subcutaneous glucose, reached a maximum value after 15 min, then the current decreased reaching the basal value after about one hour. After the end of the experiment (150 min), the electrode response decreased to below the initial value; this could be explained as the insulinic response of the animal to the rapid glucose load.

During this time the sensitivity of the biosensor was checked and we noticed only a slight decrease of less than 5% after the three hours of monitoring.

During the experiment, blood aliquots were taken from the ear of the rabbit and analysed in the clinical laboratory for glucose. The concentrations are plotted in Figure 3.

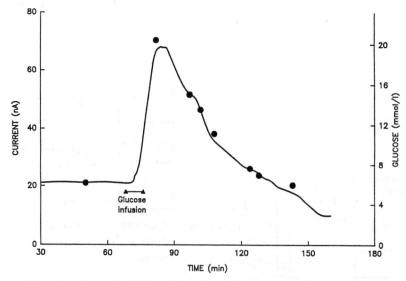

Figure 3 *In vivo glucose load with a 3 kg rabbit. The values plotted are the clinical values of blood taken out during the experiment. (Correlation coefficient r = 0.993)*

Assuming that the constant value of current before the glucose load was proportional to the concentration measured in the blood, we can continuously monitor the glucose concentration during the experiment (right side of the plot). Therefore, we assume a linear relation between the biosensor output (current) and glucose concentration in the range of interest (one point calibration). In the plot of Figure 3, we inserted also the values of the control procedure and it can be seen how this system follows, with high accuracy, the concentration value of blood in rabbits. The response time of the overall system is 2–3 min and the stability of the sensitivity of the biosensor is perfectly acceptable.

6 Human Volunteers

In Figure 4 we report 15 experiments on human volunteers submitted to glucose oral load; we follow the change in the blood glucose concentration by continuous monitoring of the subcutaneous glucose value and by normal clinical methods, taking blood samples every 30 min. The microdialysis fibre was inserted in a forearm and blood samples for clinical analysis were taken from the other arm.

The glucose value of the blood was determined by clinical standard procedures. In only four cases, volunteers 5, 6, 8, and 13, the microdialysis probe had to be replaced during the course of the experiment, due to accidental breakage or malfunctioning of the probe. This is evident from an interruption of the continuous line.

To correlate the continuous monitoring of the current output with the glucose value, we assume that the current value before the load corresponds to the glucose value of the first blood sample (one point calibration). In a few experiments, volunteers 1, 3, 8, and 9, the glucose value for calibration has been chosen in the flat part of the curve of each experiment (generally after the peak value, at the end of the experiment). This was necessary because in such an experiment, the blood samples for the 'control level' of glucose were analysed at the end of the experiment and we feared that glycolysis reduced the value of glucose in the first samples.

The *in vivo* and clinical results of the experiments reported as numbers 2, 14, and 15, did not perfectly correlate (Figure 4), but the correlation increases (Figure 5) if we assume a delay between the blood glucose value and subcutaneous values, of 30 min. A delay was in fact expected from the physiological point of view between the glucose peak of the blood and that of subcutaneous tissue (interstitial fluid). However we noticed it only with a few patients.

In our experiments the microdialysis probe was located in the forearm or in the arm, but other sampling sites could be more suitable for a better correlation with the glucose values in blood.

This has been considered in a similar experiment[12] where up to 18 min of delay was calculated between glucose blood and subcutaneous value.

In another three experiments (1, 11, and 12) a low correlation was shown; the trend of glucose in blood seems at a glance dissimilar or opposite to the

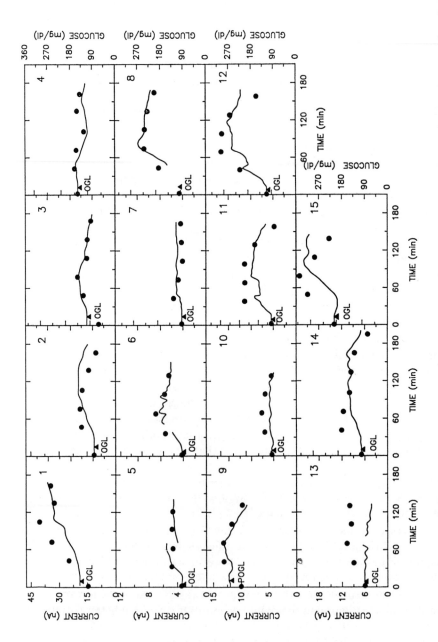

Figure 4 *Comparison of biosensor output with a subcutaneous microdialysis probe and blood glucose concentrations for 15 volunteers after glucose oral load realized by drinking 75 g of glucose in water at a time indicated by the arrow in each graph*

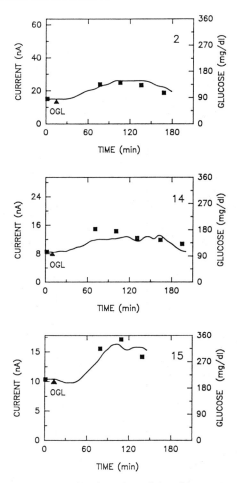

Figure 5 *The experiments 2, 14, and 15 have been delayed in respect of their glucose value in blood of 30 min. Better agreement was then obtained*

subcutaneous values trend and the introduction of a delay for subcutaneous values does not increase the correlation.

The results of experiments 3, 4, 7, 8, 9, and 10 seem to correlate much better and give the hope for a real portable subcutaneous glucose monitoring. We can notice that in some experiments, the glucose values taken from blood and subcutaneously did not reach a real peak after the glucose ingestion.

We should consider, however, that in such a test, the blood glucose profile can depend on three major effects: (i) the rate of glucose absorbance in the intestine; (ii) the volume of distribution of glucose, and (iii) the rate of glycolysis; this is mainly due to the insulin response of the patient.

Such effects depend on many physiological variables, such as: (i) the age of the patient; (ii) the diet of the patients during the days preceding the test; and even (iii) the time when the test is performed.

Therefore, it is quite normal to obtain large variations in the physiological response to the intake of the same amount of glucose.

The size of the apparatus does not permit experiments longer than 3–4 h and a system will soon be devised to extend the experiments.

During the experiments conducted, it became apparent that control of sensitivity was not necessary over a period of up to 3–4 h. Variations in the sensitivity of the glucose probe were less than 10%, which is fairly acceptable for continuous monitoring. One of the effects of the position of the microdialysis probe was involuntary muscle contraction near the sampling point. Often this contraction (generally when the blood sampling was taken for clinical analysis) led to a sudden variation of the continuous monitoring due, in our opinion, to a collapse of the subcutaneous fibre.

This was evident in experiment 6, where a series of small current peaks are evident on the continuous monitoring trace.

In another experiment (number 15) the patient was affected slightly by Parkinson disease and we obtained a series of unwanted peaks which have been smoothed out in the reported continuous monitoring.

7 Conclusions

Microdialysis, coupled with a suitable flow cell equipped with a glucose biosensor, permits real *in vivo* sampling and continuous monitoring of glucose during a glucose load experiment. One point calibration is perfectly acceptable with this approach.

Studies on the sampling site reveal new problems, but continuous glucose monitoring can be performed on the basis of the principle described.

Acknowledgement

This research was supported by the CNR Target Project on Biotechnology and Bioinstrumentation.

References

1. 'Advances in Biosensors', ed. A. P. F. Turner, Supplement 1; Chemical Sensors for *in vivo* monitoring, JAI Press Ltd., 1993.
2. '*In vivo* Chemical Sensors: Recent developments', eds. S. J. Alcock and A. P. F. Turner, Cranfield Press, 1993.
3. G. G. Guilbault and M. Mascini, 'Uses of Immobilized Biological Compounds', Kluwer Academic Publications, 1993, 115.
4. M. Mascini, D. Moscone, and G. Palleschi, 'Bioinstrumentation', Butterworths, 1990, p. 1429.
5. M. Schichiri, R. Kawamori, and Y. Yamasaki, 'Bioinstrumentation', Butterworths, 1990, p. 1501.
6. P. K. Kissinger, *J. Chromatogr.*, 1989, **488**, 31.
7. D. Moscone, M. Pasini, and M. Mascini, *Talanta*, 1992, **39**, 1039.
8. M. Mascini and F. Mazzei, *Anal. Chim. Acta*, 1987, **192**, 9.

9. P. J. Taylor, E. Kmetec, and J. M. Johnson, *Anal. Chem.*, 1977, **49**, 789.

10. G. Palleschi, M. A. Nabi Rahni, G. J. Lubrano, J. N. Ngwainbi, and G. G. Guilbault, *Anal. Biochem.*, 1986, **159**, 114.

11. D. Moscone and M. Mascini, *Ann. Biol. Clin.*, 1992, **50**, 323.

12. C. Meyerhoff, F. Bischof, F. Sternberg, H. Zier, and E. F. Pfeiffer, *Diabetologia*, 1992, **35**, 1087.

Chromatography and Capillary Electrophoresis in Biomedicine

E. Jellum

INSTITUTE OF CLINICAL BIOCHEMISTRY, RIKSHOSPITALET, 0027 OSLO, NORWAY

1 Introduction

Several human diseases often lead to characteristic alterations in the highly complex pattern of metabolites both of exogenous and endogenous origin in the body fluids, cells, and tissues. Multicomponent analytical techniques, including chromatography and electrophoresis, are suitable to detect diagnostically important changes in the biochemical 'profiles' thus produced. The profiling techniques currently used in our laboratory include gas chromatography, gas chromatography–mass spectrometry, high performance liquid chromatography with computerized diode-array detector, automated amino acid analysis, and capillary electrophoresis with laser-induced fluorescence detection.[1]

The whole system is used for problem solving in biomedicine, *e.g.* for the diagnosis of metabolic disorders, for monitoring the effect of dietary treatment, for monitoring the level of drugs, for studies related to surgical problems, for detecting markers of malignant disease, and for evaluating risk factors in cancer development.

Some applications of our multicomponent analytical system will be demonstrated herein.

2 Metabolic Disorders

This is a field where various chromatographic and mass spectrometric methods have proved to be the most valuable diagnostic tools.[2–5] Many centres in different parts of the world therefore use such methods to identify characteristic metabolites which occur in the body fluids as a consequence of an enzyme deficiency.

Selection of patients is based on a number of clinical symptoms or 'warning signals' typical of metabolic disease. These signals may be progressive disease, metabolic acidosis, recurrent vomiting, peculiar smells from the body and urine, neurological symptoms of unknown etiology, and indications of inheritance (similar cases in the family).

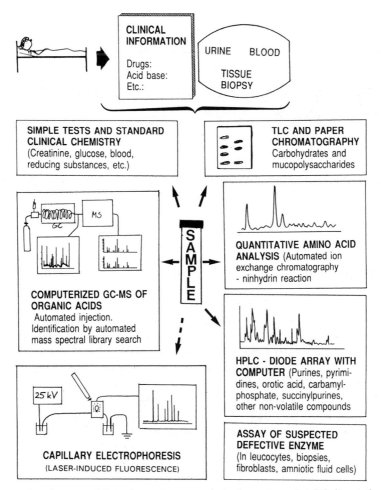

Figure 1 *Flow diagram of the multicomponent analytical system used at Rikshospitalet, Oslo*

After submission of a sample, preferably urine, to our laboratory the specimen is subjected to systematic analysis as illustrated in Figure 1. Certain dip-stick tests and simple clinical chemistry analyses are carried out first. Then TLC and paper chromatography are performed to determine mucopolysaccharides and carbohydrates (for diagnosis of mucopolysaccharidoses and disorders of carbohydrate metabolism), followed by quantitative amino acid analysis to diagnose aminoacidopathies. GC–MS with automated sample injection and peak identification by computerized mass spectral library search are used to obtain the organic acid profiles, thus revealing possible organic acidurias (over 50 such diseases are known). HPLC with a computerized diode-array detector determines, *e.g.* the content of succinylpurines (excreted in a rare genetic type of autism), and orotic acid which is important in the differentiation of disorders of the urea-cycle. Although capillary electrophoresis (CE) is not yet used in the daily routine, we have recently demonstrated that this method is particularly suitable for diagnosis of disorders related to cysteine, homocysteine, and glutathione.[6]

The analytical system outlined above has the potential to diagnose about half of the approximately 250 currently known different metabolic diseases. Follow-up studies, *e.g.* monitoring the effect of treatment and, in a few instances, prenatal diagnosis by direct analysis of cell-free amniotic fluid, are also possible by means of the methods discussed above. In these instances the analytical procedures have to be modified to obtain quantitative data, *e.g.* by using special, often stable isotope-labelled internal standards in the HPLC- and GC-analyses, and by operating the MS in the selected ion monitoring mode.

Figures 2 and 3 show some typical chromatographic profiles selected from our daily routine. Figure 2 (top) is the urinary organic acid profile of a patient with methyl malonic acidemia, compared to a control urine (bottom). Figure 3 (top) shows the profile from a patient with severe ketoacidosis, and Figure 3 (lower panel) illustrates the diagnosis of hyperoxaluria.

Recent progress in gene-technology makes it possible to diagnose more and more metabolic disorders at the gene level. The question may then be raised whether modern DNA-technology will make chromatography and mass spectrometry obsolete for diagnostic purposes. One should realize, however, that DNA-technology is only suitable for diagnosis in situations where there is a known genetic disease within a given family. The DNA methods are therefore particularly appropriate for prenatal diagnosis where one knows which type of gene defect to look for. In contrast, the chromatographic methods can diagnose close to half of all metabolic disorders recognized today without knowing anything about the underlying cause of the disease.

Therefore, one can safely postulate that the chromatographic profiling techniques will continue to be most helpful tools for diagnosis and studies of metabolic diseases for many years to come. These methods cannot be replaced by DNA-technology, which, on the other hand, will become increasingly more important for prenatal and confirmatory diagnosis.

3 Chromatography and Problem-solving in Surgery

Trans-cervical Resection of the Endometrium in Women with Bleeding Disturbances

A new surgical technique, now used in many hospitals, has been developed as an alternative to the removal of the uterus (hysterectomi) in women with bleeding disturbances. The method involves resection of the endometrium by insertion of a special instrument through the cervical canal. During this operation an electrically inert solution containing 1.5% of the amino acid glycine is infused to irrigate the uterine cavity. About 25 min are required for surgery, and the patients are conveniently cured from their bleeding problem. There are remarkably few complications, except that about one in three patients experience nausea and vomiting several hours after surgery.[7] This discomfort is not related to the anaesthesia given, and must be due to other, *e.g.* metabolic factors.

(a)

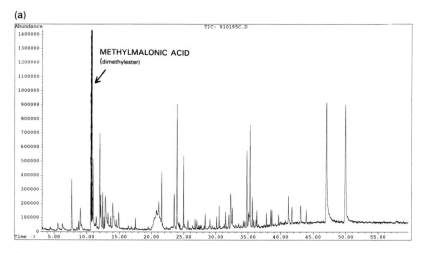

(b)

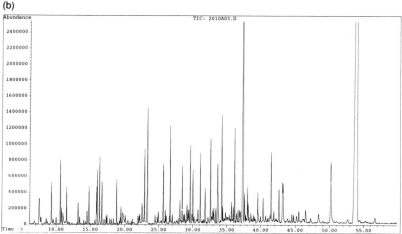

Figure 2 *Organic acid profile of urine from a patient with methylmalonic acidemia* (a) *and from a control person* (b). *The samples were acidified, extracted with diethylether, and methylated with diazomethane before analysis. The separation was achieved in a Hewlett–Packard 5970 GC–MS system with an automated sample injection system (HP 5890 GC with HP 7673A 100 sample injector) and a HP 300 data system. The SP-1000 fused silica capillary column was programmed from 80–200 °C at a rate of 4° min*$^{-1}$

In an attempt to approach the problem from the chromatographic and biochemical viewpoint, serum samples from 30 patients undergoing trans-cervical resection were analysed, using the methods outlined above. Figure 4 (top) shows the amino acid profile of serum from a patient who experienced no problems following the operation. Figure 4 (bottom) shows that large amounts of glycine were present in another patient who suffered from post-surgical discomfort with nausea and vomiting. In fact, it turned out that *all* patients with nausea had absorbed large quantities of glycine from the irrigation solution, resulting in extremely high serum levels of this amino acid.

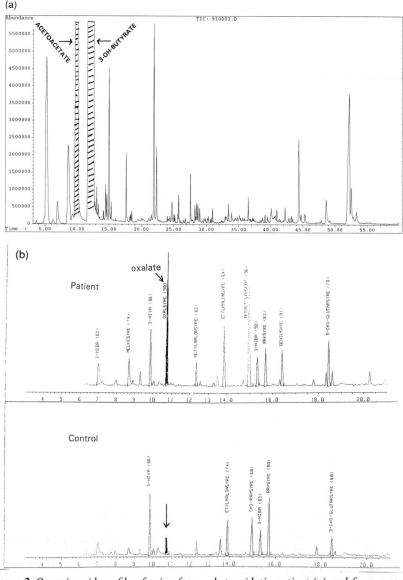

Figure 3 *Organic acid profile of urine from a ketoacidotic patient (a) and from a control and a patient with primary hyperoxaluria, type 1 (b). Analytical procedure as described in Figure 2*

Glycine is known to be toxic in high doses, and with serum values up to 100 times higher than normal, it is clear that the patients had become intoxicated with glycine. The reason why only 30% of the patients had enhanced absorption of glycine leading to the toxic effects, remains unclear. The investigation (to be published in detail elsewhere, see reference 7) also suggests that alternative irrigation solutions should be sought for use during trans-cervical surgery.

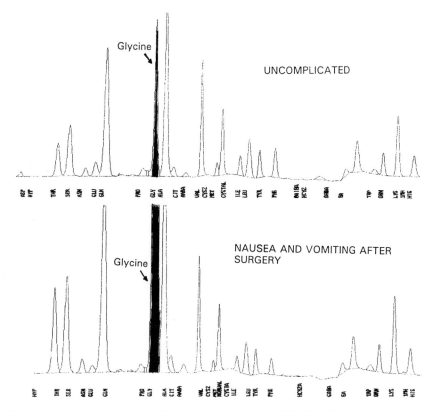

Figure 4 *Serum amino acid profiles of two patients with bleeding disturbance, undergoing trans-cervical surgery. A Biotronic automatic amino acid analyser (Biotronic, Maintal, Germany) fitted with an ion exchange chromatography column and classical post-column ninhydrin detection and a Nelson Turbochrom data system were used*

Capillary Electrophoresis of Taurine in Heart Muscle Biopsies

Recent reports[8-10] have caused a growing interest in the aminosulfonic acid, taurine, because of its possible role in the development of cardiomyopathy. It now seems clear that this compound is an essential amino acid in cats. Taurine depleted cats develop retinal degeneration, cardiomyopathy, altered white-cell function, and abnormal growth and development.[10] Evidence for a taurine-deficiency cardiomyopathy in humans is also beginning to appear,[9] and there seems to be a link between decreased taurine concentration in the myocardium and decreased myocardial mechanical function, at least in the cat.[8] Treatment of advanced cardiomyopathy in humans may involve heart transplantation, which is carried out routinely in our hospital. In view of the possible link between cardiomyopathy and lack of taurine in the heart muscle, there is a need for a method to analyse taurine in small biopsies from the myocardium.[6] HPCE may be a suitable method for this purpose. A variable wavelength fluorescence spectrophotometer detector, using a Xenon UV-lamp as light source, was used in our first investigations.[6]

Our CE-instrument has now been completely redesigned, and consists of a commercial injection system (Lauer Labs., The Netherlands) and a home-made laser-induced fluorescence detector. The latter detector utilizes a He–Cd laser (Liconics, USA) and the laser light is directed on to the fused silica column by means of fibre optics. The emitted light is passed via suitable optical lenses and filters into a photomultipler detector.

Figure 5 shows the separation of amino compounds, including taurine, in a minute quantity of a human heart biopsy (about 0.1 mg tissue). The sample had been treated with CBQCA[11] to yield highly fluorescent derivatives suitable for excitation with the He–Cd laser.

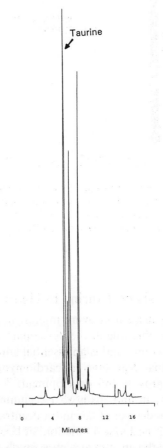

Figure 5 *Capillary Electrophoresis (CE)-separation of amino compounds in a human heart biopsy. 0.1 mg of tissue was derivatized with CBQCA[11] before analysis. The capillary column was 30 cm long, i.d. 50 μm. A 0.05 M sodiumphosphate buffer, pH 7.4 was used and the applied voltage was 20 kV. The home-made CE instrument is described in the text*

A quantitative HPCE method which requires only sub-milligram tissue is now used to study patients with cardiomyopathy and other myocardial failures.

4 Chromatography (HPLC and GC) in Studies on Risk of Breast Cancer: Use of the JANUS Serum Bank[12]

This study was based on blood samples provided by the JANUS serum bank in Norway. The bank was initiated in 1973 and now (October 1993) comprises 424 938 serum samples consolidated from 293 692 donors. The specimens are stored at $-25\,°C$. From 1–13 consecutive samples are available from each donor. Up to October 1993 about 14 000 of the donors had developed some form of cancer. Frozen serum samples, collected from a few months to 19 years prior to clinical recognition of their disease, are thus available for research purposes. The principle aim of the JANUS-project is to search in the premorbid sera for chemical, biochemical, immunological, or other changes that might be indicative of early stages of cancer development.[12]

One of the many projects utilizing the large serum bank was concerned with the role of polyunsaturated fatty acids in serum phospholipids and risk of breast cancer. In this study sera from 87 women who developed breast cancer subsequent to blood donation (cases) were compared with samples from 235 women who were free of any diagnosed cancer (controls), of similar age, and whose sera had similar storage time as the cases. HPLC was used to isolate the phospholipid fractions, which were transmethylated before determination of the fatty acid profile by capillary GC.

The results showed that there was an inverse relation between linoleic acid (C18:2, *n*-6) and risk of breast cancer, but this association was restricted to women who were 55 years and younger.

A conclusion to this study was that high levels of linoleic acid (18:2, *n*-6), as measured in serum phospholipids, seems to offer a protection against breast cancer in pre- and perimenopausal age.[13] Fatty acids of the *n*-3 series had, surprisingly, no association with risk of breast cancer.

5 Concluding Remarks

The multicomponent analytical system outlined herein has been gradually assembled over the past two decades and used as a problem-solver in the biomedical field. It has been considered all these years that new advances in analytical chemistry and instrumentation should be closely watched and whenever possible, exploited as potential tools in biomedicine. Thus when GC–MS and HPLC were introduced in the late 1960s and applied to, *e.g.* studies on metabolic disorders, many new diseases of this type were soon discovered.[2-5] Today these chromatographic methods are recognized as indispensible diagnostic tools, particularly in the field of metabolic disorders. Also in other fields, *e.g.* as exemplified in this report, chromatography may yield new information on biomedical problems.

Today we have available another technique with considerable potential, namely capillary electrophoresis (CE). The well known advantages of CE are simplicity, versatility, rapid analysis time, high separation efficiency, and small sample volume requirement. Particularly the latter fact opens up new possibilities to analyse small tissue biopsies and small amounts of serum and blood cells down to the level of single cells.[14] It is predicted that CE with its high sensitivity detection, *e.g.* using laser-induced fluorescence, may turn out to be another useful problem-solving technique in modern biomedicine.

References

1. E. Jellum and A. K. Thorsrud, *J. Chromatogr.*, 1989, **488**, 105.
2. E. Jellum, *J. Chromatogr.*, 1977, **143**, 427.
3. S. I. Goodman and S. P. Markey, 'Laboratory and Research Methods in Biology and Medicine', Alan R. Liss, New York, 1981, Vol. 6.
4. R. A. Chalmers and A. M. Lawson, 'Organic Acids in Man', Chapman and Hall, New York, 1982.
5. Z. Deyl and C. C. Sweeley (eds.), *J. Chromatogr.*, 1986, **379**, 1.
6. E. Jellum, A. K. Thorsrud, and E. Time, *J. Chromatogr.*, 1991, **559**, 455.
7. O. Istre, E. Jellum, K. Skajaa, and A. Forman, *Br. J. Gynaecol.* (submitted for publication).
8. P. D. Pion, M. D. Kittleson, Q. R. Rogers, and J. G. Morris, *Science*, 1987, **237**, 764.
9. A. Tenaglia and R. Cody, *Am. J. Cardiol.*, 1988, **62**, 136.
10. R. L. Hamlin and C. A. Buffington, *Vet. Clin. N. Am. Small Anim. Prac.*, 1989, **19**, 527.
11. J. P. Liu, Y. Z. Hsieh, D. Wiesler, and M. Novotny, *Anal. Chem.*, 1991, **63**, 408.
12. E. Jellum, Aa. Andersen, H. Ørjasaeter, O. P. Foss, L. Theodorsen, and P. Lund-Larsen, *Biochim. Clinica*, 1987, **11**, 191.
13. L. Vatten, K. S. Bjerve, Aa. Andersen, and E. Jellum, *Eur. J. Cancer*, 1993, **29A**, 532.
14. R. T. Kennedy, M. D. Oates, B. R. Cooper, B. Nickerson, and J. W. Jorgenson, *Science*, 1989, **246**, 57.

Magnesium: Clinical Significance and Analytical Determination

A. Hulanicki,[1] A. Lewenstam,[2] and M. Maj-Zurawska[1]

[1] DEPARTMENT OF CHEMISTRY, UNIVERSITY OF WARSAW, POLAND
[2] DEPARTMENT OF ANALYTICAL CHEMISTRY, ABO AKADEMI UNIVERSITY, TURKU, FINLAND

Summary

Magnesium plays a significant role in metabolic and physiological processes and its deficiency is responsible for numerous diseases. In this respect the knowledge of magnesium level is important. The determination of ionized magnesium is possible using an ion-selective electrode with a neutral carrier ionophore ETH 5220, in an automated computerized system which takes into account the selectivity of the electrode and its kinetic behaviour. Results for patients with myocardial infarct and short bowel syndrome indicate applications of such measurements.

The elemental composition of the human body has been known for decades. Magnesium is the fourth most abundant metallic element with its average content ranging to $0.5\,g\,kg^{-1}$, which corresponds to a total amount of $30-40\,g$ for an adult human. The majority of magnesium is present in the bones (53%), with muscles (27%) and soft tissues (19.2%) containing most of the rest. Only a small percentage is present in blood (0.8%), of which 0.3% corresponds to serum and 0.5% to erythrocytes. The content of magnesium in various body compartments is given in Figure 1. The mobility of blood and relative ease of its sampling and subsequent measurement make this compartment attractive for clinical determination.

Magnesium exhibits a well documented physiological and metabolic role in numerous processes,[1,2] however it is still a subject of interest and intensive study, probably because of its multifunctionality, which makes such studies difficult. The best known role of magnesium is as a cofactor in more than 300 enzymatic reactions, involving, e.g. ATPases, creatine kinase, alkaline phosphatase, and enolase. Magnesium influences the enzyme activity by ligand binding, by

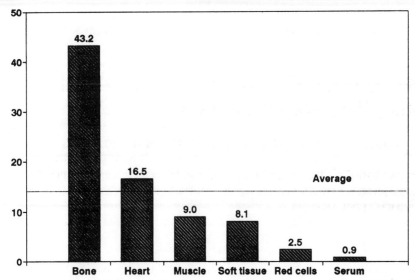

Figure 1 *Content of magnesium in various body compartments (in mmol kg⁻¹). The horizontal line represents the average magnesium content in the human body*

binding the activity sites of an enzyme, changing its conformation, and promoting aggregation of multienzyme complexes.

The role of magnesium is extremely important as an activator for the transmembrane electrolyte flux. The membrane sites available for calcium are affected by the ambient magnesium concentration. Magnesium may penetrate the membrane by diffusion, but its release is an energy-requiring process influencing the intracellular free calcium level. The electrical properties of membranes and their permeability characteristics are affected by magnesium concentration. A transmembrane potential results from the simultaneous transport of magnesium and potassium against calcium ions.

The numerous functions of magnesium are related to the intracellular/ extracellular concentration ratio of magnesium and the total decrease of magnesium content in the body is responsible for numerous acute and chronic changes. Among them should be mentioned disturbances of electrolyte levels, vascular diseases (hypertension, arteriosclerosis, *etc.*), cardiac disturbances (electrocardiographic changes, sudden cardiac death, *etc.*), neuromuscular, and neuropsychiatric diseases (apathy, delirium, seizures, *etc.*). Hypomagneseamia may be caused by both poor magnesium intake and increased magnesium losses. The mechanisms governing those processes have not been fully elucidated.

The simplest way of measuring the magnesium content of the body is by analysis of blood or its fractions. It is, however, certain that this does not reflect properly the status of magnesium in the body. The magnesium content in serum is not directly correlated with any other pool of magnesium except the interstitial fluid. Similarly, the magnesium level in erythrocytes is not correlated with its content in muscle tissue. Sampling and analysis of this compartment is however difficult. It has been indicated that highly informative results can be obtained by analysis of blood leucocytes.[2] This may be a new challenge for analysts.

Magnesium is present in the body in various speciation forms. In serum about two-thirds of magnesium exists as free hydrated ions, most of the remaining magnesium (27%) is bound to proteins, phospholipids, nucleotides, and ATP. Complexing of magnesium with phosphate groups of nucleic acids has structural effects as DNA molecules become more compact and gain stability through magnesium bridges. Only a small percentage of magnesium (8%) is bound to low molecular mass ligands such as carbonate, lactate, oxalate, *etc.*

In many biochemical reactions magnesium acts as a calcium antagonist.[3] Despite its slightly smaller ionic radius, magnesium forms complexes of similar stability with ligands having oxygen donor atoms. Introduction of nitrogen donors reduces selectivity for calcium over magnesium. The side reaction coefficient, $\alpha_{Ca(Mg)}$, for calcium complexation in the presence of magnesium

$$\alpha_{Ca(Mg)} = 1 + [Mg^{2+}]K_{Mg} \tag{1}$$

where $[Mg^{2+}]$ is the free magnesium concentration, and K_{Mg} the stability constant of the magnesium complex, significantly decreases the stability of calcium complexes. It must also be pointed out that not only thermodynamics but also kinetics must be taken into account. As the calcium complexes are more labile than magnesium ones, the rate constant for exchange of water in calcium aqua ions for other ligands is mainly limited by their diffusion rate. In the case of magnesium, the limiting factor is the release of water molecules from the magnesium aqua ions. As a result, the overall rate constant for magnesium is several orders of magnitude smaller than that for calcium.

Not only calcium influences the complexation, and in consequence, the speciation, of magnesium. As ligands which may complex magnesium in the body fluids are relatively strong bases, the hydrogen ion concentration influences the position of the metal–ligand equilibrium. For the same sample analysed at a different pH, the free magnesium ion concentration has a different value. The normal pH value of blood and blood serum is 7.4, but a change in the carbon dioxide saturation influences directly the pH of the sample. To avoid this, the sample should be prepared cautiously, *e.g.* in anaerobic conditions. Errors may occur at preliminary sample treatment or even at the sampling stage. To obtain comparable results, all measurements of 'ionized magnesium' are normalized to pH 7.4 using the simple Siggaard–Anderson formula[4]:

$$NiMg = iMg + 10^{A(7.4 - pH)} \tag{2}$$

The parameter A has been found[5] equal to -0.10 for magnesium, and NiMg and iMg represent the concentrations of 'normalized ionized magnesium' and 'measured ionized magnesium' respectively, at the indicated pH. Measurements made at a pH differing by less than 0.12 from 7.4 introduce a relative error smaller than 2.5% (Figure 2). The error caused by the approximate value of the parameter A, within the limits -0.05 to -0.15, is also negligible.

At present, the most obvious solution for measuring the ionized magnesium is ion-selective electrode potentiometry. Several electrodes sensitive to magnesium

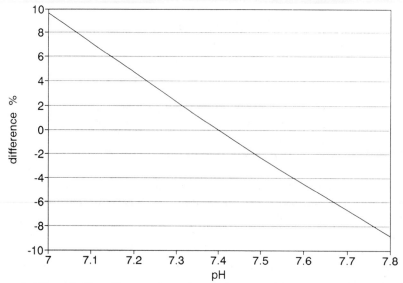

Figure 2 *Effect of pH on determination of ionized magnesium in serum*

have been developed during the last 25 years. The initial ion exchanger electrodes with dialkylphosphate membranes were replaced by electrodes with neutral carrier membranes, having mainly the β-diketone group. Later other ionophores were used, but nearly all of them originate from the research team of the late Professor Wilhelm Simon in Zurich (ETH) and at this point a tribute should be paid to his achievements in the area of ion-selective electrodes.

The principal requirement for such electrodes is their selectivity towards ions which are present in blood or serum samples. The selectivity of an electrode is quantitatively expressed by the selectivity coefficient, k_{XY}^{pot}, which is a parameter in the Nikolski–Eisenman equation:

$$E = \text{constant} + RT/nF \left[\ln(a_X + k_{XY}^{pot} a_Y^{z/n})\right] \tag{3}$$

where E is the potentiometric analytical signal, a_X and a_Y are activities of the analyte and interferent ions, respectively, n and z are the charges of these ions, and R, T, and F have their usual physiochemical meanings. The larger the value of the selectivity coefficient, the more significant the interference. Taking into account the maximal possible concentration of the interferent and the minimal possible concentration of the analyte in the extracellular measurements, the selectivity of such electrodes should be expressed by coefficients not larger than[6]: $\log k_{MgNa}^{pot} = -4.5$; $\log k_{MgK}^{pot} = -0.6$; $\log k_{MgCa}^{pot} = -2.2$; and $\log k_{MgH}^{pot} = 9.6$; to allow for a relative error not larger than 1%.

The ionophore ETH 5220[7] (Figure 3), with optimal composition of the polymeric membrane, only partially satisfies those requirements, as the selectivity coefficients are: $\log k_{MgNa}^{pot} = -3.0$; $\log k_{MgK}^{pot} = -2.15$; $\log k_{MgCa}^{pot} = -0.20$; and $\log k_{MgH}^{pot} = 1.5$. As the experimental values of $\log k_{MgNa}^{pot}$ and $\log k_{MgCa}^{pot}$ are larger

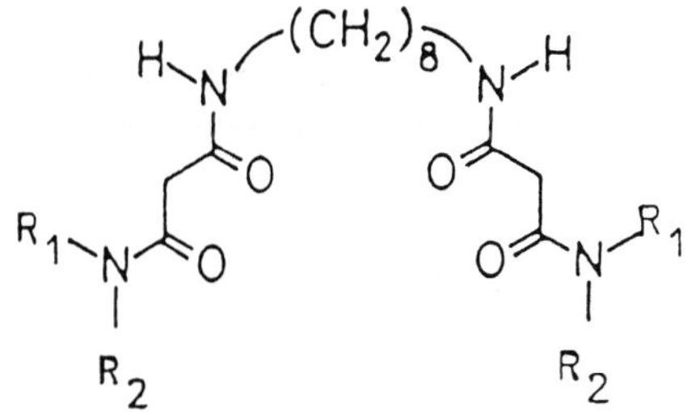

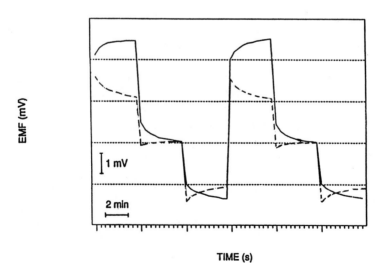

Figure 3 *Magnesium ionophore ETH 5220*

than those calculated for the limiting case, a significant interference from calcium, as its content is usually at least twice that of magnesium, and a small effect from sodium, in spite of its high level, can be expected. Those predictions correspond to measurements made under total equilibrium conditions. However, in automated measurements, in particular, the kinetics should be taken into account.[8] The time dependence of the ETH 5220 electrode response shows a different behaviour for calcium and for magnesium (Figure 4). The shape of the magnesium response is typical for ion-selective electrodes, reaching slowly the equilibrium potential caused by a finite rate of magnesium ion transfer to the membrane surface and/or slow diffusion in the membrane. The calcium response shows a non-monotonic transient, of a character observed also for some solid-membrane electrodes. Such overshoots indicate a fast calcium exchange process at the membrane concurrent with the temporary distortion of the calcium ion concentration in the vicinity of the membrane. Such behaviour results from

TIME (s)

Figure 4 *Time response for calcium and magnesium of the ETH 5220 electrode. Background solution (in mmol l^{-1}): NaCl 140; KCl 4.5; CaCl$_2$ 1.25; MgCl$_2$ 0.75 in pH TYES buffer. The solid curve corresponds respectively to 1.5; 0.75; 0.5 Mg^{2+}, the dashed curve to 175; 1.25; 0.75 Ca^{2+}*

the different rates of dehydration of magnesium and calcium, the latter apparently being higher. This causes the kinetic discrimination of calcium and magnesium ions, a phenomenon which plays an important role in the membrane interface processes in biological systems.

Such kinetic effects influence the electrode selectivity making it dependent on time and on the Mg^{2+}/Ca^{2+} concentration ratio. On this basis, by simultaneous measurement of the response of the calcium and magnesium electrodes (and sodium—because of selectivity interferences) it was possible to determine in an undisturbed way ionized magnesium in serum samples.[9] It must be stressed that such a result could be obtained only due to contemporary achievements in computerized analysers.

In spite of the fact that the analytical result depends not only on the reproducibility of the magnesium electrode readout alone, but also on the electrodes measuring the interferents (Ca, Na), the precision of measurements is satisfactory, as was checked on standards and on Quality Control Serum (Table 1). More difficult is the validation of accuracy as there is not an adequate reference method. The previously used procedures[10] were based on separation of magnesium high molecular mass complexes by ultrafiltration or dialysis, determination of total magnesium content in the ultrafiltrate, and calculation of the concentration of ionized magnesium on the basis of known stability constants of low molecular mass complexes. The last step of this procedure gives good results as was checked by comparing the ionized magnesium concentration with that obtained from total content and stability constants evaluated by coulometric titrations with high accuracy in comparable conditions. This was carried out for several ligands: acetate, lactate, oxalate, citrate, and phosphate[11] (Table 2).

The main disadvantage of measurements with the separation step is connected with the long time of determination and difficulties connected with the necessity of working in conditions preventing loss of carbon dioxide, in strictly temperature controlled conditions. The loss of carbon dioxide increases the pH, which in turn stimulates an equilibrium shift towards formation of complexes and decreases the concentration of 'ionized magnesium'. Such multistep operations may be the source of operational errors. Thus the introduction of potentiometric measurements is a significant improvement in clinical analysis.

To obtain diagnostically meaningful results, it is indispensable to measure both the ionized and total magnesium concentrations. This is a relatively simple

Table 1 *Precision of Ionized Magnesium Determination in Clinical Analyser KONE*

	Taken/ nmol l^{-1}	Found Mean/ mmol l^{-1}	SD/ mmol l^{-1}	RSD/ %	N**
High standard	1.200	1.153	0.026	2.3	22
Low standard	0.300	0.284	0.010	3.5	22
Quality Control Serum	*	0.605	0.025	4.1	12

*The ionized magnesium is not certified in serum at present.   ** Number of measurement.

Table 2 *Error of Total Magnesium Determination Obtained from Measurements of Ionized Magnesium and Calculation Based on Known Stability Constant (β) of Magnesium–Ligand Complexes. Total Mg Concentration 0.100 mmol$\,l^{-1}$, Ligand Concentrations 1.00 mmol$\,l^{-1}$, Temp. 37 °C*

Ligand	Stability Constant β	Total Mg/mmol$\,l^{-1}$	Error/%
Acetate	3.80	0.1010	+1.10
Lactate	4.36	0.0936	−6.4
Oxalate	2.45×10^2	0.0946	−5.4
Citrate	1.74×10^3	0.1020	+2.0
Phosphate	7.94×10^2	0.1030	+3.0

task and can be carried out by several techniques. A common procedure is based on molecular absorption spectrometry after complexation of magnesium with colour forming specific reagents. Calcium should be considered as a potential interferent but there are several convenient procedures which permit determination free of interferences.[12]

Another commonly used technique is atomic absorption (rarely emission) spectrometry. Usually flame atomization is applied because the magnesium concentration is mostly above the determination limit of this technique. Phosphate interferes but the effect may be eliminated by addition of lanthanum salts. The standard deviation of such procedures is close to or even better than 2%.[13] The use of electrothermal atomization is more expensive and needs larger dilution of samples. Nevertheless it may be used when only very small samples, having a lower content of magnesium, are available. Good results with electrothermal atomization were obtained in determining magnesium in leucocytes, as flame atomization gives negative errors in this case. This may be caused by the fact that in the flame, magnesium present in cell walls or organelles is not completely atomized, and the determination refers to cytosol only.[14,15]

Ion-selective electrode measurements have been used in combination with any of the procedures for total magnesium determination in several clinical instances.[5,16] Healthy adults were taken as the reference group and the results were compared with those for patients with myocardial infarct (Figure 5) and with short bowel syndrome. Significant differences were observed in concentration of both ionized and total magnesium and they were found to be diagnostically interesting. In particular, when short bowel syndrome is accompanied by symptomatic tetanus, magnesium therapy is necessary because of the significant decrease in the concentration of ionized magnesium.

When comparing those results the striking fact is the large scatter of results for both total and ionized magnesium. For healthy adults the range of ionized magnesium was found from 0.45 to 0.70 mmol$\,l^{-1}$, with maximum at 0.57–0.59. This is in accordance with literature values. Nevertheless, such scatter, obtained with relatively precise analytical tools, needs consideration. The multifunctionality of magnesium suggests that the investigated population could be subject to various diseases influencing the magnesium level (cardiac cases, cardiopulmonary

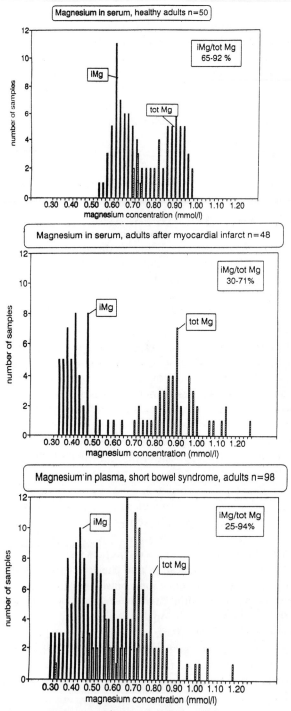

Figure 5 *Results of total and ionized magnesium determination in serum for healthy adults, after myocardial infarct and with short bowel syndrome*[16]

bypass, abnormal pregnancy, renal transplantation recipients, diabetics, asthmatics, *etc.*). There are also several possible sources of errors connected mainly with preanalytical events.[9] Samples taken after physical exercise show a reduced magnesium level.[17] Even muscular action during sampling may have an influence. Blood or serum in contact with siliconized surfaces (tubes) may show up to 10% higher values of ionized magnesium.[18] Addition of anticoagulant may also effect the magnesium level.[5] All those facts indicate that closer and broader cooperation of analysts and clinicians is still necessary.

Acknowledgement

This paper was prepared within the frame of Grant KBN 2P303–087–04.

References

1. M. F. Ryan, *Ann. Clin. Biochem.*, 1991, **28**, 19.
2. R. J. Elin, *Clin. Chem.*, 1987, **33**, 1965.
3. A. K. Campbell, 'Intracellular Calcium—Its Universal Role as Regulator', John Wiley, Chichester, 1983.
4. J. Thode, N. Fogh-Andersen, and O. Siggaard-Andersen, *Scand. J. Clin. Lab. Invest.*, 1985, **45**, 131.
5. M. Maj-Zurawska, M. Pertkiewicz, D. Drygieniec, A. Lewenstam, and A. Hulanicki, 'Proceedings IV European Congress on Magnesium', Giessen, 21–23 September 1992.
6. U. Oesch, D. Ammann, and W. Simon, *Clin. Chem.*, 1986, **32**, 1448.
7. M. Rouilly, B. Rusterholz, U. Spichiger, and W. Simon, *Clin. Chem.*, 1990, **36**, 466.
8. M. Maj-Zurawska and A. Lewenstam, *Anal. Chim. Acta*, 1990, **236**, 331.
9. A. Lewenstam, M. Maj-Zurawska, N. Blomqvist, and J. Ost, *Clin. Chem. Enzym. Comms.*, 1993, **5**, 95.
10. M. Speich, B. Bousquet, and G. Nicolas, *Clin. Chem.*, 1981, **27**, 246.
11. S. Głab, M. Maj-Zurawska, P. Łukomski, A. Hulanicki, and A. Lewenstam, *Anal. Chim. Acta*, 1993, **273**, 493.
12. Z. Marczenko, 'Separation and Spectrophotometric Determination of Elements', Ellis Horwood, Chichester, 1986.
13. G. N. Bowers and J. Pybus, in 'Standard Methods of Clinical Chemistry', ed. S. Meites, Vol. 7, Academic Press, New York, 1972, p. 143.
14. R. J. Elin and J. M. Hosseini, *Clin. Chem.*, 1985, **31**, 377.
15. A. Hulanicki and B. Godlewska (in preparation).
16. M. Maj-Zurawska, A. Hulanicki, D. Drygieniec, M. Pertkiewicz, M. Krokowski, A. Zebrowski, and A. Lewenstam, *Electroanalysis*, 1993, **5**, 713.
17. D. Boehmer, S. W. Golf, and P. E. Nowacki, 'Abstracts of IV European Congress on Magnesium', Giessen, 21–23 September 1992, p. 16.
18. Ch. Sachs, C. Ritter, M. Ghairamani, C. Kindermans, M. Deciaux, and H. Marsoner, 'Abstracts of IV European Congress on Magnesium', Giessen, 21–23 September 1992, p. 5.

Investigations of Human Metabolism of Trace Elements Using Stable Isotopes

P. J. Aggett[1] and J. E. Whitley[2]

[1] THE INSTITUTE OF FOOD RESEARCH, NORWICH RESEARCH PARK, NORWICH NR4 7UA, UK
[2] THE SCOTTISH UNIVERSITIES RESEARCH AND REACTOR CENTRE, EAST KILBRIDE G75 0QU, UK

Summary

Although the biochemical importance of trace elements such as zinc, iron, copper, and selenium is well known, there is a need to know more about: (i) the ideal dietary intakes; (ii) how metal species and dietary food matrices influence the efficiency with which trace elements are absorbed; (iii) the processes by which the elements are absorbed, distributed around the body, stored, and excreted; and (iv) how these metabolic processes change with disease states and adapt with incipient deficiencies and toxicities. One should then be able to determine physiological and dietary requirements for such elements and also be more able to diagnose the early stages of toxicity and deficiency. This review discusses the use of stable isotope tracers to investigate these topics.

1 Systemic Metabolism of Trace Elements

It is known that systemic control of trace element distribution, usage, and homeostasis is effected by exploiting their oxidation states and affinities for organic molecules to create selective physicochemical compartments which together achieve specific pathways that deliver the elements to their functional operative sites in appropriate forms and concentrations. The accumulative specificity of these pathways is important because if each step of the metabolic sequence, for the cationic trace metals, is examined in isolation, numerous interactions amongst the elements can be detected.

In this 'metabolic' context, the essential trace elements can be regarded as forming three groups:

(i) Cation-forming elements such as iron, zinc, manganese, and copper. These are transferred and utilized as inorganic ions; they need specific carriers to transfer them across lipid membranes and to maintain their solubility at physiological pH within the intracellular and extracellular environment. Their homeostasis is controlled mainly by the gastrointestinal tract and liver.

(ii) Anion-forming elements such as iodine, molybdenum, selenium, chromium, and, possibly, fluoride. These have a greater ability to cross lipid membranes spontaneously and are more water soluble at the physiological pH. The ions have a highly efficient gastrointestinal uptake and transfer, and their systemic utilization and compartmentalization is effected by manipulating their many oxidation states. Their homeostasis is dependent upon renal excretion.

(iii) Other elements, such as cobalt (in vitamin B_{12}), molybdenum, and, possibly, chromium, which have many oxidation states and are metabolized and used in organic complexes. Furthermore, it is noteworthy that iron is taken up by the gut mucosa more efficiently in the organic haem complex than in an inorganic state; the haem is then broken down by mucosal haem oxidase and the iron which is released then forms a common metabolic pool with inorganic iron.

The above knowledge is imperfect and our confidence in our estimation of human requirements for trace metals is limited by an inadequate insight of the interplay between the many systemic, luminal, and dietary factors which affect the intestinal absorption and systemic use of trace elements (Table 1). This efficiency with which a dietary constituent is absorbed and utilized in the body is known as 'bioavailability': this term reflects all the processes and factors

Table 1 *Factors Influencing the Intestinal Uptake and Transfer of Trace Elements*

Systemic factors

Anabolic demands
Growth in infancy and childhood
Pregnancy and lactation
Post-catabolic states
Endocrine effects
Infection and stress
Specific systemic reserves of metal
Genetic influence, inborn errors of metabolism
Nutritional status for other nutrients

Luminal and dietary

Chemical form and oxidation state of element in the diet
Presence of:
    antagonistic ligands (phosphate, carbonate, tannates, polyphenols, oxalate)
    facilitatory ligands (ascorbate (for iron), carboxylic acids, some sugars, amino acids, fatty acids)
    competing metals
Intestinal redox state
Luminal redox state

mentioned above. An understanding of how dietary and systemic factors influence bioavailability and the systemic fate of a nutrient depends on methods using labels. In general, tracer studies are designed to answer questions related either to determining an element's bioavailability or to monitoring how the body uses, conserves, or excretes elements once they have been absorbed (*i.e.* taken up by and transferred across the gut into the body). This outline will focus on iron, copper, zinc, and selenium; but other studies have used stable tracers of calcium, magnesium, and molybdenum.[1-3]

The use of any isotopic tracer assumes that the added tracer exchanges freely with, and behaves identically to, the natural elemental pool being investigated. Radio isotopes have long been used as tracers. They have the advantage of being assayed with ease and precision and with suitably high specific activities truly trace amounts of label can be used. Additionally, with sensitive whole body counters, the rates of tracer accumulation in and loss from the body, can be monitored easily and imagers can show the distribution in the body. However, the perceived and genuine hazards of ionizing radiation are now increasingly precluding the use of radio isotopes in metabolic studies; particularly those in children and women in their reproductive years. Such considerations also limit the use of radio isotopes for serial repetitive studies, and for multi-elemental studies, in which imprecision arising from difficulties in discriminating overlapping energy spectra is an additional significant handicap.

Low natural abundance stable isotopes are free from radiation hazards, and can be used for serial and simultaneous multi-element studies and have become used extensively in metabolic studies. They do present some disadvantages however. One cannot assay them as easily as radio isotopes and whatever analytical technique is used, pretreatment is normally involved to isolate them from interfering matrices and other elements. The analytical equipment is expensive as are the isotopes themselves. Furthermore, the commercial availability and purity of enriched stable isotopes are sometimes unreliable.

Another significant disadvantage with the use of low natural abundance stable isotopes is that it is often difficult to use them in amounts small enough to avoid significantly increasing the mass of the labelled pool and, as a result, changing its metabolism.

On the other hand, because stable isotopes do not decay, the material for analysis can be stored for batch analysis which improves quality control and economy. Additionally, in the long-term, the ability to perform repetitive multi-element and safe studies reduces the considerable effort and resource involved in recruiting human volunteers.

The isotopic composition of iron, copper, zinc, and selenium are shown in Table 2.[1-3] It can be seen that suitable low abundance isotopes are available for iron (Fe-54, Fe-57, Fe-58), zinc (Zn-67, Zn-70), and selenium (Se-74) but not for copper (Cu-63, Cu-65). Not surprisingly the least naturally abundant isotopes are particularly expensive; representative costs are shown in Table 3. All these factors are considered in designing studies. Unfortunately, the high abundance of the copper isotopes severely limits their use and our understanding of the metabolism and bioavailability of copper.

Table 2 *Isotopic Composition of Iron, Copper, Zinc, and Selenium*

	Atomic Weight	Natural Abundance		Atomic Weight	Natural Abundance
Copper	63	69.2	Iron	54	5.8
	65	30.8		56	91.8
				57	2.2
				58	0.3
Zinc	64	48.63	Selenium	74	0.88
	66	27.90		76	9.0
	67	4.10		77	7.7
	68	18.75		78	23.5
	70	0.62		80	49.6
				82	9.4

Table 3 *Representative Costs of Stable Isotopes*

Element	Isotope	Enrichment/atom %	Cost/£ mg^{-1}
Iron	54	95–97	10–20
	57	80–97	10–28
	58	65–93	100–300
Copper	63	99	<2
	65	91–99	2–5
Zinc	67	76–94	12–28
	68	37–99	2–6
	70	65–88	80–300
Selenium	74	31–99	60–520
	76	74–99	4–25
	77	68–94	15–20
	82	73–99	14–35

The practical problems in planning investigations are therefore; cost-effectiveness, consideration of whether the label being used is physiologically relevant with respect to the amount being used, its form, or species, and its ability to equilibrate with the native pool of element; and the analytical methods and computation.[1-3]

2 Analytical Methods

The earliest study using a stable isotope label involved neutron activation analysis (NAA) of Fe-58. NAA has the obvious disadvantage of needing a nuclear reactor and the use of stable isotopes with irradiation daughter products with decay characteristics suitable for determination. This technique has a variable sensitivity and precision (1–10%) which, with its lack of versatility, disadvantaged it compared with the mass spectrometric techniques which have been increasingly used since 1987.[1] Of these, thermal ionization mass spectrometry

(TI-MS) is widely accepted as giving the best precision and sensitivity. However, for some types of investigation, inductively coupled plasma mass spectrometry (ICP-MS) and fast atom bombardment (FAB-MS) are equally valuable.[1-4] TI-MS is most appropriate for analysis of small samples such as plasma and blood components when high sensitivity and precision are paramount,[5] whereas ICP-MS is useful for dietary and faecal analysis in which larger amounts of isotopes are usually present, and for multi-elemental isotope comparison and ratio-determination analyses. Additionally the faster throughput of samples with ICP-MS compared with that of FAB-MS lends it some advantage. A combination of gas chromatography and electron-impact MS has found most use for determination of Se.

Table 4 *Methods for Mass Spectrometric Stable Isotope Analysis of Metals*

Method	Sensitivity/ μg	Relative precision
Thermal ionization MS		
magnetic sector	1–10	<0.1%
quadrupole	1–10	0.1–1%
Inductively Coupled Plasma MS		
quadrupole	1–20	0.3–1%
Fast Atom Bombardment MS	1–10	0.3–2%
Gas chromatography–Electron Ionization MS	<1	0.1–10%

Intrinsic and Extrinsic Labelling

A fundamental issue in all tracer studies is knowing how best to insert the label or tag into the dietary or metabolic pool being investigated.[1-3] Extrinsic labelling is achieved by adding the label to the pool being studied. Thus, with a foodstuff, if the pool is in solution the tag is added and allowed to equilibrate and then administered. Equilibration with the dietary pool is much more difficult with dry foods, but none the less protocols are more convenient if the isotope can be added to a meal or foodstuff immediately before it is eaten. This approach raises concern about the representivity of the label for the native element both for speciation and chemical form and for perturbations which the label may induce in the size and behaviour of the pool.

These problems can be avoided by biologically incorporating the label in the foodstuff.[1-3,6] This is intrinsic labelling. Thus isotopes can be supplied to plants by hydroponic cultivation or, less physiologically, injections into their leaves, or to animals by mouth or by direct injection into the circulation or into the body. This is an expensive approach because the incorporation of the isotope into an edible part of the animal or plant is often quite low, e.g. that of Zn-68 and Zn-70 into chicken was 2–3%[7] and of Zn-70 into soy bean seeds was 18%.[8]

Fortunately, several studies have shown an acceptable similarity between extrinsic and intrinsic tags. For example the absorption of extrinsic and intrinsic

Zn labels from infant formulas is similar.[9-10] However a 10–20% lower absorption of Zn from extrinsically labelled chicken compared with the intrinsic label has been reported,[11] but this difference might have resulted from analytical difficulties.[9] Intrinsic and extrinsic labels of copper also appear to behave similarly in bioavailability studies.[12] Some differences have been noted between the absorption of extrinsic and intrinsic Se labels from chicken muscle. These probably arose because the intrinsic selenium had become incorporated into seleno-analogues of the amino acids methionine and cysteine. These intrinsic amino acids would be metabolized by the pathways used by the corresponding sulfur amino acids, whereas the extrinsic label, selenite, would be handled as an anion.[13] On the basis of our understanding of the absorption of inorganic iron and organic (haem) iron, similar differences would be expected if such contrasting labels of this element were used in comparative studies. These two examples emphasize the importance of selecting appropriate chemical forms of labels for metabolic studies. However, for most studies involving inorganic species it would seem that extrinsic labels might prove satisfactory, providing adequate time is allowed for full exchange with the native element and as long as one remembers the confounding possibilities when interpreting results.

Experimental Approaches to Monitoring Trace Element Bioavailability and Metabolism

These can be classified as studies involving (a) the metabolic balance or faecal monitoring procedures, (b) tissue incorporation or retention, and (c) a kinetic analysis based on monitoring the metabolic fate of single or simultaneously administered oral and intravenous labels.[2,3]

Metabolic Balance or Faecal Monitoring

A traditional approach used to estimate the 'absorption' of a trace metal such as Zn, Cu, Se, and Fe has been to measure, over a period of time (usually 3–7 days), the dietary intake and the faecal output, and to take the difference as being the intestinal uptake or loss of the element. This is really only the apparent net absorption. The disappearance of the nutrient between the mouth and anus does not necessarily mean that the substance has been truly absorbed into the body; it may have been retained in the intestinal mucosa. Furthermore the gut itself may secrete or excrete the element into its lumen via biliary secretions, pancreatic juice or mucosal secretion, or cellular sloughing. This endogenous element enters the faecal pool and gives an impression of a lower overall absorption value for the exogenous substance. This difficulty can to a certain extent be overcome by using isotopic labels of the exogenous elements and determining their fractional or 'luminal disappearance' between the dietary and faecal pools. By measuring the total amounts of element and isotopic tag in these pools the relative contributions to the faecal pool of exogenous and endogenous elements can be calculated and their changes with physiological states or different dietary intakes can be calculated, as has been done for zinc and copper.[14,15] Some workers have followed the more straightforward approach of just measuring the relative loss of dietary labels during intestinal transit.[16,17]

An alternative approach to the faecal monitoring procedure has been to label the endogenous pool of element and to monitor the appearance of this tracer in the stools.[5] The metabolic balance approach can be refined further by measuring also the label in urine. This is of more value for anionic elements such as Se, which are absorbed efficiently and excreted in the urine, than for cations (Zn, Cu, and Fe), the body burdens of which are controlled predominantly by the gut rather than by the kidneys.[1-3]

The mathematical formulae for analysing the data are given in the various references cited.

Tissue Incorporation

This approach is of use if the element of interest is incorporated efficiently in a specific tissue which is easy to sample. Thus it is particularly valuable for studies of the bioavailability of iron. An isotope of iron can be given orally and its incorporation into newly synthesized haemoglobin can be measured.[18,19] A protocol has now evolved which by taking a blood sample, usually 14 days after the isotope was administered, standardizes the incorporation of one isotope of iron given in food against the incorporation of another given in a standard solution of iron(II) sulfate.[20] This corrects for the considerable inter-individual variation in the intestinal uptake of iron and enables the precise effect of different foodstuffs to be ascertained. This technique has been used often in infant populations in which there are considerable problems with iron deficiency anaemia. However it is an expensive procedure and has not been applied very much to studies on adults.

Kinetic Analysis and the Use of Oral and Simultaneous Intravenous Administration of Isotopes

The approaches are currently being developed. If an isotope is given orally its appearance in the plasma can be assessed by taking, at frequent intervals, several plasma samples (over 4–6 hours) for analysis. The values can be plotted against time and the resultant curve and its shape is the product of the rate of entry of the isotope from the intestine into the plasma, and of the rate of disappearance from the plasma pool of the isotope caused by its uptake into the body tissue.[21,22] The latter process can be assessed separately by using a different isotope given intravenously to prepare a similar plot of the isotopic content in plasma. The two curves can then be assessed by convolution analysis to provide specific information about the rates of appearance and disappearance and, from the associated curve shapes, there can be derived some concept of the size and variation of different metabolic pools[23,24] during different physiological states and during systemic adaptation to different intakes of the element under investigation.

This approach assumes that isotopes given orally or intravenously are metabolized similarly. This may not be the case because orally administered elements may be bound to different carriers in the plasma than those with which the intravenous isotope binds and because orally given isotopes are more likely to be processed or taken up by liver before entering the systemic circulation. The metabolic implications of these metabolic differences have not yet been fully assessed. Even so, this is a potentially powerful technique for studying changes in the metabolism of elements[21,22] and it has recently been extended by the exploitation of the observation that the urinary content of enriched isotopes match those of plasma 24 hours after their administration. This enables longer duration studies of mineral metabolism; this method was applied first to studies of Ca metabolism but is now being applied to Zn.[2]

3 Conclusion

This overview demonstrates the increasing use of low natural abundance stable isotopes in the investigation of mineral metabolism. There are many practical problems associated with their use and analysis, and their expense has limited their application in some areas such as studies in adults. Undoubtedly, we will have to assess our ideas and protocols as the practical problems and their metabolic implications become better appreciated, but none the less, the use of such isotopes will certainly refine our understanding of the way the body uses elements such as Zn, Cu, Fe, and Se and other essential elements, we will also be able to determine our dietary requirements for these nutrients and to find ways of detecting more efficiently early deficiency and toxicity states.

References

1. J. R. Turnlund, *Crit. Rev. Food Sci. Nutr.*, 1991, **30**, 387.
2. B-M. Sandström, S. J. Fairweather-Tait, R. F. Hurrell, and W. Van Dokkum, *EC FLAIR Report Inter. J. Vit. Nutr. Res.*, 1994, in press.
3. R. F. Hurrell, L. Davidson, and P. Kastenmayer, 'A Handbook of Isotopic Studies of Mineral Metabolism', International Atomic Energy Agency, Vienna, 1994, in press.
4. J. Eagles, S. J. Fairweather-Tait, D. E. Portwood, R. Self, A. Gotz, and K. G. Heumann, *Anal. Chem.*, 1989, **61**, 1023.
5. M. J. Jackson, R. Giugliano, L. G. Giugliano, E. F. Oliveria, R. Shrimpton, and I. G. Swainbank, *Br. J. Nutr.*, 1988, **59**, 193.
6. C. M. Weaver, *Crit. Rev. Food Sci. Nutr.*, 1985, **23**, 75.
7. M. Janghorbani, B. T. G. Ting, and V. R. Young, *Br. J. Nutr.*, 1981, **46**, 395.
8. M. Janghorbani, C. M. Weaver, B. T. G. Ting, and V. R. Young, *J. Nutr.*, 1983, **113**, 973.
9. R. E. Serfass, E. E. Ziegler, B. B. Edwards, and R. S. Houk, *J. Nutr.*, 1989, **119**, 1661.
10. C. B. Egan, F. G. Smith, R. S. Houk, and R. E. Serfass, *Am. J. Clin. Nutr.*, 1991, **53**, 547.
11. M. Janghorbani, N. W. Istfan, J. O. Pagonnes, F. H. Steinke, and V. R. Young, *Am. J. Clin. Nutr.*, 1982, **36**, 537.
12. P. E. Johnson, M. A. Stuart, J. R. Hunt, L. Mullen, and T. S. Stark, *J. Nutr.*, 1988, **118**, 1522.
13. M. J. Christensen, M. Janghorbani, F. H. Steinke, N. W. Istfan, and V. R. Young, *Br. J. Nutr.*, 1983, **50**, 42.
14. C. M. Taylor, J. R. Bacon, P. J. Aggett, and I. Bremner, *Am. J. Clin. Nutr.*, 1991, **53**, 755.

15. J. R. Turnlund, W. R. Keyes, H. L. Anderson, and L. L. Acord, *Am. J. Clin. Nutr.*, 1989, **49**, 870.
16. N. W. Istfan, M. Janghorbani, and V. R. Young, *Am. J. Clin. Nutr.*, 1983, **38**, 187.
17. R. A. Ehrenkranz, P. A. Gettner, C. M. Nelli, E. A. Sherwonit, P. A. Williams, B. T. G. Ting, and M. Janghorbani, *Pediatr. Res.*, 1989, **26**, 298.
18. M. Janghorbani, B. T. Ting, and S. J. Forman, *Am. J. Hematol.*, 1986, **21**, 277.
19. S. J. Fomon, M. Janghorbani, B. T. G. Ting, E. E. Zieger, R. Rogers, S. E. Nelson, L. S. Ostedgaard, and B. B. Edwards, *Pediatr. Res.*, 1988, **24**, 20.
20. P. Kastenmayer, L. Davidsson, P. Galan, F. Cherouvrier, S. Hercberg, and R. F. Hurrell, *Br. J. Nutr.*, 1993, in press.
21. M. E. Wastney, I. G. Gökmen, R. L. Aamodt, W. F. Rumble, G. E. Gordon, and R. I. Henkin, *Am. J. Physiol.*, 1991, **260**, R134.
22. N. M. Lowe, A. Green, J. Rhodes, M. Lombard, R. Jalan, and M. J. Jackson, *Clin. Sci.*, 1993, **84**, 113.
23. R. A. Shipley and R. E. Clark, 'Tracer methods for *in vivo* kinetics', Academic Press, London, 1972.
24. C. Cobelli, G. Toffolo, D. M. Bier, and R. Nosadini, *Am. J. Physiol.*, 1987, **253**, E551.

Recent Novel Applications of Antibodies in Analysis: Production and Applications of Iodinated Conjugates for Use in Antibody Screening, Isolation, and Detection Procedures

R. O'Kennedy,[1] J. Bator,[2] C. L. Reading,[2] O. Nolan,[1] N. Quinlan,[1] D. Cahill,[1] and M. W. Thomas[2]

[1] SCHOOL OF BIOLOGICAL SCIENCES, DUBLIN CITY UNIVERSITY, GLASNEVIN, DUBLIN 9, IRELAND
[2] DEPARTMENT OF HAEMATOLOGY AND TUMOUR BIOLOGY, M. D. ANDERSON CANCER CENTRE, 1515 HOLCOMBE BOULEVARD, HOUSTON, TEXAS 77030, USA

Summary

Antibody and antibody-derivatives have major potential for analytical applications. New developments in production methods mean that they can be more easily exploited. Novel reagents incorporating defined antigens can expedite this approach. Conjugates between iodinated Bolton–Hunter (IBH) reagent and horseradish peroxidase (HRP), alkaline phosphatase, streptavidin peroxidase, avidin, fluorescein isothiocyanate (FITC)–human serum albumin (HSA) complex, HSA–HRP complex, and peroxidase anti-peroxidase (PAP) complex were produced. They were used to screen for monoclonal antibodies reacting with IBH and bifunctional antibodies reacting with IBH and tumour cells. Quadromas secreting such bifunctional antibodies were successfully detected using (a) FITC–HSA–IBH and flow cytometry and (b) IBH–HSA–HRP and IBH–PAP. They could also be isolated and partially purified by affinity chromatography on IBH–HSA–Sepharose 4B. The production and applications of these conjugates for screening and isolation of bifunctional antibodies are described. These and similar conjugates have potential for use in novel immunoassays for antigen detection on cells or in single step plate-based immunoassays. They may also have potential for use in drug analysis and with biosensors.

1 Introduction

Antibodies and antibody derivatives offer ideal reagents for use in analysis. With the advent of fusion and genetic approaches and the greater efficiency of chemical methods of antibody production, it is now feasible to construct very specific antibody-based structures.[1] There is now considerable interest in the production of bifunctional antibodies, that can bind to the surface of a tumour cell and to a therapeutic agent simultaneously (see Figure 1), for use in immunodiagnosis and therapy.[1-4] Based on this concept we have produced bifunctional antibodies which bind to tumour-associated antigens and to a hapten.[1-3] The hapten selected was 3-(3'-iodo-4'-hydroxy)-phenyl propionic

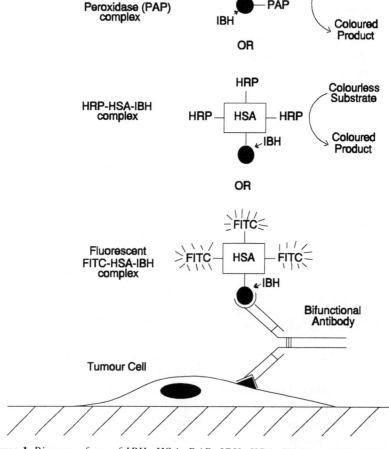

Figure 1 *Diagram of use of IBH–HSA–PAP, IBH–HSA–HRP, or IBH–HSA–FITC complexes for the detection of bifunctional antibodies. FITC is fluorescent and can be detected by flow cytometry. The peroxidase enzyme can be detected when using IBH–HSA–HRP/PAP as it gives a coloured product following reaction with the substrate*

acid. It forms the labelling group of iodinated Bolton–Hunter (IBH) reagent and is available as the *N*-hydroxysuccinamide derivative[5] and binds mainly to the ε-NH$_2$ group of lysine.[6] This hapten can be directly labelled with ^{131}I or ^{125}I and then attached to a protein, which in turn can be conjugated with a drug, a toxin, or a chemotherapeutic agent. The localization of this haptenated complex on the surface of a tumour cell can lead to its detection and/or destruction.[1-3] We have also developed a microassay that can detect IBH without the use of radioactivity[7] and we have also successfully used IBH-labelled antibodies for immunoassays.[8] Using a combination of these technologies it is now possible to develop novel immunoassays using IBH and bifunctional antibodies.

Monoclonal antibodies to IBH were produced using spleen cells from mice immunized with IBH–albumin.[1-3] The hybridomas secreting these antibodies were fused in turn with hybridomas secreting anti-tumour-antibodies (113F1) to produce quadromas. These quadromas secrete bifunctional antibodies reacting with IBH and with tumour associated-antigens, consisting of a 200/100/80/40 Kd complex, which are present on the surface of MCF7 cells, a human mammary carcinoma cell line.

This paper describes the production and evaluation of a range of IBH conjugates for use with these bifunctional antibodies. Several approaches were used:

  (i) direct conjugation of IBH to enzymes such as HRP and alkaline phosphatase;
  (ii) conjugation of IBH to proteins conjugated with either FITC or HRP; and
  (iii) conjugation of IBH to an antibody–enzyme complex such as PAP.

The use of IBH–HSA–Sepharose 4B for the isolation and partial purification of bifunctional antibodies (113F1 $\times$ IBH) was also investigated. Important discoveries derived from this research were the feasibility of production and use of the conjugates described and the effects of iodination on conjugate activity.

As shown in Figure 1, the conjugates that we describe for use with bifunctional antibodies need to have a reporter molecule (*e.g.* FITC or HRP) conjugated to an antigen which reacts with one arm of the antibody. Such conjugates can be used to detect bifunctional antibody activity during screening at the initial production stage. Thereafter, they can be key reagents in assays using bifunctional antibodies.

2 Materials and Methods

Materials

Sephadex G-25, avidin, biotin, horseradish peroxidase (HRP), human placental alkaline phosphatase, iminobiotin–agarose, and FITC–Celite were obtained from the Sigma Chemical Company (St. Louis, MO). *N,N'*-Dimethylformamide (DMF) and HRP-labelled streptavidin were obtained from the Fisher Chemical Co. (Orangeburg, NY). Bolton–Hunter reagent was obtained from Aldrich

Chemical Co. (Milwaukee, WI). Peroxidase anti-peroxidase and HRP-conjugated goat anti-mouse antibody were obtained from Cappel. Ultrogel AcA 44, molecular weight fractionation range 10 000–130 000, was obtained from LKB.

Preparation of Mono-iodinated Bolton–Hunter Reagent (IBHR) and its Conjugation to Proteins

IBHR was prepared as previously described.[7] It was conjugated to protein as follows. IBHR was dissolved at 4 mg ml^{-1} in benzene, containing approximately 4% (w/v) DMF. One milligram of IBHR was dried down under nitrogen and dissolved in 200 µl DMF. One millilitre of a 1 mg ml^{-1} solution of protein (peroxidase, alkaline phosphatase, streptavidin–peroxidase, avidin, *etc.*), or protein conjugate, was added to the IBHR solution with thorough mixing. For larger scale preparations the ratios of IBHR and protein were maintained. The reaction was left at 4 °C for at least two hours. The solution was then centrifuged to remove any precipitate formed (1 min at 13 000 r.p.m.). The conjugate was then purified by gel filtration.

Production and Use of IBH–HSA–FITC

A 0.5 ml volume of 0.8 M NaHCO$_3$ and 50 mg FITC–Celite were added to 0.5 mg HSA (2 mg ml^{-1}) in 0.01 M PBS, pH 7.4. The solution was incubated at 4 °C for three hours, with constant mixing. Precipitated material was removed by centrifugation. Unbound FITC was separated from the HSA–FITC conjugate by gel filtration on Sephadex G-25-150 (10 mm i.d. × 300 mm) in 0.1 M NaHCO$_3$. The first eluted peak (conjugate) was exhaustively dialysed against 0.01 M, PBS, pH 9.0. IBH reagent (1 mg ml^{-1} protein) in DMF (0.5 ml) was then added to the HSA–FITC conjugate and left for two hours at 4 °C. Any precipitate formed was removed by centrifugation and the supernatant applied to Ultrogel AcA 44 (16 mm i.d. × 400 mm). The column was eluted with 0.01 M PBS, pH 9.0, at a flow rate of 1 ml min^{-1} and 2 ml fractions were collected. The absorbances of fractions at 280 and 493 nm were recorded. The presence of IBH was also determined.[7]

Production of IBH–HSA–HRP

HRP was conjugated to HSA by a modification of the method of Nakane and Kawaoi.[9] It was dialysed against PBS, treated with IBH, fractionated on Ultrogel AcA 44, and analysed, as previously described.

Production of IBH–PAP

IBH was directly conjugated to PAP and the conjugate isolated by gel filtration as above; 1.1 ml fractions were collected.

Additional Iodination Techniques

HRP and alkaline phosphatase were also iodinated by the ICl method of Contreras et al.[10] and the iodinated enzymes fractionated by gel filtration on Ultrogel AcA 44.

Cell Culture

Bifunctional antibodies (IgG$_3$) were produced by fusion of a hybridoma secreting anti-IBH antibodies with a hybridoma secreting 113F1 antibody.[1-3] Hybridomas and quadromas were grown in SDMEM (Gibco) and 2% rabbit serum.[11]

MCF7 cells were grown in SDMEM with 2% rabbit serum, harvested by trypsinization, washed in PBS, and resuspended at a concentration of 1×10^6 cells ml^{-1}. Viability was estimated by trypan blue dye exclusion. For flow analysis, the cells were resuspended and washed in 0.01 M PBS, pH 7.4, containing 3% foetal calf serum and 0.01% (w/v) azide.

Flow Cytometry

Flow cytometry was performed on a Coulter EPICS C profile analyser. MCF7 cells were first incubated with media or antibody solutions, washed, and 50 µl FITC-labelled goat anti-mouse antibody (50 µg ml^{-1}) or IBH–HSA–FITC added. Incubations were for 30 min at 4 °C with occasional mixing. Samples were also pre-incubated with IBH–HSA–FITC and then added to MCF7 cell suspensions for 1 h. Following washing and viability testing, samples were fixed in paraformaldehyde (1%, v/v). Cells were used only if viability before fixing was greater than 95%.

Preparation of Antibody-coated Plates and ELISA

Microtitre plates with 96 wells (Nunc Immunoplate, Irvine Scientific, Santa Ana, CA) were precoated with 100 µl goat anti-mouse antibody (10 µg ml^{-1}) diluted in 50 mM sodium bicarbonate buffer, pH 9.6, and incubated overnight at 4 °C. They were then washed five times in PBS, pH 7.2, containing 0.05% (v/v) Tween-20 and once in PBS alone. The plates were blocked by incubation with 1% (w/v) BSA–1% (v/v) polyvinyl pyrrolidone (PVP) in 0.01 M PBS, pH 9, for one hour at room temperature. After washing, 100 µl of the test solutions were added and incubated in the wells for at least 30–60 min at room temperature. The plates were rewashed as before. A 100 µl volume of HRP-labelled goat anti-mouse antibody (diluted to 5–10 µg ml^{-1} in PBS, pH 8, containing 1% (w/v) BSA and 1% (v/v) PVP) or 100 µl of IBH–protein–enzyme conjugates (e.g. IBH–HSA–HRP, IBH–PAP, IBH–HRP, or IBH–alkaline phosphatase, etc.) was added. The substrate used for HRP was 100 µl ABTS (2,2'-azi-no-bis(3-ethylbenzthiazoline-6-sulfonic acid, Sigma, 400 µM) and H$_2$O$_2$ (0.3% v/v) in 0.1 M citrate buffer, pH 4. The substrate used for alkaline phosphatase was p-nitro-phenyl phosphate in 0.1 M Tris–0.1 M NaCl with 50 mM MgCl$_2$, pH 8.0. Absorbances were measured at 414 nm on a Titertek plate Reader (Flow Laboratories).

Affinity Chromatography

(a) *Avidin*: An iminobiotin–agarose column (1 or 5 ml plastic syringes) was used to isolate avidin after conjugation with IBH. After addition of avidin the column was washed with four column volumes of 50 mM carbonate buffer, pH 11, containing 1 M NaCl. Avidin was eluted using 50 mM acetate buffer, pH 4.0, containing 0.1 M NaCl and 1 ml fractions were collected. Avidin was estimated by the method of Green.[12] Conjugation of IBH was also carried out directly on avidin bound to iminobiotin–agarose by the method previously described in this paper. Successful conjugation was achieved.

(b) *Use of IBH–HSA–Sepharose to isolate anti-IBH monoclonal and bifunctional antibodies*: Sepharose 4B resin was activated by cyanogen bromide treatment.[13] It was then conjugated with HSA (approximately 2.78 ± 0.15 mg HSA ml^{-1} of resin; mean $\pm$ SD of three determinations). The immobilized HSA was then washed extensively (0.01 M PBS, pH 9.0) and conjugated with IBH (1 mg mg^{-1} HSA). Conjugation was confirmed by the detection of iodide.[7] Free IBH was removed by extensive washing with PBS. HSA-resin was conjugated with IBH just before use. IBH/HSA-resin was stored in PBS, pH 7.4, with 0.1% (w/v) azide.

Samples containing anti-IBH antibody or 113F1 × IBH bifunctional antibody were applied to the affinity column and left for 1–2 h at room temperature. The column was then washed with 20 volumes PBS, pH 7.7, and 1 ml fractions collected. Antibody was eluted using a gradient consisting of equal volumes 0.01 M PBS, pH 7.7, and 0.2 N acetic acid containing 3 M NaCl.[14] Absorbances at 280 nm were recorded. Fractions were tested for the presence of protein (absorbance at 280 nm), pH, and the ability to bind to microtitre plates coated with IBH–albumin. Antibody binding was detected using HRP-labelled goat anti-mouse antibody. Fractions were also tested for antibody using flow cytometry (see section on Flow Cytometry).

3 Results

IBH was initially conjugated to HSA. As shown in Table 1, the average number of IBH molecules conjugated per mole of HSA was 13 and this is similar to values reported in the literature.[6] Good levels of conjugation were also achieved with ovalbumin. Such conjugates were useful for immunization to produce anti-IBH antibodies and for subsequent detection of both monoclonal and bifunctional anti-IBH antibodies.

In order to produce a conjugate for the detection of bifunctional antibodies, binding to IBH and tumour cells, a fluorescent IBH complex was developed. This was achieved by conjugating IBH to HSA and subsequently labelling with FITC (see Figure 1). The isolation of this complex by gel filtration is shown in Figure 2a. It was also found that it was possible to reverse the order of conjugation and still produce the required IBH–HSA–FITC conjugate. This conjugate was then successfully used to detect binding of the 113F1 × IBH bifunctional antibody to MCF7 cells by flow cytometry (Figure 3). Here only

Table 1 *IBH Conjugation of Proteins*

Protein	μmole IBH/ μmole Protein	Comment
Ovalbumin	2–4	Good for screening and immunization
Human serum albumin (HSA)	13	Good for screening and immunization
Horseradish peroxidase	1–2	Lysine not fully available
Alkaline phosphatase	—	Low levels of conjugation
Streptavidin-peroxidase	1–2	Insufficient lysine residues
Avidin	variable	Binding to biotin inhibited or reduced
Goat anti-mouse IgG	17–23	Good conjugation
HSA–HRP complex	ND	Used for screening: background high
Peroxidase anti-peroxidase (PAP)	ND	Good for screening; low background on dilution

IBH = iodinated Bolton–Hunter reagent.
ND = not determined.

the bifunctional antibody gives significant binding of IBH–HSA–FITC complex to the MCF7. It is important to ensure that adequate controls are used and that there are very few dead cells present.

In order to develop reagents for use in enzyme-based assays IBH was conjugated to HRP and alkaline phosphatase (Table 1). With HRP a low level of conjugation was achieved, possibly due to the low level of lysine residues available. Conjugation of alkaline phosphatase with IBH resulted in a reduction in enzyme activity. Iodination of both enzymes, using the ICl method,[10] also failed to produce useful reagents (results not shown).

The potential use of IBH–streptavidin conjugates was also examined (Table 1). With streptavidin peroxidase a low level of IBH conjugation was achieved. This is probably associated with the relatively small number of available lysine residues in streptavidin. In contrast to streptavidin, avidin has lysine residues available for conjugation. However, direct conjugation of IBH to avidin greatly reduced its ability to bind to biotin. To overcome this problem avidin was bound to an iminobiotin–agarose column. This ensured that some of the binding sites on avidin were protected. Avidin was then conjugated *in situ* with IBH. However, the resultant avidin–IBH conjugate did not prove efficient for use in enzyme-based assays for detecting antibody binding activity.

Two IBH conjugates were found to be useful for ELISA, *i.e.* IBH–HSA–HRP, and IBH–PAP. Figure 2b shows the isolation of IBH–HSA–HRP by gel filtration. The early-eluting peak contains HRP activity. It also had bound IBH. The ability of this conjugate to detect anti-IBH monoclonal and bifunctional antibodies was examined as shown in Table 2. Pre-incubation of the present anti-IBH antibody or the bifunctional antibody (113F1 × IBH) with the

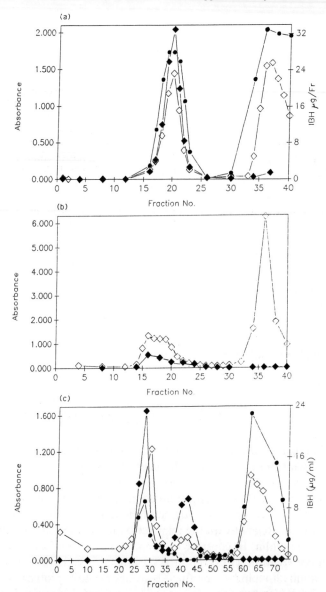

Figure 2 *Isolation of IBH conjugates on Ultrogel AcA 44 (16 mm i.d., c 400 i.d.).*
(a) Isolation of IBH–HSA–FITC: The eluent was PBS, pH 9, the flow rate was
1 ml min⁻¹ and 2 ml fractions were collected. The absorbances at 280 nm
(◇ protein/IBH), 493 nm (◆, FITC), and the levels of IBH (●, μg/fraction) are shown
(b) Isolation of HRP–HSA–IBH: The eluent was 0.01 M PBS, pH 9.0, the flow
rate was 1 ml min⁻¹ and 2 ml fractions were collected. The absorbances at 280 nm
(◇ protein/IBH), and 403 nm (◆, peroxidase) are shown
(c) Isolation of IBH–PAP: The eluent was 0.01 M PBS, pH 9.0, total flow rate
was 1 ml min⁻¹ and 1.1 ml fractions were collected. The absorbances at 280 nm
(◇ protein), 493 nm (◆, peroxidase), and the levels of IBH (●, μg ml⁻¹) are shown

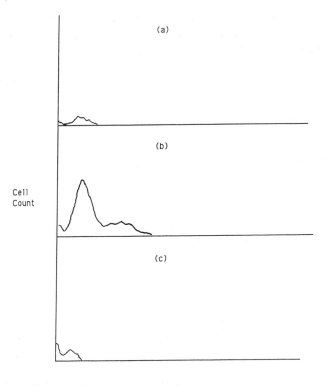

Figure 3 *Use of flow cytometry and IBH–HSA–FITC for detection of bifunctional antibodies, 113F1 × IBH, reacting with MCF7 tumour cells. (a) IBH–HSA–FITC; (b) IBH–HSA–FITC and 113F1 × IBH bifunctional antibody; and (c) IBH–HSA–FITC and 113F1, only*

IBH–HSA–HRP conjugate prior to addition to the plate gave higher values than simultaneous incubation on the plate. This may be due to the fact that the pre-incubation step facilitates greater interaction with those antibodies in solution. On the plate the presence of the goat anti-mouse capture antibody may reduce the degree of interaction occurring, as only captured antibodies that are orientated in a manner such that their active sites may bind the IBH–HSA–HRP complex will be detected. The conjugate could be diluted 1:100 to 1:1000 and still detect anti-IBH monoclonal antibodies. The detection of the bifunctional anti-IBH activity was difficult to achieve below dilutions of 1:100 of the conjugate with supernatants from quadromas containing both monoclonal and bifunctional antibodies as the latter may be present in very low concentrations. Therefore, preincubation of the bifunctional supernatant and the IBH–HSA–HRP conjugate may be very useful and important under these conditions. Extensive blocking and washing procedures were necessary to minimize background effects. The results shown in Table 2 demonstrate the use of this conjugate for screening for the presence of bifunctional IBH × 113F1

Table 2 *Testing IBH–HSA–HRP Conjugates on Microtitre Plates Precoated with Goat Anti-mouse Antibody for their Ability to Specifically Detect Monoclonal and Bifunctional Antibodies*

Test Antibodies ($\pm$ Conjugates)	Incubation Conditions	A_{414}
SDMEM (no conjugate)	A	0.155
SDMEM + IBH–HSA–HRP	A	0.339
	B	0.385
	C	0.357
113F1 + IBH–HSA–HRP	A	0.354
	B	0.358
	C	0.321
Anti-IBH + IBH–HSA–HRP	A	0.404
	B	0.840
	C	0.832
113F1 × IBH + IBH–HSA–HRP (undiluted)	A	1.089
	B	0.933
	C	1.047
113F1 × IBH	A	0.385
	B	0.534
	C	0.566

Incubation conditions: A: Simultaneous incubation for test solution $\pm$ conjugate for 3 h at room temperature on microtitre plate.
B: Preincubate test solution and conjugate overnight and then transfer to plate for 2 h at room temperature.
C: Preincubate test solution and conjugate for 1 h at room temperature and add to plate for 3 h.
All plates were precoated with goat anti-mouse antibody as described in Methods. A_{414} is absorbance at 414 nm.

antibody. The results obtained were confirmed by use of flow cytometry with IBH–HSA–FITC and MCF7 cells.

The isolation of IBH–PAP is shown in Figure 2c. The first peak contains peroxidase and IBH and retained measurable peroxidase activity at a 1:10 000 dilution. For detection of anti-IBH monoclonal antibodies it could be used at dilutions of 1:1000–1:2000. For the detection of bifunctional activity, dilutions of 1:100–1:500 were optimal. The detection of antibodies using this reagent is shown in Table 3. The reagent should be diluted as much as possible to minimize effects due to high backgrounds which may be associated with the use of PAP complexes. Such high backgrounds may result from non-specific binding of the large PAP complex to the plate.

As shown in Figure 4 IBH–HSA–Sepharose-4B could be successfully used to isolate 113F1 × IBH bifunctional antibody by affinity chromatography. Peak III from this separation was shown to have bifunctional activity when screened with IBH–HSA–FITC and MCF7 cells (Figure 5). It also was capable of binding IBH–PAP as measured by an enzyme assay on microtitre plates (Table 3). Fractions from this peak also bound to MCF7 cells when tested by a cell suspension ELISA method. Thus, the material in this peak clearly had

Table 3 *Use of IBH–PAP for the Detection of Bifunctional Antibodies Bound to Microtitre Plates Coated with Goat Anti-mouse Antibody*

Test solution	A_{414}
PBS only	0.142
SDMEM + IBH–PAP	0.383
Unpurified 113F1 × IBH + IBH–PAP	0.778
Affinity purified 113F1 × IBH + IBH–PAP	1.016

A_{414} is the absorbance at 414 nm.
113F1 × IBH is the bifunctional antibody which was isolated by affinity chromatography as described in Methods.

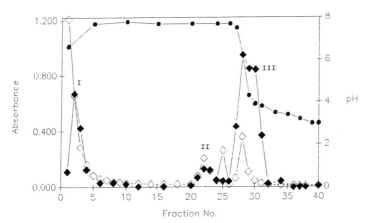

Figure 4 *Affinity chromatography of bifunctional antibody 113F1 × IBH on IBH–HSA– Sepharose 4B (1 ml column). The column was washed with 20 ml 0.01 M PBS, pH 7.7, after addition of the sample. A linear gradient of 10 ml 0.01 M PBS, pH 7.7 and 10 ml 0.2 N acetic acid with 3 M NaCl was then applied; 1 ml fractions were collected. The absorbance at 280 nm (◇), the pH (●), and the binding of column fractions to IBH–albumin (◆) attached to the walls of microtitre plates were determined. The latter was measured by ELISA using a HRP-labelled goat anti-mouse antibody. The HRP substrate used was ABTS and the absorbances were measured at 414 nm. The relative absorbances at 414 (◆) are shown. Protein peaks I, II, and III are shown*

bifunctional activity although monoclonal anti-IBH antibody could also be present. Further purification can be achieved by adsorbing the antibody preparations from this peak with MCF7 cells or the appropriate antigen.

4 Discussion

The use of bifunctional antibodies for analytical purposes requires the availability of reagents which can be used in a variety of formats. One step assay procedures

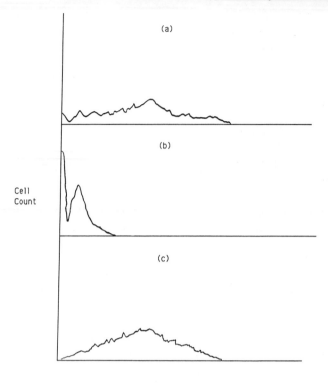

Fluorescence

Figure 5 *Flow cytometry of peak III from affinity chromatography of bifunctional antibody 113F1 × IBH on IBH−HSA−Sepharose. The presence of bifunctional antibody was confirmed using MCF7 cells and IBH−HSA−FITC (a). The negative control consisted of media and IBH−HSA−FITC (b). Cell binding by the antibody was also confirmed using FITC-labelled goat anti-mouse antibody (c)*

(where all reagents are added together) may be one approach where bifunctional antibodies have advantages over other more classical systems (*e.g.* multi-step sandwich assays using monoclonal or polyclonal antibodies). To maximize the utility of such approaches we envisaged the use of a hapten as shown in Figure 1. The hapten, *e.g.* IBH, may be directly conjugated to the label or it may be linked to a carrier (*e.g.* HSA) which is then labelled. Hence, a variety of labels may be used, *e.g.* radioactive, fluorescent, or enzyme-based, thus making many different detection formats possible. In addition, drugs, toxins, or other therapeutic agents could be linked to the carrier further extending its potential applications.

Several workers have produced bifunctional antibodies which directly bind to an enzyme such as HRP, β-galactosidase, or urease.[15−18] We have produced such a system where one arm of the bifunctional antibody binds to HRP while the other binds an anti-leukaemic cell antibody.[19] However, this approach is limited in relation to the detection formats available and specific antibodies to each enzyme used are required. In addition, they then have to be made into bifunctional antibodies. We have also produced bifunctional antibodies, by chemical means, which bind to a carrier (HSA) with one arm while the other arm has specificity for leukaemic cells. In this case, the carrier may be labelled with HRP or FITC. The former has been used successfully in microplate-based assays.[20]

The work described shows that the level of conjugation of proteins with IBH is quite variable. This is mainly due to the number of free-ε-NH_2 groups in the protein, but from the level of conjugation achieved, it would appear that many of the potential conjugation sites may be unavailable. This is probably due to the three dimensional structure of the protein. High levels of conjugation are necessary to produce good conjugates for immunization and monoclonal/bifunctional antibody screening and it was found that HSA was an ideal carrier for IBH for such applications.

The production of conjugates for use as reagents in bifunctional antibody-based assays also requires a good level of conjugation to be achieved. This is at least partly due to the fact that in any given bifunctional antibody molecule only one of the arms will be available for binding to the hapten. It was found from the work described that IBH steptavidin-peroxidase was not suitable for bifunctional antibody screening. The IBH–HSA–HRP conjugate was capable of being used for both antibody screening and in assay systems. The IBH–HSA–FITC conjugate could also be used successfully for screening for bifunctional antibody (113F1 × IBH) binding to MCF7 cells using flow cytometry as shown in Figure 3.

IBH–PAP conjugates were produced in an attempt to develop reagents with greater assay sensitivity and to eliminate the use of HSA as a carrier. While this reagent was more sensitive than IBH–HSA–HRP, problems were encountered with higher levels of non-specific background binding. Those could be reduced when high dilutions of the reagent were used.

Attempts to produce active IBH–avidin conjugates were unsuccessful. It would appear that conjugation with IBH may deactivate the binding sites on avidin. *In situ* conjugation of bound avidin did not overcome this problem. This may be due to the fact that amino acid residues capable of reacting with IBH are involved in the binding site of avidin.[21] Previous work has shown that modification of lysine residues with 1-fluoro-2,3-dinitrobenzene results in total inhibition of biotin binding.[22] It might, however, be possible to use IBH-labelled biotin derivatives with avidin-conjugated enzymes.[23]

While we have concentrated on the use of IBH as the hapten, similar reagents to those described in this paper could also be used for screening for bifunctional antibody binding to drugs.[1,3] In this case, the drug acts as the hapten and it is conjugated to carriers such as those described. At present, we are developing such a system for the analysis of anti-cancer drugs. There are many other systems (*e.g.* sensors) which could utilize conjugates such as those we have developed in conjunction with bifunctional antibodies to produce novel methods of analysis.[24] We are currently investigating such applications.

Acknowledgements

We wish to thank Polycell, Dublin City University, BioResearch Ireland, and the Cancer Research Advancement Board for financial support and Ms. Barbara Drew for help with the preparation of this manuscript.

References

1. R. O'Kennedy and P. Roben, 'Essays in Biochemistry', Portland Press, London, 1991, **26**, Chapter 6, p. 59.
2. C. Reading, J. Bator, and R. O'Kennedy, 'American Association of Blood Banks', Arlington, VA, 1989, **6**, Chapter 6, p. 145.
3. O. Nolan and R. O'Kennedy, *Biochim. Biophys. Acta*, 1990, **1040**, 1.
4. C. Milstein and A. C. Cuello, *Nature (London)*, 1983, **305**, 537.
5. A. E. Bolton and M. W. Hunter, *Biochem. J.*, 1973, **133**, 529.
6. L. C. Knight and M. J. Welch, *Biochim. Biophys. Acta*, 1978, **534**, 185.
7. R. O'Kennedy, J. M. Bator, and C. Reading, *Anal. Biochem.*, 1989, **179**, 138.
8. R. O'Kennedy and P. Keating, *Anal. Biochem.*, 1991, **194**, 345.
9. P. K. Nakane and A. Kawaoi, *J. Histochem. Cytochem.*, 1974, **22**, 1084.
10. M. A. Contreras, W. F. Bale, and L. L. Spar, *Methods Enzymol.*, 1983, **92**, 277.
11. C. L. Reading, *J. Immunol. Method*, 1982, **53**, 261.
12. N. M. Green, *Methods Enzymol.*, 1970, **18(A)**, 418.
13. S. C. March, I. Parikh, and P. Cuatrecasas, *Anal. Biochem.*, 1974, **60**, 149.
14. J. C. Jaton, D. Ch. Brandt, and P. Vassal, in 'Immunological Methods', eds. I. Leftkovits and B. Pernis, Academic Press, NY, 1979, p. 54.
15. C. Milstein and A. C. Cuello, *Immunol. Today*, 1984, **5**, 299.
16. H. Tada, Y. Toyoda, and S. Iwasa, *Hybridoma*, 1989, **8**, 73.
17. G. Györogy, A. Gandolfi, G. Paradasi, E. Rolleri, E. Klasen, V. Dessi, R. Strom, and F. Celada, *J. Immunol. Method.*, 1989, **123**, 131.
18. M. Takahash and S. A. Fuller, *Clin. Chem.*, 1988, **34/9**, 1693.
19. N. Quinlan and R. O'Kennedy, Abstract No. We48, 21st FEBS Meeting, Dublin, 11–14th August, 1992.
20. O. Nolan and R. O'Kennedy, 'Generation of Antibodies by Cell and Gene Immortalisation', The Year in Immunology, Karger, Basle, 1993, **7**, p. 81.
21. G. Gitlin, E. A. Bayer, and M. Wilchek, *Biochem. J.*, 1988, **250**, 291.
22. G. Gitlin, E. A. Bayer, and M. Wilchek, *Biochem. J.*, 1987, **242**, 923.
23. M. Wilchek and E. A. Bayer, *Anal. Biochem.*, 1988, **171**, 1.
24. H. Paulus, *Behring Inst. Mitt.*, 1985, **78**, 118.

The Role of Analytical Methods in Drug Development

S. Görög, M. Gazdag, B. Herényi, P. Horváth, P. Kemenes-Bakos, and K. Mihályfi

CHEMICAL WORKS GEDEON RICHTER LTD., POB 27, H-1475, BUDAPEST, HUNGARY

1 Introduction

Analytical methods play dominant roles in practically all phases of drug research and development.[1] The main fields of activities of analytical chemists working in this area are summarized in Table 1.

Four of the various areas of analytical research included in Table 1, namely:

(i) the development of specific assay methods for bulk drugs and their formulations with emphasis on stability assays,

(ii) impurity profiling (isolation, identification, and quantitation of impurities in bulk drugs);

(iii) chiral separations including the estimation of enantiomeric and diastereomeric impurities; and

(iv) estimation of the degradation kinetics of drugs under stressed conditions

will be discussed in detail and illustrated using examples from the drugs developed by the Chemical Works of Gedeon Richter Ltd., Budapest.

The drugs included in this review are phlogosam [sodium dihydrogen disulfosalicylate samarium (III)], mazipredone [21-N-(N'-methylpiperazinyl)-11β, 17α-dihydroxy-1,4-pregnadiene-3,20-dione hydrochloride], tolperisone [2-methyl-1-(4-methylphenyl)-3-(1-piperidinyl)-propan-1-one hydrochloride], vinpocetine (3α, 16α-eburnamenine-14-carboxylic acid ethyl ester), pipecuronium bromide [2β, 16β-bis-(N'-dimethyl-1-piperazinyl)-3α, 17β-diacetoxy-5α-androstane dibromide], flumecinol (3-trifluoromethyl-α-ethyl-benzhydrol), and two experimental drugs being under clinical examination: thymotrinan (L-Arg-L-Lys-L-Asp) and posatirelin (L-6-keto-piperidine-2-carbonyl-L-leucyl-L-prolinamide).

2 Specific Assay Methods

In the course of the development of the first two of the drugs listed above, about 30 years ago, spectrophotometric methods played a major role in the

Table 1 *Some Important Fields of Activities of Analytical Chemists in Drug Research and Development*

Synthesis of Drugs	Spectral, chromatographic, *etc.* characterization of the intermediates. In-process control.
Bulk Materials	Spectral characterization (UV, IR, NMR, MS, ORD, CD, *etc.*). *X*-ray (powder, single-crystal). Chromatographic characterization (TLC, GLC, HPLC, *etc.*). Chiral chromatography. Elemental analysis; Functional group analysis. Specific assay methods. Thermochemical characterization (DTG, DTA, DSC). Morphology, polymorphism, particle size distribution. Impurity profiling. Solvent residues. Identification and quantification of degradation products. Stability studies (long-term, accelerated, severe test; solid phase, solutions). Degradation kinetics.
Formulations	Analytical aspects of preformulation studies. Specific assay methods. Content uniformity tests. Dissolution and dissolution rate testing of solid dosage forms (sustained release formulations). Stability tests (as above).
Biological Samples	Bioavailability, bioequivalence studies. Analytical aspects of pharmacokinetic and metabolic studies. Pharmacokinetics of chiral drugs. Therapeutic drug monitoring.
Method Validation Automation	

determination of the active ingredient content of bulk drugs and in their formulation. The specificity of the assay methods was ensured by the application of sufficiently selective chemical reactions.

For example, several colorimetric reagents were available to be considered for the assay of phlogosam ointment and gargle on the basis of their samarium content.[2] None, however, was suitable for the determination of samarium in the presence of ethylenediamine tetraacetic acid which was a constituent of the formulations. This was also the case with arsenazo-III, among others, which is one of the most frequently used colorimetric reagents for rare earths. The reason for this is, that under the usual conditions of its use (aqueous solution, pH 2–2.5), EDTA forms a more stable complex with an equivalent quantity of samarium(III) than does the colorimetric reagents. The problem was solved on the basis of an unusually strong solvent effect. The pH-profile of the complex formation between samarium(III) and arsenazo-III is dramatically shifted if the solvent water is replaced by the 4:1 v/v mixture of methanol and water as is seen in Figure 1. In this solvent mixture not only was a more than twofold increase of the sensitivity achieved but the pH optimum was also shifted to 1–1.5. Thus in 0.1 M hydrochloric acid, where the samarium(III)–EDTA complex decomposes, the highly selective and sensitive determination of samarium

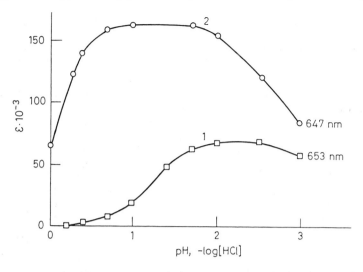

Figure 1 *pH-profile of the formation of the samarium(III)–arsenazo-III complex from phlogosam and arsenazo-III (absorbance at the absorption maxima as a function of hydrochloric acid concentration and pH, respectively). Concentration: phlogosam [sodium dihydrogen disulfosalicylate samarium(III)], $2.5 \times 10^{-6}\,M\,l^{-1}$; arsenazo-III, $8 \times 10^{-5}\,M\,l^{-1}$. Solvent: (1) water; (2) 4:1 v/v mixture of methanol and water*

content of the formulations could be carried out even in the presence of EDTA.[3]

The determination of the water-soluble prednisolone derivative, mazipredone was based on the oxidation of its α-aminoketone moiety by mercury(II)chloride. 21-Hydroxy corticosteroids do not react with this reagent. The selectivity of this reaction could be exploited for various purposes. Even the determination of the weight of the mercury(I)chloride formed enabled an indirect gravimetric stability assay to be carried out since the hydrolytic and oxidative degradation products of mazipredone do not react with mercury(II)chloride.[4] When the reaction is carried out on TLC plates followed by spraying the plate with ammonia solution, mazipredone and other 21-aminocorticoids appear as grey spots as a consequence of the reaction with ammonia to form mercury(II) amidochloride and metallic mercury.[5] As is shown by the data in Figure 2, the 20-keto-21-aldehyde forming during the reaction is the basis of a spectrophotometric assay using 1,2-diamino-4,5-dimethylbenzene as the reagent to form the spectrophotometrically active quinoxaline derivative.[6] The applicability of this method can also be extended to 21-hydroxy corticoids, if copper(II)-acetate is used as the oxidizing agent in the first step.[7,8] It is interesting to note that the two-step reaction can be successfully used for their precolumn derivatization in the HPLC determination of urinary corticosteroids. As is also shown in Figure 2 1,2-diamino-4,5-methylenedioxybenzene was used in this case as the second reagent leading to a fluorimetrically active quinoxaline derivative.[9]

Since the introduction of high performance liquid chromatography into the pharmaceutical analysis this technique has become the most important method

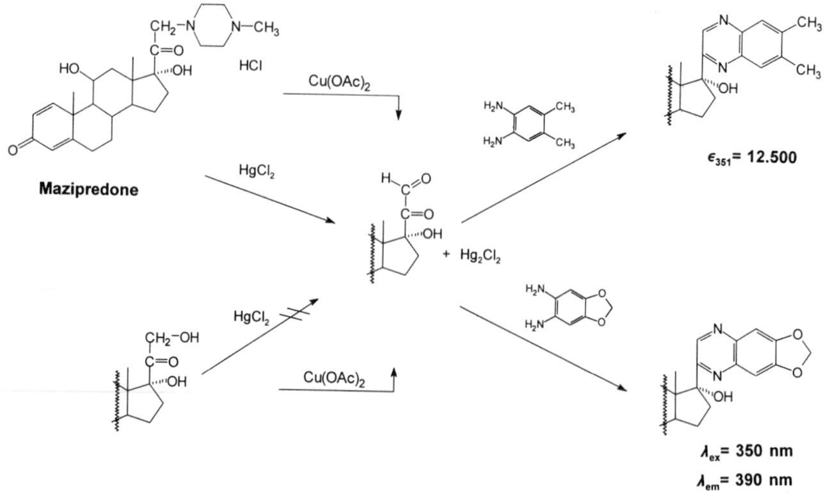

Figure 2 *Reaction scheme of the selective oxidation of mazipredone and 21-hydroxy-corticosteroids followed by condensation to spectrophotometrically and fluorimetrically active quinoxaline derivatives*

for the assay of bulk drugs and in their formulations. For example, the HPLC methods to be discussed in the section on impurity profiling and stability testing of pipecuronium bromide, thymotrinan, and posatirelin, respectively, were successfully used for this purpose too.

3 Impurity Profiling

Various strategies are in use for the impurity profiling of drugs, among them are on-line and off-line combinations of various chromatographic and spectroscopic techniques.[10-19] In the case of flumecinol the GC–MS technique using packed[14] and capillary[16] columns supplemented by HPLC–diode-array UV[17] have proved ideal tools for the identification and quantification of the numerous impurities.

The analytical protocols for the impurity profiling of a particular drug have to be revised from time to time to fulfil the increasing demands as regards specificity, sensitivity, ruggedness, and also the economy of the methods for routine use. In the case of pipecuronium bromide, for example, where in addition to a thin-layer densitometric method,[21] various HPLC methods are available for the detection and quantification of the impurities.[20-23] Among them a method used routinely involves a silica column and an eluent containing high concentrations of ammonium chloride, carbonate, and free ammonia.[20-23] To avoid the use of this aggressive eluent which decreases the lifetime of the column and the pump, an ion-pairing system with a neutral eluent containing perchlorate anion to form the ion pair was recently described.[24] Work is still in progress, illustrated by the chromatogram of pipecuronium bromide spiked with possible

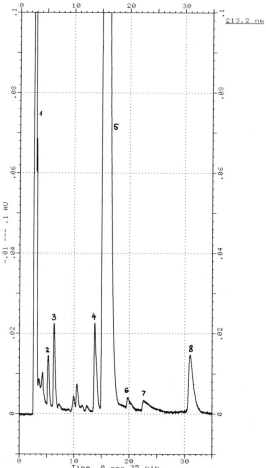

Figure 3 *HPLC separation of the potential impurities of pipecuronium bromide*
Column: Hypersil BDS (Shandon); C-8; 5 μm; 250 × 4.6 mm.
Eluent: 0.2 M NaClO₄ in acetonitrile–water 32:68 v/v; pH 6.8 at 1 ml min⁻¹.
UV detector: 213 nm.
Key: (1) bromide; (2) 3,17-bis-desacetyl analog; (3) 3-desacetyl analog; (4)
17-desacetyl analog; (5) pipecuronium; (6) monoquaternary 2β-(N′-methyl-1-
piperazinyl) analog; (7) monoquaternary 16β-(N′-methyl-1-piperazinyl)
analog; (8) 16β-(2′(3′)-dehydro-N′-dimethyl-1-piperazinyl) analog

Pipecuronium bromide

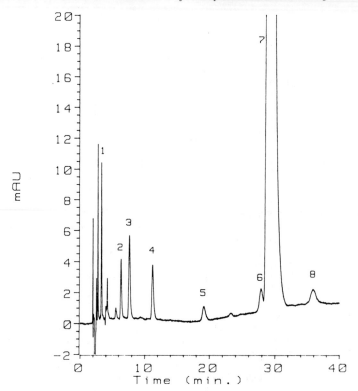

Figure 4 *HPLC separation of the potential impurities of thymotrinan*
Column: Ultrasphere IP (Beckman); C-18; 5 μm; 250 × 4.6 mm.
Eluent: 0.02 M NaH_2PO_4 + 0.03 M sodium hexane sulfonate in methanol–water
33:67 v/v; pH 2.86 at 1 ml min[-1].
UV detector: 210 nm.
Key: (1) cyclo-Lys-Asp; (2) Lys-Asp; (3) (Ac-Arg)-Lys-Asp; (4) cyclo-Arg-Lys;
(5) i-BuOCO-Lys-Asp; (6) L-Arg-L-Lys-D-Asp (diastereomer of thymo-
trinan); (7) thymotrinan (L-Arg-L-Lys-L-Asp); (8) Arg-Lys

and known impurities is shown in Figure 3. As it is seen, using the above mentioned ion pairing reagent in a reversed phase system, the separation of the isomeric monoacetates (3 and 4) and monoquaternary derivatives (6 and 7) is excellent. The highly UV-active impurity (8), an oxidative degradation product, merits special mention. The identification and quantitative determination of this impurity have been achieved by the joint application of HPLC, diode-array UV, and NMR spectroscopies.[19]

A further example is the impurity profiling of thymotrinan. The HPLC chromatogram in an ion-pair system of a sample spiked with its possible and real impurities is given in Figure 4. Table 2 gives the R_f data of the compounds collected using TLC systems routinely used for the purity test of the bulk material.

Table 2 *Thin-layer Chromatographic Separation of the Potential Impurities of Thymotrinan (R$_f$ Values)*

	System-1[1]	System-2[2]
Thymotrinan (Arg-Lys-Asp)	0.25	0.27
Arg-Lys-D-Asp	0.25	0.27
cyclo-Arg-Lys	0.10	0.06
Arg-Lys	0.28	0.11
Lys-Asp	0.30	0.48
(Ac-Arg)-Lys-Asp	0.35	0.39
cyclo-Lys-Asp	0.47	0.57
i-BuOCO-Lys-Asp	0.55	0.65

[1] *Method 1*

Chromatoplate: HPTLC Silica gel (Merck 5631). At least one day before the chromatographic run the plate is prerun in a 300:265:240:220 v/v mixture of n-butanol–pyridine–glacial acetic acid–water.

Mobile phase: n-butanol–pyridine–glacial acetic acid–water 300:341:151:208 v/v. Running distance: 10 cm.

Visualization: chlorine–tolidine reaction. Semi-quantitative evaluation. Detection limit 50 ng thymotrinan.

[2] *Method 2*

Chromatoplate: as above.

Mobile phase: n-propanol–12.5% ammonia solution 1:1 v/v. Running distance: 10 cm.

Visualization: the plate is dipped into 0.2% methanolic ninhydrin and heated until maximum colour intensity is reached.

Quantification: Densitometry at 500 nm in reflection mode. Detection limit 200 ng thymotrinan. Linearity range 0.2–2 µg.

4 Chiral Separation

The estimation of the optical purity by chromatographic methods of chiral drugs used therapeutically as the pure enantiomers is one of the most important tasks in contemporary pharmaceutical analysis. This applies also to chiral drugs such as tolperisone and flumecinol used as racemates, since the drug authorities require the pure enantiomers to be separated and examined.

The α$_1$-glycoprotein column (Chiral AGP—ChromTech AB) has proved suitable for the separation of the enantiomers of tolperisone (see Figure 5). It is interesting to note if the organic modifier is changed from acetonitrile to 2-propanol no separation is achieved. Using this column the enantiomers of flumecinol (and several other α-ethyl benzhydrols) were also successfully separated.[25] For flumecinol the separation factor α$_{S/R}$ is 1.34 using a phosphate buffer of pH 7 containing 10% v/v 2-propanol as the eluent. Even better separation (α$_{R/S}$ = 1.53) was achieved using a Chiralcel-OD column (Daicel) in which the chiral selector is dimethylphenylcarbamoylated cellulose coated on silica and the eluent was 98:2 v/v mixture of hexane and 2-propanol. Since the efficiency of this column was much better, it has been successfully used for the simultaneous separation of the enantiomers of flumecinol and its chiral impurities as well as the separation of the achiral impurities.[17]

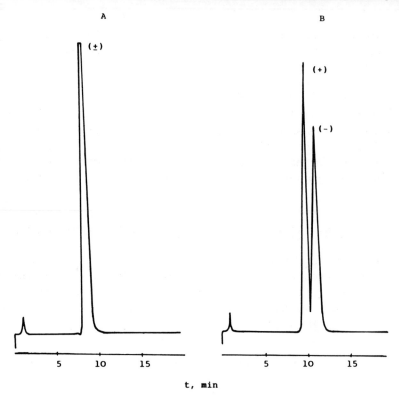

Figure 5 *HPLC separation of the enantiomers of tolperisone [2-methyl-1-(4-methyl-phenyl)-(1-piperidinyl)-propan-1-one]*
Column: Chiral-AGP (ChromTech) 100 × 4.6 mm.
Eluent (A) 0.1 M phosphate buffer (pH 7)–2-propanol 85:15 v/v; (B) 0.1 M phosphate buffer (pH 7)–acetonitrile 85:15 v/v at 1 ml min^{-1}.
UV detector: 245 nm.

In the case of drugs with two chiral centres, such as vinpocetine, two enantiomeric pairs of the two diastereomers, altogether four stereoisomers have to be separated. This is achievable using a Chiral AGP column if higher pH (7.73) and gradient elution of 2-propanol from 18 to 35% v/v are used.[26] Since potential stereoisomeric impurities [3(R), 16(R); 3(S), 16(R); 3(R), 16(S)] are eluted after vinpocetine [3(S), 16(S) derivative] this method is not suitable for the estimation of traces of stereoisomeric impurities in vinpocetine. For this purpose the method based on dynamic formation of diastereomeric adducts using achiral cyanopropylsilica column and an eluent containing the chiral additive (+)-camphorsulfonic acid, described by Szepesi, Gazdag, and Iváncsics[27] is suitable. This method ensures the elution of the enantiomeric impurity [3(R), 16(R)] before the main peak allowing its determination down to the 0.2% level.

If the aim is the separation of the diastereomers only in the case of the analysis of drugs with two or more chiral centres neither chiral column nor chiral additive in the eluent is necessary but the conditions of the achiral chromatography have to be carefully optimized. For example, in the course of the impurity profiling of thymotrinan, its diastereomeric L-Arg-L-Lys-D-Asp impurity was separated from the main compound (L-Arg-L-Lys-L-Asp) only at or above pH 9 when an RP-18 column and, as the eluent, phosphate buffer was used: $\alpha_{LLD/LLL}$ at pH 9 is 1.50.[28,29] Because the aggressive nature of this eluent destroys the column rather rapidly further research was carried out to find another chromatographic system for the separation. As demonstrated in Figure 4 in the preceding section the recently developed ion-pairing system with its acidic eluent is also suitable for this purpose. Another advantage of this system was that the diastereometric impurity eluted before the main peak.

5 Estimation of Degradation Pathways and Kinetics

The description of the degradation pathways and their kinetics in acidic and alkaline media under stressed conditions is an important part of the documentation required by most drug registration authorities. The selectivity and the ruggedness of the analytical (mainly HPLC) methods used are prerequisites for obtaining reliable data for this purpose.

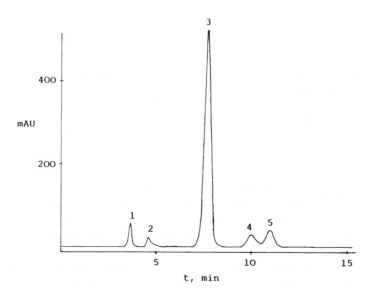

Figure 6 *HPLC separation of the acidic and basic degradation products of posatirelin [L-6-ketopiperidine-2-carbonyl-L-leucyl-L-prolinamide]*
Column: Nucleosil 10 RP-18 (Macherey-Nagel/BST); 250 × 4 mm.
Eluent: 0.05 M phosphate buffer (pH 3)–methanol 65:35 at 1 ml min⁻¹.
UV detector: 215 nm.
Key: (1) H-L-Aad-L-Leu-L-Pro-NH₂; (2) H-L-Aad-L-Leu-L-Pro-OH; (3) posatirelin (L-Kpc-L-Leu-L-Pro-NH₂); (4) D-Kpc-L-Leu-L-Pro-NH₂; (5) L-Kpc-L-Leu-L-Pro-OH

As an example the investigation of posatireline will be outlined. Preliminary studies revealed that the main degradation products in acidic and alkaline media were H-Aad-Leu-Pro-NH$_2$, H-Aad-Leu-Pro-OH, Kpc-Leu-Pro-OH, and D-Kpc-Leu-Pro-NH$_2$. A typical chromatogram showing the separation of these from the main component is shown in Figure 6.

The scheme of the degradation in acidic and alkaline media is summarized in Figure 7.

The Arrhenius plots of the degradation reactions in 0.01 M hydrochloric acid are shown in Figure 8.

The main reaction was shown to be the hydrolytic splitting of the ketopiperidine ring to form H-Aad-Leu-Pro-NH$_2$. At elevated temperatures the hydrolysis of the prolinamide moiety has also to be taken into consideration.

The Arrhenius plot of the reactions in 0.01 M sodium hydroxide is shown in Figure 9. Generally speaking, the posatireline is less stable in alkaline medium than in acidic solution. It is interesting to note that the main degradation reaction at room temperature is the hydrolytic splitting of the ketopiperidine ring just as in acidic medium whilst at elevated temperature epimerization at the ketopiperidine moiety is the main reaction (the two Arrhenius plots intersect).

References

1. C. E. Hammer, (ed.), 'Drug Development', CRC Press, Boca Raton, 1986.
2. K. Burger, 'Organic Reagents in Metal Analysis', Akadémiai Kiadó, Budapest and Pergamon Press, Oxford, 1973.
3. S. Görög and J. Sütõ, *Acta Pharm. Hung.*, 1981, **51**, 217.
4. S. Görög, Z. Tuba, and I. Egyed, *Analyst (London)*, 1969, **94**, 1044.
5. S. Görög and Gy. Hajós, *J. Chromatogr.*, 1969, **43**, 541.
6. S. Görög and G. Szepesi, *Analyst (London)*, 1972, **97**, 59.
7. S. Görög and G. Szepesi, *Anal. Chem.*, 1972, **44**, 1079.
8. G. Szepesi and S. Görög, *Boll. Chim. Farm.*, 1975, **114**, 98.
9. T. Yoshitake, S. Hara, M. Yamaguchi, M. Nakamura, Y. Ohkura, and S. Görög, *J. Chromatogr.*, 1989, **489**, 364.
10. S. Görög, *Pharmacon (Seoul)*, 1991, **21**, 190.
11. S. Görög, B. Herényí, and É. Csizér, *Acta Chim. Hung.*, 1956, **122**, 251.
12. S. Görög and B. Herényi, *J. Chromatogr.*, 1987, **400**, 177.
13. S. Görög, A. Laukó, B. Herényi, A. Georgakis, É. Csizér, G. Balogh, Gy. Gálik, S. Mahó, and Z. Tuba, *Chromatographia*, 1988, **26**, 316.
14. S. Görög, A. Laukó, and B. Herényi, *J. Pharm. Biomed. Anal.*, 1988, **6**, 697.
15. S. Görög, 'Estimation of impurity profiles', ed. S. Görög, in 'Steroid Analysis in the Pharmaceutical Industry', Ellis Horwood, Chichester, 1989, pp. 200–206.
16. A. Laukó, É. Csizér, and S. Görög, Proceedings of the 13th International Symposium on Capillary Chromatography, Vol. II, ed. P. Sandra, 1991, pp. 1548.
17. S. Görög, B. Herényi, and M. Rényei, *J. Pharm. Biomed. Anal.*, 1992, **10**, 831.
18. A. Laukó, É. Csizér, and S. Görög, *Analyst (London)*, 1993, **118**, 609.
19. S. Görög, G. Balogh, and M. Gazdag, *J. Pharm. Biomed. Anal.*, 1991, **9**, 829.

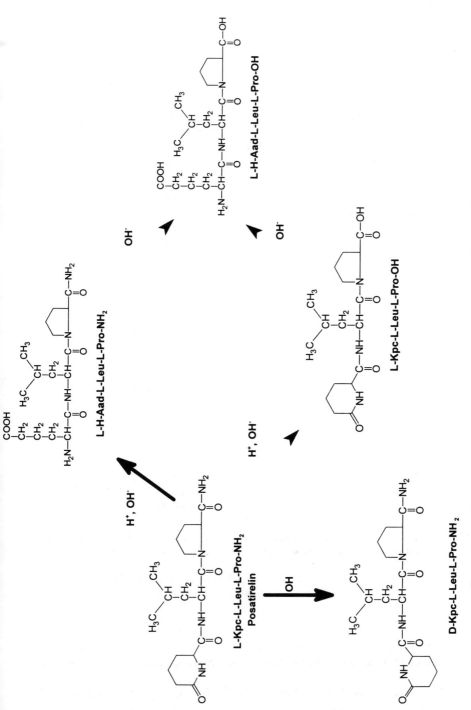

Figure 7 *Reaction scheme of the degradation of posatirelin in acidic and alkaline media*

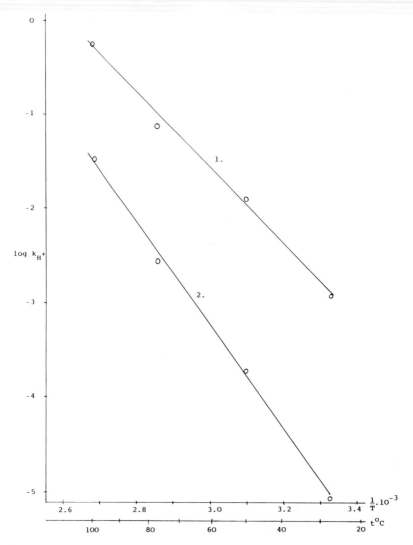

Figure 8 *Arrhenius plots of the degradation reactions of posatirelin in 0.01 M hydrochloric acid (logarithm of the pseudo first order rate constant, k_H–in h^{-1} vs $1/T$). $c_{posatirelin} = 3 \times 10^{-3}$ M.*
Key: (1) posatirelin $\rightarrow$ L-H-Aad-L-Leu-L-Pro-NH$_2$ reaction; (2) posatirelin $\rightarrow$ L-Kpc-L-Leu-L-Pro-OH reaction

20. M. Gazdag, G. Szepesi, K. Varsányi-Riedl, and Z. Tuba, 'Advances in Steroid Analysis 1984', ed. S. Görög, Akadémial Kiadó, Budapest, and Elsevier, Amsterdam, 1985, p. 431.
21. M. Gazdag, G. Szepesi, K. Varsányi-Riedl., Z. Végh, and Zs. Pap-Sziklay, *J. Chromatogr.*, 1985, **328**, 279.
22. M. Gazdag, K. Varsányi-Riedl, and G. Szepesi, *J. Chromatogr.*, 1985, **347**, 284.

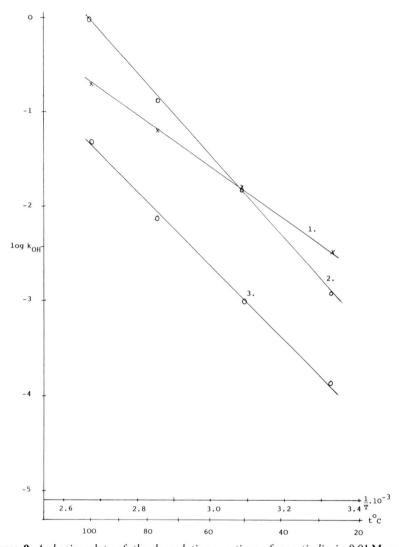

Figure 9 *Arrhenius plots of the degradation reactions of posatirelin in 0.01 M sodium hydroxide (logarithm of the pseudo-first order rate constant, k_{OH} – in hour^{-1} vs. $1/T$). $c_{posatirelin} = 3 \times 10^{-3}$ M.*
Key: (1) posatirelin → L-H-Aad-L-Leu-L-Pro-NH$_2$ reaction; (2) posatirelin → D-Kpc-L-Leu-L-Pro-NH$_2$ reaction; (3) posatirelin → L-Kpc-L-Leu-L-Pro-OH reaction

23. G. Szepesi, M. Gazdag, and K. Mihályfi, *J. Chromatogr.*, 1989, **464**, 265.
24. M. Gazdag, M. Babják, P. Kemenes-Bakos, and S. Görög, *J. Chromatogr.*, 1991, **550**, 639.
25. S. Görög and B. Herényi, *J. Pharm. Biomed. Anal.*, 1990, **8**, 837.
26. B. Herényi and S. Görög, *J. Chromatogr.*, 1992, **592**, 297.
27. G. Szepesi, M. Gazdag, and R. Iváncsics, *J. Chromatogr.*, 1982, **244**, 33.

28. S. Görög, B. Herényi, O. Nyéki, I. Schön, and L. Kisfaludy, *J. Chromatogr.*, 1988, **452**, 317.
29. B. Herényi, S. Görög, O. Nyéki, I. Schön, and L. Kisfaludy, *Chromatographia*, 1990, **29**, 395.

Subject Index